STRUCTURAL ANALYSIS

A Classical and Matrix Approach

Second Edition

STRUCTURAL ANALYSIS
A Classical and Matrix Approach

Second Edition

Jack C. McCormac
Clemson University

James K. Nelson, Jr.
Clemson University

 **ADDISON-WESLEY**

An imprint of Addison Wesley Longman, Inc.

Reading, Massachusetts • Menlo Park, California • New York • Harlow, England
Don Mills, Ontario • Sydney • Mexico City • Madrid • Amsterdam

Sponsoring Editor: Michael Slaughter
Project Coordinator: Elizabeth Fresen/Cathy Wacaser
Art Development Editor: Vita Jay
Design Administrator: Jess Schaal
Text and Cover Design: Jess Schaal
Cover Illustration/Photo: Jeff Goldberg/ESTO
Photo Research: Nina Page
Production Administrator: Randee Wire
Compositor: Interactive Composition Corporation
Cover Printer: Phoenix Color Corporation

Library of Congress Cataloging-in-Publication Data

McCormac, Jack C.
    Structural analysis: A classical and matrix approach / Jack C.
McCormac, James K. Nelson. — 2nd ed.
      p. cm.
    Includes biblographical references and index.
    ISBN 0-673-99753-7
    1. Structural analysis (Engineering)   I. Nelson, James K.
II. Title.
TA645.M38 1996
624.1'71—dc20   96-23304
                                  CIP

ISBN 0-673-99753-7

96 97 98 99—DOC—9 8 7 6 5 4 3 2 1

This Book Is Dedicated to Our Dear Friend, the Late Rudolf E. Elling

Contents

PREFACE   *xiv*

PART ONE
STATICALLY DETERMINATE STRUCTURES   **1**

CHAPTER 1
Introduction   **3**

1.1   Structural Analysis and Design   3
1.2   History of Structural Analysis   4
1.3   Basic Principles of Structural Analysis   7
1.4   Structural Components and Systems   9
1.5   Structural Forces   12
1.6   Structural Idealization (Line Diagrams)   14
1.7   Calculation Accuracy   15
1.8   Checks on Problems   16
1.9   Impact of Computers on Structural Analysis   17

CHAPTER 2
Structural Loads   **18**

2.1   Introduction   18
2.2   Specifications and Building Codes   19
2.3   Types of Structural Loads   22
2.4   Dead Loads   23
2.5   Live Loads   24
2.6   Live Load Impact Factors   25
2.7   Live Loads on Roofs   26
2.8   Rain Loads   27

2.9   Wind Loads   29
2.10   Snow Loads   43
2.11   Other Loads   46
          Problems   50

CHAPTER 3
System Loading and Behavior   **53**

3.1   Introduction   53
3.2   Tributary Areas   54
3.3   Live Load Reduction   59
3.4   Loading Conditions for Allowable Stress Design   61
3.5   Loading Conditions for Strength Design   63
3.6   Placing Loads on the Structure   64
3.7   Concept of the Force Envelope   68
          Problems   69

CHAPTER 4
Reactions   **71**

4.1   Equilibrium   71
4.2   Moving Bodies   72
4.3   Calculation of Unknowns   72
4.4   Types of Support   72
4.5   Stability, Determinacy, and Indeterminacy   74
4.6   Geometric Instability   76
4.7   Sign Convention   77
4.8   Horizontal and Vertical Components   78
4.9   Free-Body Diagrams   78
4.10   Reactions by Proportions   78
4.11   Reactions Calculated by Equations of Statics   79
4.12   Principle of Superposition   82
4.13   The Simple Cantilever   82
4.14   Cantilevered Structures   84
4.15   Reaction Calculations for Cantilevered Structures   85
4.16   Arches   86
4.17   Three-Hinged Arches   87
4.18   Uses of Arches and Cantilevered Structures   92
4.19   Cables   93
          Problems   98

CHAPTER 5
Shear and Moment Diagrams   **110**

5.1   Introduction   110
5.2   Shear Diagrams   112
5.3   Moment Diagrams   113
5.4   Relations Among Loads, Shears, and Bending Moments   113
5.5   Moment Diagrams Drawn from Shear Diagrams   115
5.6   Shear and Moment Diagrams for Statically Determinate Frames   122
       Problems   125

CHAPTER 6
Introduction to Plane or Two-Dimensional Trusses   **136**

6.1   General   136
6.2   Assumptions for Truss Analysis   137
6.3   Effect of Assumptions   138
6.4   Truss Notation   138
6.5   Roof Trusses   139
6.6   Bridge Trusses   141
6.7   Arrangement of Truss Members   143
6.8   Statical Determinacy of Trusses   144
6.9   Use of Sections   147
6.10   Horizontal and Vertical Components   147
6.11   Arrow Convention   148
6.12   Method of Joints   149
       Problems   152

CHAPTER 7
Plane Trusses, Continued   **161**

7.1   Method of Moments   161
7.2   Forces in Members Cut by Sections   161
7.3   Application of the Method of Moments   163
7.4   Method of Shears   169
7.5   Zero-Force Members   172
7.6   When Assumptions Are Not Correct   173
7.7   Simple Trusses   174
7.8   Compound Trusses   175
7.9   The Zero-Load Test   175
7.10   Complex Trusses   177
7.11   Stability   181
7.12   Equations of Condition   183
7.13   Computer Example   186
       Problems   189

CHAPTER 8
Three-Dimensional or Space Trusses   **202**

8.1   General   202
8.2   Basic Principles   203
8.3   Statics Equations   204
8.4   Stability of Space Trusses   205
8.5   Special Theorems Applying to Space Trusses   206
8.6   Types of Support   207
8.7   Illustrative Examples   209
8.8   More Complicated Space Trusses   213
8.9   Simultaneous-Equation Analysis   218
8.10   Computer Example   222
        Problems   224

CHAPTER 9
Influence Lines   **227**

9.1   Introduction   227
9.2   The Influence Line Defined   228
9.3   Influence Lines for Simple Beam Reactions   229
9.4   Influence Lines for Simple Beam Shears   229
9.5   Influence Lines for Simple Beam Moments   231
9.6   Qualitative Influence Lines   232
9.7   Uses of Influence Lines; Concentrated Loads   235
9.8   Uses of Influence Lines; Uniform Loads   237
9.9   Common Simple Beam Formulas from Influence Lines   238
9.10   Placing Live Loads to Cause Maximum Values Using Influence
        Lines   238
9.11   Placing Live Loads to Cause Maximum Values Based on Maximum
        Curvature   241
9.12   Influence Lines for Trusses   242
9.13   Arrangement of Bridge Floor Systems   242
9.14   Influence Lines for Truss Reactions   244
9.15   Influence Lines for Member Forces of Parallel-Chord Trusses   245
9.16   Influence Lines for Members of Nonparallel-Chord Trusses   247
9.17   Influence Lines for K Truss   248
9.18   Determination of Maximum Forces   249
9.19   Counters in Bridge Trusses   251
9.20   Live Loads for Highway Bridges   253
9.21   Live Loads for Railway Bridges   256
9.22   Impact Loadings   257
9.23   Maximum Values for Moving Loads   258
        Problems   261

CHAPTER 10
Deflection and Angle Changes—Geometric Methods   **271**

10.1   Introduction   271
10.2   Sketching Deformed Shapes of Structures   272
10.3   Reasons for Computing Deflections   273
10.4   The Moment-Area Theorems   275
10.5   Application of the Moment-Area Theorems   278
10.6   The Method of Elastic Weights   285
10.7   Application of the Method of Elastic Weights   286
10.8   Limitations of the Elastic-Weight Method   290
10.9   Conjugate-Beam Method   291
10.10   Summary of Beam Relations   293
10.11   Application of the Conjugate Method to Beams   294
10.12   Long-Term Deflections   297
10.13   Application of the Conjugate Method to Frames   297
        Problems   300

CHAPTER 11
Deflection and Angle Changes—Energy Methods   **308**

11.1   Introduction to Energy Methods   308
11.2   Conservation of Energy Principle   308
11.3   Virtual Work or Complementary Virtual Work Method   309
11.4   Truss Deflections by Virtual Work   311
11.5   Application of Virtual Work to Trusses   314
11.6   Deflections of Beams and Frames by Virtual Work   317
11.7   Example Problems for Beams and Frames   319
11.8   Rotations or Angle Changes by Virtual Work   325
11.9   Maxwell's Law of Reciprocal Deflections   328
11.10   Introduction to Castigliano's Theorems   329
11.11   Castigliano's Second Theorem   330
        Problems   335

CHAPTER 12
Introduction to Statically Indeterminate Structures   **343**

12.1   General   343
12.2   Continuous Structures   344
12.3   Advantages of Statically Indeterminate Structures   346
12.4   Disadvantages of Statically Indeterminate Structures   348
12.5   Looking Ahead   349

PART TWO
STATICALLY INDETERMINATE STRUCTURES—
CLASSICAL METHODS   351

CHAPTER 13
Force Methods of Analyzing Statically Indeterminate Structures   **353**

13.1   Methods of Analyzing Statically Indeterminate Structures   353
13.2   Beams and Frames With One Redundant   354
13.3   Beams and Frames With Two or More Redundants   363
13.4   Support Settlement   365
13.5   Analysis of Externally Redundant Trusses   367
13.6   Analysis of Internally Redundant Trusses   372
13.7   Analysis of Trusses Redundant Internally and Externally   374
13.8   Temperature Changes, Shrinkage, Fabrication Errors, and So On   376
13.9   Computer Example   378
        Problems   380

CHAPTER 14
Influence Lines for Statically Indeterminate Structures   **389**

14.1   Influence Lines for Statically Indeterminate Beams   389
14.2   Qualitative Influence Lines   394
14.3   Influence Lines for Statically Indeterminate Trusses   397
14.4   Influence Lines Using SABLE   402
        Problems   402

CHAPTER 15
Castigliano's Theorems and the Three-Moment Theorem   **405**

15.1   Castigliano's Second Theorem   405
15.2   Castigliano's First Theorem—The Method of Least Work   414
15.3   The Three-Moment Theorem   416
15.4   Development of the Theorem   417
15.5   Application of the Three-Moment Theorem   419
        Problems   425

CHAPTER 16
Slope Deflection—A Displacement Method of Analysis   **430**

16.1   Introduction   430
16.2   Derivation of Slope-Deflection Equations   431
16.3   Application of Slope-Deflection Equations to Continuous Beams   433
16.4   Analysis of Frames—No Sidesway   439
16.5   Analysis of Frames With Sidesway   441
16.6   Analysis of Frames With Sloping Legs   447
        Problems   448

PART THREE
STATICALLY INDETERMINATE STRUCTURES—
MODERN METHODS   453

CHAPTER 17
Approximate Analysis of Statically Indeterminate Structures   **455**

17.1   Introduction   455
17.2   Trusses With Two Diagonals in Each Panel   456
17.3   Continuous Beams   457
17.4   Analysis of Building Frames for Vertical Loads   462
17.5   Analysis of Portal Frames   465
17.6   Lateral Bracing for Bridges   466
17.7   Analysis of Mill Buildings   468
17.8   Analysis of Building Frames for Lateral Loads   471
17.9   Moment Distribution   480
17.10   Analysis of Vierendeel "Trusses"   480
        Problems   482

CHAPTER 18
Moment Distribution for Beams   **488**

18.1   General   488
18.2   Introduction   489
18.3   Basic Relations   490
18.4   Definitions   492
18.5   Sign Convention   493
18.6   Fixed-End Moments for Various Loads   495
18.7   Application of Moment Distribution   495
18.8   Modification of Stiffness for Simple Ends   499
18.9   Shear and Moment Diagrams   501
18.10   Computer Solutions   503
        Problems   505

CHAPTER 19
Moment Distribution for Frames   **510**

19.1   Frames With Sidesway Prevented   510
19.2   Frames With Sidesway   512
19.3   Sidesway Moments   514
19.4   Frames With Sloping Legs   521
19.5   Multistory Frames   526
19.6   Computer Example   530
        Problems   532

CHAPTER 20
Introduction to Matrix Methods   **539**

20.1   Reasons for Matrix Methods   539
20.2   Use of Matrix Methods   540
20.3   Force and Displacement Representations   541
20.4   Some Necessary Definitions   542
20.5   The Fundamental Concept   543
20.6   Systems With Several Elements   545
20.7   Bars Instead of Springs   548
20.8   Solution for a Truss   548
20.9   System Matrices Using Strain Energy   552
20.10   Looking Ahead   554
            Problems   554

CHAPTER 21
More About Matrix Methods   **556**

21.1   Introduction   556
21.2   Definition of Coordinate Systems   557
21.3   The Elemental Stiffness Relationship   560
21.4   Truss Element Matrices   560
21.5   Beam Element Matrices   562
21.6   Transformation to Global Coordinates   565
21.7   Assembling the Global Stiffness Matrix   565
21.8   Loads Acting on the System   566
21.9   Computing Final Beam End Forces   568
21.10   Putting It All Together   568
            Problems   572

APPENDIX A
The Catenary Equation   **574**

APPENDIX B
Matrix Algebra   **579**

B.1   Introduction   579
B.2   Matrix Definitions and Properties   579
B.3   Special Matrix Types   580
B.4   Determinant of a Square Matrix   581
B.5   Adjoint Matrix   583
B.6   Matrix Arithmetic   583
B.7   Gauss Method of Solving Simultaneous Equations   588
B.8   Special Topics   590

GLOSSARY   595
INDEX   599

Preface

The purpose of this book is to introduce the elementary fundamentals of structural analysis for beams, trusses, and frames. Sufficient information is included so that students may develop a thorough understanding of both statically determinate and statically indeterminate structures.

In this second edition, the book has been divided into three parts. Part I is devoted almost entirely to statically determinate structures, while Parts II and III address statically indeterminate structures.

Several of the classical methods of analyzing statically indeterminate structures are presented in Part II. Due to time constraints, some instructors might prefer to skip some or all of this portion of the book. Though the methods included are of considerable historical interest, they almost never are used in actual structural practice today. Also, it is not necessary to study these classical methods in order to be able to understand and apply the modern methods of analysis presented in Part III. Nevertheless, a study of Part II (particularly Chapter 13, concerning the method of consistent distortions) can provide the student with a good understanding of the behavior of statically indeterminate structures.

The increasing availability of digital computers over the past few decades has completely changed the practical application of structural analysis. Instead of applying specialized classical methods, engineers generally use computer programs prepared with the broad and comprehensive matrix methods. Enclosed at the back of this book is a diskette containing a computer program entitled SABLE (Structural Analysis and Behavior for Learning Engineering). With SABLE, almost all of the exercise problems included in this book can be solved.

Modern tools of analysis, such as computer software, enable the engineer to analyze both small and large structures with rapidity. Unfortunately, however, the mathematical efficiency of such analyses tends to hide the basic principles of structural analysis. It is for this reason that considerable space is used herein to present the elementary methods of analysis. It is hoped that students using this textbook will obtain an understanding and fundamental "feel" for the behavior of structures that they might not gain if such methods were omitted from their study of structural analysis.

Finally, to enhance the development of "feel" in student engineers, the authors have emphasized the sketching of the deformed shapes of structures under load. In addition, the program SABLE enables users to apply loads to various types of structures and to observe their deformed shapes on the computer screen.

The authors thank the following persons who reviewed this edition.

Fouad H. Fouad, University of Alabama at Birmingham
Hany J. Farran, California State Polytechnic University—Pomona
Fouad Fanous, Iowa State University
S. D. Rajan, Arizona State University
Carl E. Kurt, University of Kansas
Roberto A. Osegueda, University of Texas at El Paso
Yook-Kong Yong, Rutgers University

The authors also are grateful to Nina Kristeva, John Murden, Roberto Osegueda, David Rosowsky, Scott Schiff, Bala Sockalingam, Peter Sparks, Yook-Kong Yong, and the civil engineering class of 1995 at The Citadel for their review and contributions to the computer program SABLE.

Jack C. McCormac
James K. Nelson, Jr.

PART ONE

STATICALLY DETERMINATE STRUCTURES

Chapter 1

Introduction

1.1 STRUCTURAL ANALYSIS AND DESIGN

The application of loads to structures causes those structures to deform. As a result, various forces are produced in the structures. The calculations of the magnitudes of these forces and deformations is referred to as *structural analysis*. It is an extremely important topic to humankind. Indeed, almost every branch of technology becomes involved at some time or another with questions concerning strength and deformation.

Structural design includes the arrangement and proportioning of structures and their parts so that they will satisfactorily support the loads to which they may be subjected. In detail, structural design involves the following: the general layout of structures; studies of the possible structural forms or types that may provide feasible solutions; consideration of loading conditions; preliminary structural analyses and designs of the possible solutions; the selection of a solution; and the final structural analysis and design of the structure, including the preparation of design drawings.

This book is devoted to structural analysis, with only occasional remarks concerning the other phases of structural design. Structural analysis can be so interesting to many persons that they become completely attached to it and have the feeling that they want to become 100% involved in the subject. Although analyzing and predicting the behavior of structures and their parts is an extremely important part of structural design, it is only one of several important and interrelated steps. As a result, it is rather unusual for a person to be employed completely as a structural analyst. He or she in almost all probability will be involved in several or all phases of structural design.

It is said that Robert Louis Stevenson for a time studied structural engineering, but he apparently found the "science of stresses and strains" too dull for his lively imagination. He then went on to study law for a while before devoting the rest of his

White Bird Canyon Bridge, White Bird, Idaho. (Courtesy of the American Institute of Steel Construction, Inc.)

life to writing prose and poetry.[1] Most of us who have read *Treasure Island, Kidnapped,* or his other work would agree that the world is a better place because of his decision. Nevertheless, there are a great number of us who regard structural analysis and design as extremely interesting topics. It is hoped that this book will add to that number.

I.2 HISTORY OF STRUCTURAL ANALYSIS

Structural analysis as we know it today evolved over several thousand years. During this time many types of structures such as beams, arches, trusses, and frames were used in construction for hundreds or even thousands of years before satisfactory methods of analysis were developed for them.

Though ancient engineers showed some understanding of structural behavior (as evidenced by their successful construction of great bridges, cathedrals, sailing vessels, and so on), real progress with the theory of structural analysis occurred only in the last 150 years.

The Egyptians and other ancient builders surely had some kinds of empirical rules drawn from previous experiences for determining sizes of structural members. There is, however, no evidence that they had developed any theory of structural analysis. The Egyptian Imhotep who built the great step pyramid of Sokkara in about 3000 B.C. sometimes is referred to as the world's first structural engineer.

[1] *Proceedings of the First United States Conference on Prestressed Concrete* (Cambridge, Mass.: Massachusetts Institute of Technology, 1951), 1.

Although the Greeks built some magnificent structures, their contributions to structural theory were few and far between. Pythagoras (about 582–500 B.C.), who is said to have originated the word *mathematics,* is famous for the right angle theorem that bears his name. (This theorem actually was known by the Sumerians in about 2000 B.C.) Further, Archimedes (287–212 B.C.) developed some fundamental principles of statics and introduced the term *center of gravity*.

The Romans were outstanding builders and were very competent in using certain structural forms such as semicircular masonry arches. As did the Greeks, they, too, had little knowledge of structural analysis and made even less scientific progress in structural theory. They probably designed most of their beautiful buildings from an artistic viewpoint. Perhaps their great bridges and aqueducts were proportioned with some rules of thumb; however, if these methods of design resulted in proportions that were insufficient, the structures collapsed and no historical records were kept. Only their successes endured.

One of the greatest and most noteworthy contributions to structural analysis, as well as to all other scientific fields, was the development of the Hindu-Arabic system of numbers. Unknown Hindu mathematicians in the 2nd or 3rd centuries B.C. originated a numbering system of one to nine. In about 600 A.D. the Hindus invented the symbol *sunya* (meaning empty), which we call zero. (The Mayan Indians of Central America, however, had apparently developed the concept of zero about 300 years earlier.)[2]

In the 8th century A.D. the Arabs learned this numbering system from scientific writings of the Hindus. In the following century, a Persian mathematician wrote a book that included the system. His book was translated into Latin some years later and brought to Europe.[3] In around 1000 A.D., Pope Sylvester II decreed that the Hindu-Arabic numbers were to be used by Christians.

Before real advances could be made with structural analysis, it was necessary for the science of mechanics of materials to be developed. By the middle of the 19th century, much progress had been made in this area. A French physicist Charles Augustin Coloumb (1736–1806) and a French engineer-mathematician Louis Marie Henri Navier (1785–1836), building upon the work of numerous other investigations over hundreds of years, are said to have founded the science of mechanics of materials. Of particular significance was a textbook published by Navier in 1826, in which he discussed the strengths and deflections of beams, columns, arches, suspension bridges, and other structures.

Andrea Palladio (1518–1580), an Italian architect, is thought to have been the first person to use modern trusses. He may have revived some ancient types of Roman structures and their empirical rules for proportioning them. It was actually 1847, however, before the first rational method of analyzing jointed trusses was introduced by Squire Whipple (1804–1888). His was the first significant American contribution to structural theory. Whipple's analysis of trusses often is said to have signalled the beginning of modern structural analysis. Since that time there has been an almost continuous series of important developments in the subject.

[2] *The World Book Encyclopedia* (Chicago, IL, 1993, Book N–O), 617.

[3] See note 2 above.

Several excellent methods for calculating deflections were published in the 1860s and 1870s which further accelerated the rate of structural analysis development. Among the important investigators and their accomplishments were: James Clerk Maxwell (1831–1879) of Scotland, for the reciprocal deflection theorem in 1864; Otto Mohr (1835–1918) of Germany, for the method of elastic weights presented in 1870; Alberto Castigliano (1847–1884) of Italy, for the least work theorem in 1873; and Charles E. Greene (1842–1903) of the United States, for the moment-area theorems in 1873.

The advent of railroads gave a great deal of impetus to the development of structural analysis. It was suddenly necessary to build long span bridges capable of carrying very heavy moving loads. As a result, the computation of stresses and strains became increasingly important. Fatigue and impact stresses became serious matters. Furthermore, up until this time there had not been a great deal of need to analyze statically indeterminate structures; however, continuous span railroad bridges created the need to do so.

One method for analyzing continuous statically indeterminate beams—the three-moment theorem—was introduced in 1857 by the Frenchman B. P. E. Clapeyron (1799–1864), and was used for analyzing many railroad bridges. In the decades that followed, many other advances were made in indeterminate structural analysis based upon the recently developed deflection methods.

Otto Mohr, who worked with railroads, is said to have reworked into practical, usable form many of the theoretical developments up to his time. Particularly notable in this regard was his 1874 publication of the method of consistent distortions for analyzing statically indeterminate structures.

In the United States two great developments in statically indeterminate structure analysis were made by G. A. Maney (1888–1947) and Hardy Cross (1885–1959). In 1915 Maney presented the slope deflection method, while Cross introduced moment distribution in 1924.

In the first half of the 20th century, many complex structural problems were expressed in mathematical form, but sufficient computing power was not available for practically solving the resulting equations. This situation continued in the 1940s, when much work was done with matrices for analyzing aircraft structures. Fortunately, the development of digital computers made practical the use of equations for these and many other types of structures, including high-rise buildings.

Some particularly important historical references on the development of structural analysis are listed below in footnotes 4, 5, and 6. They document the slow but steady development of the fundamental principles involved. It seems ironic that the college student of today can learn in a few months the theories and principles of structural analysis that took humankind several thousand years to develop.

[4] J. S. Kinney, *Indeterminate Structural Analysis* (Reading, Mass.: Addison-Wesley, 1957), 1–16.

[5] S. P. Timoshenko, *History of Strength of Materials* (New York: McGraw-Hill, 1953), 1–439.

[6] H. M. Westergaard, "One Hundred Fifty Years Advance in Structural Analysis," *ASCE 94* (1930): 226–240.

Pacific Gas and Electric Company Headquarters Building, San Francisco.
(Courtesy Bethlehem Steel Corporation.)

1.3 BASIC PRINCIPLES OF STRUCTURAL ANALYSIS

Structural engineering embraces an extensive variety of structural systems. When speaking of structures, people typically think of buildings and bridges. But there are many other types of systems with which structural engineers deal, including sports and entertainment stadiums, radio and television towers, arches, storage tanks, aircraft and space structures, concrete pavements, and fabric air-filled structures. These structures can vary in size from a single member, as in the case of a light pole, to structures such as the Sears Tower in Chicago, which is over 1450 feet tall, or the Humber Estuary bridge in England, which has a suspended span that is over 4,626 feet long. (It is anticipated that in 1998 or 1999 the Akashi-Kaikyo suspension bridge in Japan will be completed, with its suspended span of approximately 6530 feet.)

In order to be able to address this wide range of sizes and types of structure, a structural engineer must have a solid understanding of the basic principles that apply to all structural systems. It is unwise to learn how to analyze a particular structure, or even a few different types of structures. Rather, it is more important to learn the fundamental principles that apply to all structural systems, regardless of their type or use. One never knows what types of problems the future holds or what type of structural system may be conceived for a particular application, but a firm understanding of basic principles will help us to analyze new structures with confidence.

The fundamental principles used in structural analysis are Sir Isaac Newton's laws of inertia and motion:

1. A body will exist in a state of rest or in a state of uniform motion in a straight line unless it is forced to change that state by forces imposed on it.
2. The rate of change of momentum of a body is equal to the net applied force.
3. For every action there is an equal and opposite reaction.

Cold-storage warehouse, Grand Junction, Colorado. (Courtesy of the American Institute of Steel Construction, Inc.)

These laws of motion can be expressed by the equation

$$\Sigma F = ma$$

In this equation ΣF is the summation of all the forces that are acting on the body, m is the mass, and a is its acceleration.

In this textbook, we will be dealing with a particular type of equilibrium called *static equilibrium,* in which the system is not accelerating. The equation of equilibrium thus becomes:

$$\Sigma F = 0$$

These structures either are not moving, as is the case for most civil engineering structures, or are moving with constant velocity, such as space vehicles in orbit. Using the principle of static equilibrium, we will be studying the forces that act on structures and methods to determine the response of structures to these forces. By response, we mean the displacement of the system and the forces that occur in each component of the system. This emphasis should provide readers with a solid foundation for advanced study and, it is hoped, convince them that structural theory is not difficult and that it is not necessary to memorize special cases.

1.4 STRUCTURAL COMPONENTS AND SYSTEMS

All structural systems are composed of components. The primary components in a structure are considered to be the following:

Ties—those members that are subjected to axial tension forces only. A tie is not loaded along its length and cannot resist flexural forces.

Struts—those members that are subjected to axial compression forces only. Like a tie, a strut is not loaded along its length and cannot resist flexural forces.

Beams and girders—those members that are subjected to flexural forces. They usually are thought of as being horizontal members that are primarily subjected to gravity forces; but there are frequent exceptions—rafters, for example.

Columns—those members that are primarily subjected to axial compression forces and may be subjected to flexural forces also. Columns usually are thought of as being vertical members, but they may be inclined.

Diaphragms—those components that are flat plates. Diaphragms have very high in-plane stiffness and are commonly used for floors and shear-resisting walls. Usually, diaphragms span between beams or columns and may themselves be stiffened with ribs to better resist out-of-plane forces.

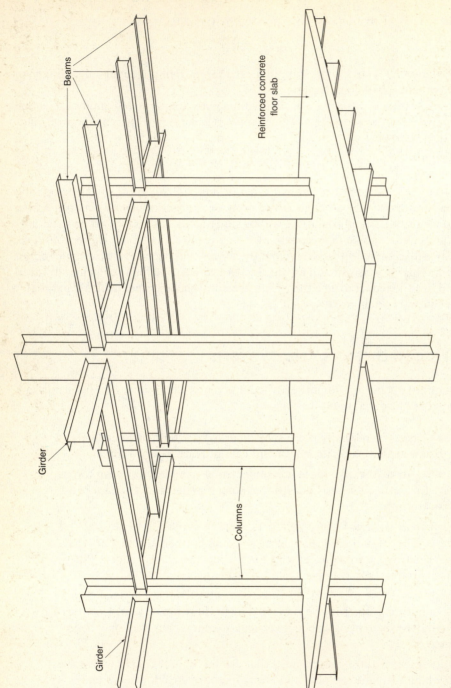

Figure 1.1 Typical building frame.

Las Vegas Convention Center. (Courtesy Bethlehem Steel Corporation.)

Structural components are assembled to form structural systems. In this textbook we will be dealing with typical framed structures. A building frame is shown in Figure 1.1. (In this figure, a girder is considered to be a large beam with smaller beams framing into it.)

A *truss* is a special type of structural frame. It is composed entirely of struts and ties—that is to say, all of its components are connected in such a manner that they are subjected only to axial forces. It is assumed that all of the external loads acting on trusses are applied at their joints and not directly to their components where they might cause bending in the truss members. An old type of bridge structure consisting of two trusses is shown in Figure 1.2. In this figure, the *stringers* are the beams that

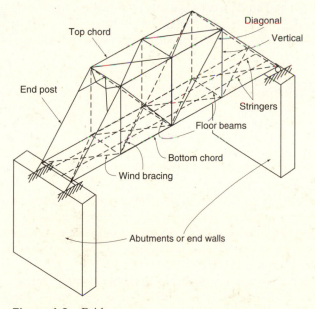

Figure 1.2   Bridge truss.

Access bridge, Renton, Washington. (Courtesy Bethlehem Steel Corporation.)

support the bridge floor and run parallel to the trusses. They are supported at their ends by the *floor beams,* which run transverse to the roadway and frame into the joints of the truss.

There are other types of structural systems. These include fabric structures (e.g., tents and outdoor arenas) and curved shell structures (e.g., dams or sports arenas such as the Astrodome). The analysis of these types of structures requires advanced principles of structural mechanics and is beyond the scope of this book.

1.5 STRUCTURAL FORCES

A structural system is acted upon by forces. Under the influence of these forces, the entire structure is assumed herein to be in a state of static equilibrium and, as a consequence, each component of the structure also is in a state of static equilibrium. The forces that act on a structure include the applied loads and the resulting reaction forces.

The applied loads are the known loads that act on a structure. They can be the result of the structure's own weight, occupancy loads, environmental loads, and so on. The reactions are the forces the supports exert on a structure. They are considered to be part of the external forces applied and are assumed to balance the other external loads on the structure.

To introduce loads and reactions, three simple structures are shown in Figure 1.3. The beam shown in part (a) of the figure is supporting a uniformly distributed gravity load and is itself supported by upward reactions at its ends. The barge in

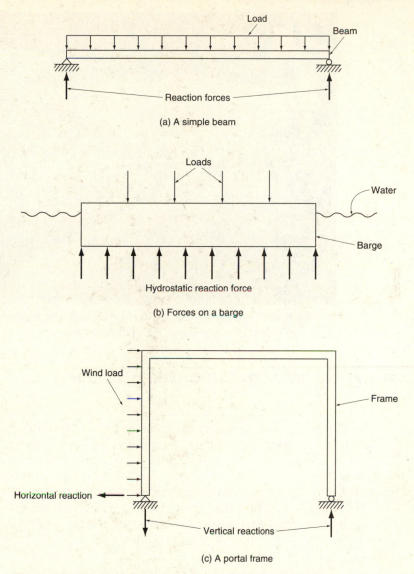

Figure 1.3 Loads and reactions for three simple structures.

part (b) of the figure is carrying a group of containers on its deck. It is in turn supported by a uniformly distributed hydrostatic pressure provided by the water beneath. Part (c) shows a building frame subjected to a lateral wind load. This load tends to overturn the structure, thus requiring an upward reaction at the right-hand support and a downward one at the left-hand support. These forces create a couple that offsets the effect of the wind force. A detailed discussion of reactions and their computation is presented in Chapter 4.

Hungry Horse Dam and Reservoir, Rocky Mountains, in northwestern Montana. (Courtesy of the Montana Travel Promotion Division.)

1.6 STRUCTURAL IDEALIZATION (LINE DIAGRAMS)

To calculate the forces in the various parts of a structure with reasonable simplicity and accuracy, it is necessary to represent the structure in a simple manner that is conducive to analysis. Structural components have width and thickness. Concentrated forces rarely act at a single point; rather, they are distributed over small areas. If these characteristics are taken into consideration in detail, however, an analysis of the structure will be very difficult, if not impossible to perform.

The process of replacing an actual structure with a simple system conducive to analysis is called *structural idealization.* Most often, structural components are represented by lines that are located along the center lines of the components. The sketch of a structure idealized in this manner usually is called a *line diagram.*

The preparation of line diagrams is shown in Figure 1.4. In part (a) of the figure, the wood beam shown supports several floor joists and in turn is supported by three

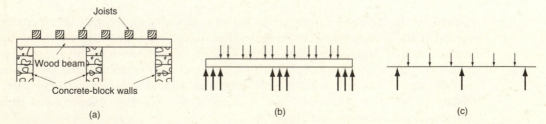

Figure 1.4   Replacing a structure and its forces with a line diagram.

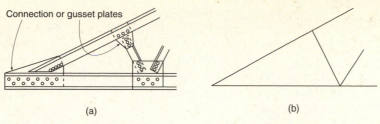

Figure 1.5   A line diagram for a portion of a steel roof truss.

concrete-block walls. The actual distribution of the forces acting on the beam is shown in part (b) of the figure. For purposes of analysis, though, we can conservatively represent the beam and its loads and reactions with the line diagram of part (c). The loaded spans are longer with the result that shears and moments are higher.

Sometimes the idealization of a structure involves assumptions about the behavior of the structure. As an example, the bolted steel roof truss of Figure 1.5(a) is considered. The joints in trusses often are made with large connection or gusset plates and, as such, can transfer moments to the ends of the members. However, experience has shown that the stresses caused by the axial forces in the members greatly exceed the stresses caused by flexural forces. As a result, for purposes of analysis we can assume that the truss consists of a set of pin-connected lines, as shown in Figure 1.5(b).

Although the use of simple line diagrams for analyzing structures will not result in perfect analyses, the results usually are quite acceptable.

Sometimes, though, there may be some doubt in the mind of the analyst as to the exact line diagram or model to be used for analyzing a particular structure. For instance, should beam lengths be clear spans between supports, or should they equal the distances center to center of those supports? Should the supports be assumed to be free to rotate under loads, should they be assumed to be completely fixed against rotation, or should they fall somewhere in between? Because of many questions such as these, it may be necessary to consider different models and perform the analysis for each one to determine the worst cases.

1.7 CALCULATION ACCURACY

A most important point that many students with their superb pocket calculators and personal computers have difficulty in understanding is that structural analysis is not an exact science for which answers can confidently be calculated to eight or more places. Computations to only three places probably are far more accurate than the estimates of material strengths and magnitudes of loads used for structural analysis and design. The common materials dealt with in structures (wood, steel, concrete, and a few others) have ultimate strengths that can only be estimated. The loads applied to structures may be known within a few hundred pounds or no better than a few thousand pounds. It therefore seems inconsistent to require force computations to more than three significant digits.

Several partly true assumptions will be made about the construction of trusses: truss members are connected with frictionless pins; the deformation of truss members under load is so slight as to cause no effect on member forces; and so forth. These deviations from actual conditions emphasize that it is of little advantage to carry structural analysis to many decimal places. Furthermore, retention of calculations to more than three significant places may be harmfully misleading, giving a fictitious sense of precision.

1.8 CHECKS ON PROBLEMS

A definite advantage of structural analysis is the possibility of making mathematical checks on the analysis by some method other than the one initially used, or by the same method from some other position on the structure. The reader should be able in nearly every situation to determine if his or her work has been done correctly.

All of us, unfortunately, have the weakness of making exasperating mistakes, and the best that can be done is to keep them to the absolute minimum. The application of the simple arithmetical checks suggested in the following chapters will eliminate many of these costly blunders. The best structural designer is not necessarily the one who makes the fewest mistakes initially, but probably is the one who discovers the largest percentage of his or her mistakes and corrects them.

Oxford Valley Mall, Langhorne, Pennsylvania. (Courtesy Bethlehem Steel Corporation.)

1.9 IMPACT OF COMPUTERS ON STRUCTURAL ANALYSIS

The availability of personal computers has drastically changed the way in which structures are analyzed and designed. In nearly every engineering school and office, computers are used to address structural problems. Though in the past computers were used much more for analysis than for design, the situation is rapidly changing as more and more design software is developed and sold commercially.

Though computers do increase productivity by expediting the work and reducing mathematical errors, they undoubtedly tend to reduce the analyst's "feel" for structures because less thought is given to behavior when an analytical model is developed and used. This can be a serious problem, particularly for young engineers with little previous experience. Unless engineers develop this "feel," computer usage may occasionally result in large mistakes. That is, completely unrealistic answers may be obtained and yet not be recognized as such.

It is interesting to note that, up to the present, the faculties at most engineering schools think that the best way to teach structural analysis is on the chalkboard, supplemented with some computer examples. Thus, in various places throughout this book students will read comments advising them of areas in which computers can be particularly helpful. Further, at the back of this book is a diskette containing a structural analysis program called SABLE (*S*tructural *A*nalysis and *B*ehavior for *L*earning *E*ngineering). This software is frequently used to solve example problems throughout the text. Note that it was prepared for IBM and IBM-compatible computers.

Chapter 2

Structural Loads

2.1 INTRODUCTION

This textbook discusses a large number of structures (beams, frames, trusses, etc.) that have all sorts of loads applied to them. In fact, the reader may very well wonder "Where in the world did they get these loads?" That very important and logical question is answered in this chapter and the next one.

Today's structural engineers probably use some kind of computer package in their work. Although the typical package will enable them to quickly analyze and design structures once the loads are established, it will provide little help in selecting the loads. Perhaps the most critical task, and one of the most difficult faced by structural engineers, is the accurate estimation of the magnitude and character of the loads that may be applied to a structure during its lifetime. No loads that may reasonably be expected to occur can be overlooked.

In this chapter various types of loads are introduced and specifications are presented with which the loads' individual magnitudes may be estimated. Our objective is to be able to answer questions such as the following: How heavy could the snow load be on a structure in Minneapolis? What maximum wind force might be expected on a hotel in Miami? How much rain load is probable for a flat roof in Houston?

The methods used for estimating loads are constantly being refined and may involve some very complicated formulas. The student should not be concerned about committing such expressions to memory, however. Rather, he or she should learn the different types of loads that may be applied to a particular type of structure and where information on estimating them is available.

The world's largest radio telescope, Green Bank, West Virginia. (Courtesy of Lincoln Electric Company.)

2.2 SPECIFICATIONS AND BUILDING CODES

The design of most structures is controlled by specifications. Even if not so controlled, however, the designer probably will refer to them as a guide. No matter how many structures a person has designed, it is impossible for him or her to have encountered every situation. By referring to specifications, he or she is making use of the best available material on the subject. Engineering specifications that are developed by various organizations present the best opinion of those organizations as to what represents good practice.

Municipal and state governments concerned with the safety of the public have established building codes with which they control the construction of various structures within their jurisdiction. These codes, which actually are laws or ordinances, specify design loads, design stresses, construction types, material quality, and other factors. They vary considerably from city to city, a fact that causes some confusion among architects and engineers.

The determination of the magnitude of loads is only a part of the structural loading problem. The structural engineer must be able to determine which loads can reasonably be expected to act concurrently on a structure. For example, would a highway bridge completely covered with ice and snow be simultaneously subjected to fast moving lines of heavily loaded trucks in every lane and a 90-mile-per-hour lateral wind, or is some lesser combination of these loads more reasonable? The topic of concurrent loads is addressed initially in Chapter 3 along with a related problem, that is, the placement of loads on a structure so the most severe conditions occur.

Quite a few organizations publish recommended practices for regional or national use. Their specifications are not legally enforceable, however, unless they are embodied in the local building code or made part of a particular contract. Among these organizations are ANSI (American National Standards Institute), ASCE (American Society of Civil Engineers), AASHTO (American Association of State Highway and Transportation Officials), and AREA (American Railway Engineering Association).

The following specifications published by the above mentioned organizations frequently are used to estimate the maximum loads to which buildings, bridges, and some other structures may be subjected during their estimated lifetimes:

Minimum Design Loads for Buildings and Other Structures, published by ASCE and referred to herein as ASCE 7-95.[1]

AASHTO LRFD Bridge Design Specifications, published by AASHTO.[2]

Specifications for Steel Railway Bridges, published by AREA.[3]

Readers should note that logical and clearly written codes are quite helpful to designers. *Furthermore, there are far fewer structural failures in areas that have good codes that are strictly enforced.*

Some people feel that specifications prevent engineers from thinking for themselves—and there may be some basis for the criticism. They say that the ancient engineers who built the great pyramids, the Parthenon, and the great Roman bridges were controlled by few specifications, which certainly is true. On the other hand, it should be said that only a few score of these great projects were built over many

---

[1] New York: ASCE, 1995.

[2] Washington, D.C.: AASHTO, 1994.

[3] Washington, D.C.: AREA, 1994.

centuries, and they were apparently built without regard to cost of material, labor, or human life. They probably were built by intuition and by certain rules of thumb developed by observing the minimum size or strength of members that would fail only under given conditions. Their probably numerous failures are not recorded in history; only their successes endured.

Today, however, there are hundreds of projects being constructed at any one time in the United States that rival in importance and magnitude the famous structures of the past. It appears that if all engineers in our country were allowed to design projects such as these without restrictions there would be many disastrous failures. *The important thing to remember about specifications, therefore, is that they are not written for the purpose of restricting engineers but for the purpose of protecting the public.*

Yet, no matter how many specifications are written, it is impossible for them to cover every possible situation. As a result, no matter which code or specification is or is not being used, the ultimate responsibility for the design of a safe structure lies with the structural engineer.

Specifications will on many occasions clearly prescribe the loads for which structures are to be designed. Despite the availability of this information, however, the designer's ingenuity and knowledge of the situation often are needed to predict the loads a particular structure will have to support in years to come. For example, over the past several decades insufficient estimates of future traffic loads by bridge designers have resulted in a great amount of replacement with wider and stronger structures.

This chapter provides an introduction to the basic types of loads with which the structural engineer needs to be familiar. Its purpose is to help the reader develop an understanding of structural loads and their behavior and to provide a foundation for estimating their magnitudes. It should not be regarded, however, as an absolutely complete essay on the subject of the loads that might be applied to any and every type of structure the engineer may design.

As building loads are those most commonly faced by designers, they are the loads most frequently referred to in this text. The basic document currently being used by a large number of structural designers for estimating the loads to be applied to buildings is the ASCE specification. It is referred to constantly in this chapter and at frequent intervals throughout the book. This specification was originally prepared and published by ANSI and referred to as the ANSI 58.1 Standard. As such, it went through several revisions. In 1988 it was taken over by ASCE and called ANSI/ASCE 7. Much information in this book is based on the 1995 edition of this specification, which now is called ASCE 7-95.

When studying the information provided in this chapter, or when reviewing any standard providing design loads, the reader is cautioned that minimum design load standards are presented. An engineer should *always* view minimum design standards with some skepticism. Although the design standards are excellent and well prepared for most situations, there may be a structural geometry or facility use for which the specified design loads are inadequate. A structural engineer should evaluate the minimum specified design loads to determine whether they are adequate for the structural system being designed.

2.3 TYPES OF STRUCTURAL LOADS

Structural loads usually are categorized by means of their character and duration. Loads commonly applied to buildings are categorized as follows:

Dead loads: Dead loads are those loads of constant magnitude that remain in one position. They include the weight of the structure under consideration, as well as any fixtures that are permanently attached to it.

Live loads: Live loads are those loads that can change in magnitude and position. They include occupancy loads, warehouse materials, construction loads, overhead service cranes, and equipment operating loads. In general, they are induced by gravity.

Environmental loads: Environmental loads are those loads caused by the environment in which the structure is located. For buildings, they are caused by rain, snow, wind, and earthquake. Strictly speaking, these are also live loads, but they are the result of the environment in which the structure is located.

South Fork Feather River Bridge in northern California, being erected by use of a 1626-ft-long cableway strung from 210-ft-high masts anchored on each side of the canyon. (Courtesy Bethlehem Steel Corporation.)

With the exception of the earthquake loads, each of these types of loads is discussed in considerable detail in the sections that follow. Earthquake or seismic loads are only briefly introduced herein, as they are a subject requiring more advanced study.

2.4 DEAD LOADS

The dead loads that must be supported by a particular structure include all of the loads that are permanently attached to that structure. Not only the weight of the structural frame, but also the weight of the walls, roofs, ceilings, stairways, and so on must be included.

Permanently attached equipment, described as "fixed service equipment" in ASCE 7-95, also is included in the dead load applied to the building. This equipment will include ventilating and air-conditioning systems, plumbing fixtures, electrical cables and support racks, and so on. Depending upon the use of the structure, kitchen equipment such as ovens and dishwashers, laundry equipment such as washers and dryers, or suspended walkways could be included in the dead load.

The dead loads acting on the structure are determined by reviewing the architectural, mechanical, and electrical drawings for the building. From these drawings, the structural engineer can estimate the size of the frame necessary for the building layout and the equipment and finish details indicated. Standard handbooks and manufacturers' specifications can be used to determine the weight of floor and ceiling finishes, equipment, and fixtures. The approximate weights of some common materials used for walls, floors, and ceilings are given in Table 2.1.

The estimates of building or other structure dead loads may have to be revised one or more times during the analysis/design process. Before a structure can be designed, it must be analyzed. Among the loads used for the first analysis are the estimates of the weights of the components of the frame. The frame is designed using the results of that analysis. Its weight can then be computed and compared with the initial weight estimate used for analysis. If there is a significant difference, the structure should be reanalyzed using a revised frame weight estimate. This cycle is repeated as often as is necessary.

TABLE 2.1 WEIGHTS OF SOME COMMON BUILDING MATERIALS

Reinforced concrete	150 pcf	2 × 12 @ 16-in. double wood floor	7 psf
Acoustical ceiling tile	1 psf	Linoleum or asphalt tile	1 psf
Suspended ceiling	2 psf	Hardwood flooring (7/8-in.)	4 psf
Plaster on concrete	5 psf	1-in cement on stone-concrete fill	32 psf
Asphalt shingles	2 psf	Movable steel partitions	4 psf
3-ply ready roofing	1 psf	Wood studs w/1/2-in. gypsum	8 psf
Mechanical duct allowance	4 psf	Clay brick wythes-4 in.	39 psf

2.5 LIVE LOADS

Live loads are those loads that can vary in magnitude and position with time. They are caused by the building being occupied, used, and maintained. Virtually all of the loads applied to a building that are not dead loads are live loads. Environmental loads, which actually are live loads by our usual definition, are listed separately by ASCE 7-95. Although environmental loads do vary with time, they are not all caused by gravity or operating conditions, as is typical with other live loads.

Some typical live loads that act on building structures are presented in Table 2.2. These loads, which are taken from Table 4-1 in ASCE 7-95, act downward and are distributed uniformly over an entire floor or roof.

TABLE 2.2   SOME TYPICAL UNIFORMLY DISTRIBUTED LIVE LOADS

Lobbies of assembly areas	100 psf	Classrooms in schools	40 psf
Dance hall and ballrooms	100 psf	Upper-floor corridors in schools	80 psf
Library reading rooms	60 psf	Stairs and exitways	100 psf
Library stack rooms	150 psf	Heavy storage warehouse	250 psf
Light manufacturing	125 psf	Retail stores—first floor	100 psf
Offices in office buildings	50 psf	Retail stores—upper floors	75 psf
Residential dwelling areas	40 psf	Walkways and elevated platforms	60 psf

Many building specifications provide concentrated loads to be considered in design. This is the situation in Section 4.3 of the ASCE 7-95, in which it states that the designer must consider the effect of certain concentrated loads as an alternative to the previously discussed uniform loads. It is, of course, their intent that the loading used for design be the one that causes the most severe stresses.

Table 4-1 of ASCE 7-95 presents the minimum concentrated loads to be considered. Some typical values from this table are shown in Table 2.3 of this chapter. The appropriate loads are to be positioned on a particular floor or roof so as to cause the greatest stresses (a topic to be discussed in detail in Chapters 3 and 9). Unless otherwise specified, each of the concentrated loads is assumed to be uniformly distributed over a 2.5 ft $\times$ 2.5 ft square area (6.25 ft^2).

TABLE 2.3   TYPICAL CONCENTRATED LIVE LOADS

Elevator machine room grating on 4-in^2	300 lb
Office floors	2000 lb
Center of stair tread on 4-in^2.	300 lb
Sidewalks	8000 lb
Accessible ceilings	200 lb

When estimating the magnitude of the live loads that may be applied to a particular structure during its lifetime, it is necessary to consider the future utilization of that structure. For example, modern office buildings often are constructed with large open spaces that may later be divided into offices and other work areas by means of partitions. These partitions may be moved, removed, or added to during the life of the structure. Building codes typically require that partition loads be considered if the floor live load is less than 80 psf, even if partitions are not shown on the drawings. (A rather common practice of structural designers is to increase the otherwise specified floor design live loads of office buildings by 15 to 20 psf to estimate the effect of the impossible-to-predict partition configurations of the future.)

The method used to establish the magnitude of the ASCE 7-95 live loads is a rather complicated process that is described in the Commentary of that specification. Among the factors contributing to a particular specified value are the mean expected load, its variation over time, the magnitude of short duration transient loads, and the reference time period—typically taken to be 50 years.

To convince the reader that the specification loads are reasonable, a brief examination of one of the specified values is considered. The example used here is the 100 psf live load specified by ASCE 7-95 for the lobbies of theaters and for assembly areas. It is desired to determine if such a load is reasonable for a crowd of people standing quite close together. It is assumed that the area in question is full of average adult males each weighing 165 pounds and each occupying an area 20 inches by 12 inches or 1.67 sq ft. The average load applied equals $165/1.67 = 98.8$ psf. As such, the 100 psf specified live load seems reasonable. It actually is on the conservative side, as it would be rather difficult to have men standing that close together over either a small or a large floor area.

2.6 LIVE LOAD IMPACT FACTORS

Impact loads are caused by the vibration and sudden stopping or dropping of moving or movable loads. It is obvious that a crate dropped on the floor of a warehouse or a truck bouncing on uneven pavement of a bridge causes greater forces than would occur if the loads were applied gently and gradually. Impact loads are equal to the difference between the magnitude of the loads actually caused and the magnitude of the loads had they been dead loads. In other words, impact loads result from the dynamic effects of a load as it is applied to a structure. For static loads, these effects are short lived and do not necessitate a dynamic structural analysis. They do, however, cause an increase in stress in the structure that must be considered. Impact loads usually are specified as percentage increases of the basic live load. Table 2.4 shows the impact percentages for buildings given in Section 4.7 of ASCE 7-95.

TABLE 2.4 LIVE LOAD IMPACT FACTORS

Elevators	100%
Motor-driven machinery	20%
Reciprocating machinery	50%
Hangers for floors and balconies	33%

2.7 LIVE LOADS ON ROOFS

The live loads that act on roofs are handled in most building codes in a little different manner than are the other building live loads. The pitch of the roof (the ratio of the rise of the roof to its span) affects the amount of load that realistically can be placed upon the roof. As the pitch increases, the amount of load that can be placed on the roof before it begins to slide off decreases. Furthermore, as the area of the roof that contributes to the load acting on a supporting component increases, it is less likely that the entire area will be loaded at any one time.

The largest roof live loads usually are caused by repair and maintenance operations that probably do not occur simultaneously over the entire roof. This is not true of the environmental snow and rain loads, however, which are considered in Sections 2.8 and 2.10 of this chapter.

In the equations presented in this section, the term *tributary area* is used. This term, which is discussed in detail in Chapter 3, is defined as the loaded area of a structure that directly contributes to the load applied to a particular member. When a building is being analyzed, it is customary for the analyst to assume that the load supported by a member is the load that is applied to its tributary area, which is assumed to extend from the member in question halfway to the adjacent members in each direction. The tributary area for a column is shown in Figure 2.1.

The basic minimum roof live load to be used in design is 20 psf, as given in Section 4.9 of ASCE 7-95, but it may be reduced depending on the size of the tributary area and the rise of the roof. The actual value to be used is determined with the expression:

$$L_r = 20R_1R_2 \geq 12 \qquad (2.1)$$

The term L_r represents the roof live load in psf of horizontal projection, while R_1 and R_2 are reduction factors. R_1 is used to account for the size of the tributary area A_t while R_2 is included to estimate the effect of the rise of the roof. The greater the tributary area or the greater the rise of the roof, the smaller will be the applicable

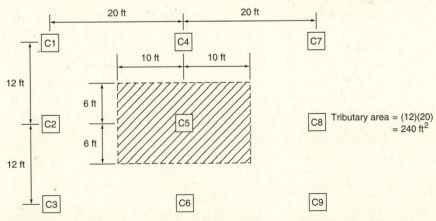

Figure 2.1   Tributary load area for Column C5.

reduction factor and the smaller the roof live load. The maximum roof live load is 20 psf and the minimum is 12 psf. Expressions for computing R_1 and R_2 follow.

$$R_1 = \begin{cases} 1 & \text{for } A_t \leq 200 \text{ sq ft} \\ 1.2 - 0.001A_t & \text{for } 200 \text{ sq ft} < A_t < 600 \text{ sq ft} \\ 0.6 & \text{for } A_t \geq 600 \text{ sq ft} \end{cases} \quad (2.2)$$

$$R_2 = \begin{cases} 1 & \text{for } F \leq 4 \\ 1.2 - 0.05F & \text{for } 4 < F < 12 \\ 0.6 & \text{for } F \geq 12 \end{cases} \quad (2.3)$$

The term F represents the number of inches of rise of the roof per foot of span. If the roof is a dome or an arch, the term F is the rise-to-span ratio of the dome or arch multiplied by 32.

2.8 RAIN LOADS

It has been claimed that almost 50 percent of the lawsuits faced by building designers are concerned with roofing systems.[4] *Ponding,* a problem with many flat roofs, is one of the common subjects of such litigation. If water accumulates more rapidly on a roof than it runs off, ponding results because the increased load causes the roof to deflect into a dish shape that can hold more water, which causes greater deflections, and so on. This process continues until equilibrium is reached or until collapse occurs. Through proper selection of loads and good design providing adequate roof stiffness, we try to avoid the latter situation. Many useful references on the subject of ponding are available.[5, 6, 7, 8]

During a rainstorm, water accumulates on a roof for two reasons. First, when rain falls, time is required for the rain to run off the roof. As such, some water will accumulate. Second, roof drains may not be even with the roof surface and/or may become clogged. Generally, roofs that have slopes of 0.25 inches per foot or greater are not susceptible to ponding, unless the roof drains become clogged and allow deep ponds to form.

In addition to ponding, another problem may occur for very large flat roofs (with perhaps an acre or more of surface area). During heavy rainstorms strong winds frequently occur. If there is a great deal of water on the roof, a strong wind may very well push a large quantity of it toward one end. The result can be a dangerous water

[4] Gary Van Ryzin, "Roof Design: Avoid Ponding by Sloping to Drain," *Civil Engineering* (New York: ASCE, January 1980), 77–81.

[5] F. J. Marino, "Ponding of Two-Way Roof System," *Engineering Journal,* AISC, 3rd quarter, no. 3 (1966), 93–100.

[6] L. B. Burgett, "Fast Check for Ponding," *Engineering Journal,* AISC, 10, no. 1 (1st quarter, 1973), 26–28.

[7] J. Chinn, "Failure of Simply-Supported Flat Roofs by Ponding of Rain," *Engineering Journal,* AISC, no. 2, 2nd quarter (1965), 38–41.

[8] J. L. Ruddy, "Ponding of Concrete Deck Floors," *Engineering Journal,* AISC, 23, no. 2 (3rd quarter, 1986), 107–115.

depth as regards the load in psf on that end of the roof. For such situations *scuppers* are sometimes used. Scuppers are large holes or tubes in walls or parapets that enable water above a certain depth to quickly drain off the roof.

Generally, two different drainage systems are provided for roofs. These normally are referred to as the primary and secondary drains.

Usually the primary system will collect the rainwater through surface drains on the roof and direct it to storm sewers. The secondary system consists of scuppers or other openings or pipes through the walls that permit the rainwater to run over the sides of the building. The inlets of the secondary drains are normally located at elevations above the inlets to the primary drains.

The secondary drainage system is used to provide adequate drainage of the roof in the event that the primary system becomes clogged or disabled in some manner. The rainwater design load then is based on the amount of water that can accumulate before the secondary drainage system becomes effective.

Determination of the water that can accumulate on a roof before runoff during a rainstorm will depend upon local conditions and the elevation of secondary drains. Section 8 of ASCE 7-95 specifies that the rain load (in psf) on an undeflected roof can be computed from:

$$R = 5.2(d_s + d_h) \qquad (2.4)$$

In this equation d_s is the depth of water (in inches) on the undeflected roof up to the inlet of the secondary drainage system when the primary drainage system is blocked. This is the static head, which can be determined from the drawings of the roof system. The term d_h is the additional depth of water on the undeflected roof above the inlet of the secondary drainage system at its design flow. This is the hydraulic head. It is dependent upon the capacity of the drains installed and the rate at which rain falls.

TABLE 2.5   FLOW RATE, Q, IN GALLONS PER MINUTE OF VARIOUS DRAINAGE SYSTEMS AT VARIOUS HYDRAULIC HEADS, d_h, IN INCHES

Drainage system	Hydraulic head d_h (in.)									
	1	2	2.5	3	3.5	4	4.5	5	7	8
4-in.-diameter drain	80	170	180							
6-in.-diameter drain	100	190	270	380	540					
8-in.-diameter drain	125	230	340	560	850	1100	1170			
6-in.-wide, channel scupper**	18	50	*	90	*	140	*	194	321	393
24-in.-wide, channel scupper	72	200	*	360	*	560	*	776	1,284	1,572
6-in.-wide, 4-in.-high, closed scupper**	18	50	*	90	*	140	*	177	231	253
24-in.-wide, 4-in.-high, closed scupper	72	200	*	360	*	560	*	708	924	1,012
6-in.-wide, 6-in.-high, closed scupper	18	50	*	90	*	140	*	194	303	343
24-in.-wide, 6-in.-high, closed scupper	72	200	*	360	*	560	*	776	1,212	1,372

*Interpolation is appropriate, including between widths of each scupper.

**Channel scuppers are open-topped (i.e., three-sided). Closed scuppers are four-sided.

From Section 8.3 of the ASCE 7-95 Commentary, the flow rate (in gallons per minute) that a particular drain must accommodate can be computed from:

$$Q = 0.0104Ai \qquad (2.5)$$

The term A is the area of the roof (in square feet) that is served by a particular drain, and i is the rainfall intensity (in inches per hour). The rainfall intensity is specified by the code that has jurisdiction in a particular area. After the flow quantity is determined, the hydraulic head can be determined from Table 2.5 (ASCE 7-95, Table C8-1) for the type of drainage system being used. If the secondary drainage system is simply run-off over the edge of the roof, the hydraulic head will equal zero.

Example 2.1 illustrates the calculation of the design rainwater load for a roof with scuppers using the ASCE 7-95 specification.

EXAMPLE 2.1 ———————————————————————————

A roof measuring 240 feet by 160 feet has 6-in. wide channel-shaped scuppers serving as secondary drains. The scuppers are 4 inches above the roof surface and are spaced 20 feet apart along the two long sides of the building. The design rainfall for this location is 3 inches per hour. What is the design roof rain load?

Solution. The area served by each scupper is: $A = 20 \times 80 = 1600$ ft². The runoff quantity for each scupper is: $Q = 0.0104(1600)(3) = 49.92$ gal/min. Referring to Table 2.5, we observe that the hydraulic head at this flow rate for the scupper used is 2 inches. The design roof load, then, is: $R = 5.2(4 + 2) = 31.2$ psf. ■

2.9 WIND LOADS

A survey of engineering literature for the past 150 years reveals many references to structural failures caused by wind. Perhaps the most infamous of these have been bridge failures such as those of the Tay Bridge in Scotland in 1879 (which caused the deaths of 75 persons) and the Tacoma Narrows Bridge (Tacoma, Washington) in 1940. But there have also been some disastrous building failures due to wind during the same period, such as that of the Union Carbide Building in Toronto in 1958. It is important to realize that a large percentage of building failures due to wind have occurred during their erection.[9]

A great deal of research has been conducted in recent years on the subject of wind loads. Nevertheless, a great deal more study is needed as the estimation of wind forces can by no means be classified as an exact science. The magnitude and duration of wind loads vary with geographical locations, the height of the structure above ground, the type of terrain around the structure, the proximity of other buildings, and the character of the wind itself.

---

[9] Wind Forces on Structures, Task Committee on Wind Forces. Committee on Loads and Stresses, Structural Division, ASCE, Final Report, *Transactions ASCE* 126, Part II (1961): 1124–1125.

Fortunately, structural engineers usually do not need to determine the exact forces at each point in a building as a function of time. The approach taken by most building codes for regularly shaped buildings is to determine a set of distributed loads that encompass or envelop the actual loads expected. These distributed loads are applied to the building in a series of design wind load cases.

The discussion about wind forces that follows is intended to be only an introduction to the subject. Furthermore, it addresses only the wind forces to be applied to the main wind-force resisting system of enclosed buildings. There are additional design criteria for wind forces applied to individual components of the building and to the *cladding* (the exterior covering of the structural parts of the building).

Section 6 of the ASCE 7-95 specification provides a rather lengthy procedure for estimating the wind pressures applied to buildings.[10] The procedure involves several factors with which we attempt to account for the terrain around the building, the importance of the building regarding human life and welfare, and of course the wind speed at the building site. Though use of the equations is rather complex, the work can be greatly simplified with the tables presented in the specification. The reader is cautioned, however, that the tables presented are for buildings of regular shapes. If a building having an irregular or unusual geometry is being considered, wind tunnel studies may be necessary. Guidelines for conducting such studies are provided in Section 6.4.3 of the ASCE 7-95 specification.

The basic form of the equation presented in the specification is:

$$p = qCG \qquad (2.6)$$

In this equation p is the estimated wind load (in psf) acting on the structure. This wind load will vary with height above the ground and with the location on the structure. The quantity q is the reference velocity pressure. It varies with height and with exposure to the wind. The aerodynamic shape factor C is dependent upon the shape and orientation of the building with respect to the direction from which the wind is blowing. Lastly, the gust response factor G is dependent upon the nature of the wind and the location of the building.

The writers of ASCE 7-95 recognized that pressure can be acting on both sides of a wall at the same time. On one side of the wall there is a force caused directly by the wind, while on the other side there is a force caused by wind-induced internal pressure in the building. The writers of the specification also recognized that the force distribution around low-rise buildings is different than around other buildings. Low-rise buildings are those that have (1) mean roof heights (h) equal to or less than 60 ft and (2) mean roof heights that are not larger than their least horizontal dimensions. Because of these considerations, two equations are presented in the specification for calculating the design wind pressure. For low-rise buildings the design wind pressure is calculated from:

$$p = q_h(GC_{pf}) - q_h(GC_{pi}) \qquad (2.7)$$

[10] K. C. Mehta, R. D. Marshall, and D. C. Perry, *Guide to the Use of Wind Load Provisions of ASCS 7-88* (New York: American Society of Civil Engineers: 1991).

For all other enclosed buildings the design wind pressure is computed from:

$$p = qGC_p - q_h(GC_{pi}) \tag{2.8}$$

These equations are a modification of the basic pressure equation. They explicitly include the effects of the internal pressure and the different pressure distribution around low-rise buildings. We will examine the terms in each of these equations separately. Before doing so, though, a few comments concerning the equations are appropriate. First, the terms to the right of the equal sign in each equation represent the external and the internal pressure acting on the wall. The first term is the external pressure and the second term is the internal pressure. Secondly, the internal pressure term in each equation is the same.

Regardless of the type of building being considered, the velocity pressure q needs to be computed. Following the specification, the basic expression for calculating the velocity pressure (in psf) at an elevation z above the ground is:

$$q_z = 0.00256K_z K_{zt} V^2 I \tag{2.9}$$

Jacobs Field, the home of the Cleveland Indians. (Courtesy of The Lincoln Electric Company.)

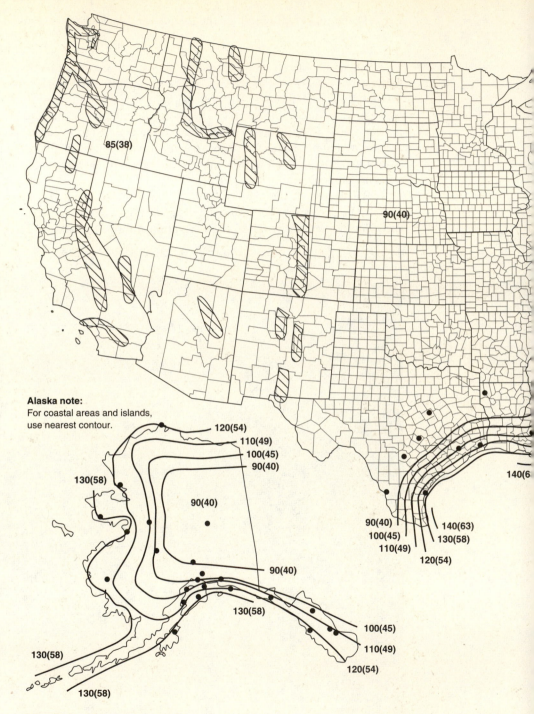

Figure 2.2 Basic wind speed.

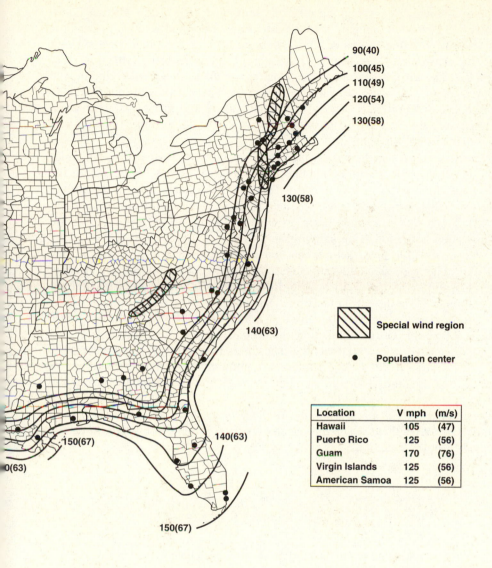

90(40)
100(45)
110(49)
120(54)
130(58)

130(58)

140(63)

150(67)

140(63)

0(63)

150(67)

	Special wind region
•	Population center

Location	V mph	(m/s)
Hawaii	105	(47)
Puerto Rico	125	(56)
Guam	170	(76)
Virgin Islands	125	(56)
American Samoa	125	(56)

Notes: 1. Values are 3-second gust speeds in miles per hour (m/s) at 33 ft (10m) above ground for Exposure C category and are associated with an annual probability of 0.02.
2. Linear interpolation between wind speed contours is permitted.
3. Islands and coastal areas shall use wind speed contour of coastal area.
4. Mountainous terrain, gorges, ocean promontories, and special wind regions shall be examined for unusual wind conditions.

The elevation at which the velocity pressure is computed depends on whether Equation 2.7 or 2.8 is used to compute the design wind pressure and whether external or internal pressure is being computed. If the pressure is being computed for a low-rise building, the elevation used is the mean roof height h. This yields the value q_h used in Equation 2.7. This also is the value of the velocity pressure used when determining the design pressure on side walls, leeward walls, and the internal pressure for buildings other than low-rise buildings. The actual height above the ground is used when computing the design pressure on the windward walls of buildings other than low-rise buildings. This yields the value q_z.

In Equation 2.9, V is the basic wind speed (in MPH) that is specified in codes for a particular location or can be obtained from figures such as Figure 2.2 on the previous page (Figure 6-1 in ASCE 7-95). The term I in Equation 2.9 is the importance factor. Values for this term are provided in Table 2.6 (taken from ASCE 7-95 Table 6.2). These categories, which are presented in Table 1.1 of ASCE 7–95, are briefly discussed in the following paragraph.

Category I is used for structures that represent a low hazard to humans if they were to fail. This type of building includes most agricultural buildings and some temporary buildings. Category III buildings are those which represent a substantial hazard to human life in the event of a failure. They include power-generating stations, schools and day care facilities with capacity greater than 250, and buildings where more than 300 people can congregate in one area. Category IV is for buildings that have been designated as essential facilities (hospitals having surgery and emergency treatment facilities, fire and police stations, communication centers required for emergency response, and so forth). All other buildings and structures are included in Category II.

The term K_z in Equation 2.9 is the *velocity pressure exposure coefficient*. It corresponds to the height at which the velocity pressure is being computed. K_z can be determined from the equations

$$K_z = \begin{cases} 2.58\left(\dfrac{15}{z_g}\right)^{2/\alpha} & \text{for } z < 15 \text{ ft} \\[2em] 2.58\left(\dfrac{z}{z_g}\right)^{2/\alpha} & \text{for } 15 \text{ ft} \le z \le z_g \end{cases} \qquad (2.10)$$

or from the data in Table 2.8 for four different exposure categories.

TABLE 2.6   IMPORTANCE FACTORS FOR WIND LOADS

Category	Importance factor I
I	0.87
II	1.00
III	1.15
IV	1.15

TABLE 2.7 EXPOSURE CATEGORY
CONSTANTS

Category	α	z_g (ft)
A	5.0	1500
B	7.0	1200
C	9.5	900
D	11.5	700

Exposure Category A is a large city center where at least 50% of the buildings are taller than 70 feet. Exposure B is for urban and suburban areas, wooded areas, and other areas with closely spaced obstructions having the size of single family dwellings or larger. Category C is open terrain with scattered obstructions that are less than 30 feet high. Category D is used for flat, unobstructed areas exposed to wind flowing over large bodies of water.

If the equations are used to compute K_z, two other terms need to be defined. These are the gradient height z_g and the term α used in the power-law representation of the wind profile. They are given in Table 2.7, which is Table 2.6 in the ASCE 7-95 Commentary.

The last term in the velocity pressure equation (2.9) that needs to be defined is K_{zt}, the topographic factor. This factor applies to buildings in Exposure Categories B,

TABLE 2.8 VELOCITY PRESSURE
COEFFICIENTS, K_z

H (ft)	Exposure category			
	A	B	C	D
0–15	0.32	0.57	0.85	1.03
20	0.36	0.62	0.90	1.08
25	0.39	0.66	0.94	1.12
30	0.42	0.70	0.98	1.16
40	0.47	0.76	1.04	1.22
50	0.52	0.81	1.09	1.27
60	0.55	0.85	1.13	1.31
70	0.59	0.89	1.17	1.34
80	0.62	0.93	1.21	1.38
90	0.65	0.96	1.24	1.40
100	0.68	0.99	1.26	1.43

C, and D. It is intended to account for the effects of wind on buildings that are located on the upper half of isolated hills or escarpments (sudden rises in terrain). When wind flows over these isolated topographic features, there is an increase in its velocity usually called speed-up. This increase can have a negative impact on the performance of the building. The topography of the area in question is considered isolated when the upwind terrain is free of such features for a distance of 1 mile or 50 times the height of the hill or escarpment, whichever is smaller.

The topographic factor is computed from:

$$K_{zt} = (1 + K_1 K_2 K_3)^2 \qquad (2.11)$$

The factors K_1, K_2, and K_3 in Equation 2.11 are defined in Figure 2.3 (ASCE 7-95, Figure 6.2). They account for the shape of the topographic features and the maximum

K_1 MULTIPLIER

H/L_h	2-D ridge	2-D escarpment	3-D axisym. hill
0.10	0.14	0.09	0.11
0.15	0.22	0.13	0.16
0.20	0.29	0.17	0.21
0.25	0.36	0.21	0.26
0.30	0.43	0.26	0.32
0.35	0.51	0.30	0.37
0.40	0.58	0.34	0.42
0.45	0.65	0.38	0.47
0.50	0.72	0.43	0.53

K_2 MULTIPLIER

x/L_h	2-D escarpment downwind of crest	All other cases
0.00	1.00	1.00
0.50	0.88	0.67
1.00	0.75	0.33
1.50	0.63	0.00
2.00	0.50	0.00
2.50	0.38	0.00
3.00	0.25	0.00
3.50	0.13	0.00
4.00	0.00	0.00

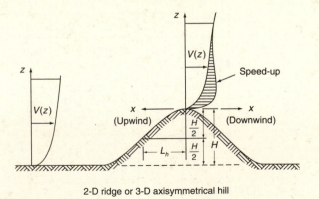

2-D ridge or 3-D axisymmetrical hill

Figure 2.3 Multipliers for obtaining topographic factor, K_{zt}.

K_3 MULTIPLIER

z/L_h	2-D ridge	2-D escarpment	3-D axisym. hill
0.00	1.00	1.00	1.00
0.10	0.74	0.78	0.67
0.20	0.55	0.61	0.45
0.30	0.41	0.47	0.30
0.40	0.30	0.37	0.20
0.50	0.22	0.29	0.14
0.60	0.17	0.22	0.09
0.70	0.12	0.17	0.06
0.80	0.09	0.14	0.04
0.90	0.07	0.11	0.03
1.00	0.05	0.08	0.02
1.50	0.01	0.02	0.00
2.00	0.00	0.00	0.00

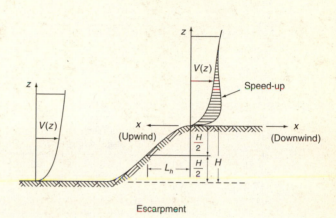

Escarpment

NOTES:

1. For values of H/L_h, x/L_h and z/L_h other than those shown, linear interpolation is permitted.
2. For $H/L_h > 0.5$, assume $H/L_h = 0.5$, and substitute $2H$ for L_h in x/L_h and z/L_h.
3. Multipliers are based on the assumption that wind approaches the hill or escarpment along the direction of maximum slope.
4. Effect of wind speed-up shall not be required to be accounted for when $H/L_h < 0.2$ or when $H < 15$ ft (4.5 m) for Exposure D, or < 30 ft (9 m) for Exposure C, or < 60 ft (18 m) for all other exposures.
5. Notation:
    H: Height of hill or escarpment relative to the upwind terrain, in feet (meters).
    L_h: Distance upwind of crest to where the difference in ground elevation is half the height of hill or escarpment, in feet (meters).
    K_1: Factor to account for shape of topographic feature and maximum speed-up effect.
    K_2: Factor to account for reduction in speed-up with distance upwind or downwind of crest.
    K_3: Factor to account for reduction in speed-up with height above local terrain.
    x: Distance (upwind or downwind) from the crest to the building site, in feet (meters).
    z: Height above local ground level, in feet (meters).

speed-up effect, the reduction in speed-up with distance upwind or downwind of the crest, and the reduction in speed-up with height above the local terrain, respectively. Under certain conditions specified in the figure, the effect of speed-up does not need to be considered. The minimum value of the topographic factor is unity.

We can now return to the remaining terms in the design pressure equations, Equations 2.7 and 2.8. These terms are the gust response factor and the aerodynamic coefficient. In ASCE 7-95 the aerodynamic shape factor is referred to as the external pressure coefficient. When applied to external pressure on buildings, other than low-rise buildings, the gust response factor G and the pressure coefficient C_p are separate terms. For these buildings, the gust factor is equal to 0.8 in Exposure Categories A and B and is equal to 0.85 in Exposure Categories C and D. The pressure coefficient is specified in Figure 2.4 (ASCE 7-95, Figure 6.3). Negative values of the pressure coefficient indicate a negative pressure or a suction. In some cases, two values of the coefficient are specified. In these cases both values, one at a time, must be used to evaluate the wind loads to determine the worst case.

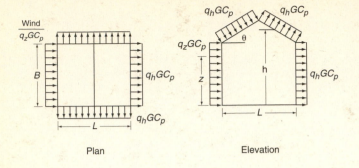

| Wind | $q_z GC_p$ | | | | $q_h GC_p$ ... (Plan) | $q_h GC_p$ ... $q_h GC_p$ ... (Elevation) |

ROOF PRESSURE COEFFICIENTS, C_p, FOR USE WITH q_h

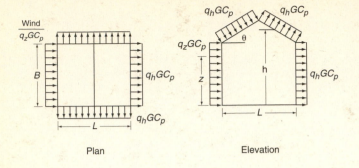

Plan   **Elevation**

WALL PRESSURE COEFFICIENTS, C_p

Surface	L/B	C_p	Use with
Windward wall	All values	0.8	q_z
Leeward wall	0–1	−0.5	q_h
	2	−0.3	
	≥ 4	−0.2	
Side walls	All values	−0.7	q_h

ROOF PRESSURE COEFFICIENTS, C_p, FOR USE WITH q_h

Wind direction	h/L	Windward angle, θ (degrees)								Leeward angle, θ (degrees)		
		10	15	20	25	30	35	45	≥ 60#	10	15	≥ 20
Normal to ridge for $\theta \ge 10°$	≤ 0.25	−0.7	−0.5	−0.3	−0.2	−0.2	0.0*			−0.3	−0.5	−0.6
			0.0*	0.2	0.3	0.3	0.4	0.5	0.01 θ			
	0.5	−0.9	−0.7	−0.4	−0.3	−0.2	−0.2	0.0*		−0.5	−0.5	−0.6
				0.0*	0.2	0.2	0.3	0.4	0.01 θ			
	≥ 1.0	−1.3**	−1.0	−0.7	−0.5	−0.3	−0.2	0.0*		−0.7	−0.6	−0.6
				0.0*	0.2	0.2	0.3	0.01 θ				

Normal to ridge for $\theta < 10°$ and parallel to ridge for all θ	≤ 0.5	Horizontal distance from windward edge	
		0 to h/2	−0.9
		h/2 to h	−0.9
		h to 2h	−0.5
		> 2h	−0.3
	≥ 1.0	0 to h/2	−1.3**
		> h/2	−0.7

* Value is provided for interpolation purposes.

** Value can be reduced linearly with area over which it is applicable as follows:

Area (sq ft)	Reduction Factor
≤ 100 (9.29 sq m)	1.0
250 (23.23 sq m)	0.9
≥ 1000 (92.9 sq m)	0.8

NOTES:

1. Plus and minus signs signify pressures acting toward and away from the surfaces, respectively.
2. Linear interpolation is permitted for values of L/B, h/L and θ other than shown. Interpolation shall be carried out only between values of the same sign. Where no value of the same sign is given, assume 0.0 for interpolation purposes.
3. Where two values of C_p are listed, this indicates that the windward roof slope is subjected to either positive or negative pressures and the roof structure shall be designed for both conditions. Interpolation for intermediate ratios of h/L in this case shall only be carried out between C_p values of like sign.
4. For monoslope roofs, entire roof surface is either a windward or a leeward surface.
5. For flexible buildings use appropriate G_f as determined by rational analysis.
6. Refer to Table 6.5 for arched roofs.
7. Notation:
   B: Horizontal dimension of building, in feet (meter), measured normal to wind direction.
   L: Horizontal dimension of building, in feet (meter), measured parallel to wind direction.
   h: Mean roof height in feet (meters), except that eave height shall be used for $\theta \le 10$ degrees.
   z: Height above ground, in feet (meters).
   G: Gust effect factor.
   q_z, q_h: Velocity pressure, in pounds per square foot (N/m^2), evaluated at respective height.
   θ: Angle of plane of roof from horizontal, in degrees.
For roof slopes greater than 80°, use $C_p = 0.8$.

Figure 2.4 External pressure coefficient, C_p, for loads on main wind-force resisting systems for enclosed or partially enclosed buildings of all heights.

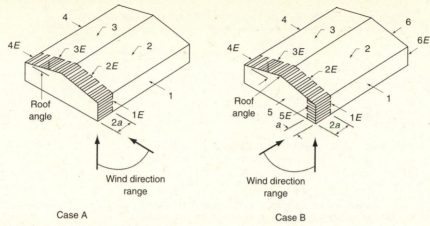

Case A   Case B

CASE A

Roof angle θ (degrees)	Building surface							
	I	2	3	4	IE	2E	3E	4E
0–5	0.40	−0.69	−0.37	−0.29	0.61	−1.07	−0.53	−0.43
20	0.53	−0.69	−0.48	−0.43	0.80	−1.07	−0.69	−0.64
30–45	0.56	0.21	−0.43	−0.37	0.69	0.27	−0.53	−0.48
90	0.56	0.56	−0.37	−0.37	0.69	0.69	−0.48	−0.48

CASE B

Roof angle θ (degrees)	Building surface											
	I	2	3	4	5	6	IE	2E	3E	4E	5E	6E
0–90	−0.45	−0.69	−0.37	−0.45	0.40	−0.29	−0.48	−1.07	−0.53	−0.48	0.61	−0.43

NOTES:

1. Case A and Case B are required as two separate loading conditions to generate the wind actions, including torsion, to be resisted by the main wind-force resisting system.
2. To obtain the critical wind actions, the building shall be rotated in 90° increments so that each corner in turn becomes the windward corner while the loading patterns in the sketches remain fixed. For the design of structural systems providing lateral resistance in the direction parallel to the ridge line, Load Case A shall be based on $\theta = 0°$.
3. Plus and minus signs signify pressures acting toward and away from the surfaces, respectively.
4. For Case A loading the following restrictions apply:
   a. The roof pressure coefficient GC_{pf}, when negative in zone 2, shall be applied in zone 2 for a distance from the edge of roof equal to 0.5 times the horizontal dimension of the building measured perpendicular to the eave line or 2.5h, whichever is less; the remainder of zone 2 extending to the ridge line shall use the pressure coefficient GC_{pf} for zone 3.
   b. Except for moment-resisting frames, the total horizontal shear shall not be less than that determined by neglecting wind forces on roof surfaces.
5. Combinations of external and internal pressures (see Table 6.4) shall be evaluated as required to obtain the most severe loadings.
6. For buildings sited within Exposure B, calculated pressures shall be multiplied by 0.85.
7. For values of θ other than those shown, linear interpolation is permitted.
8. Notation:
   a: 10 percent of least horizontal dimension or 0.4h, whichever is smaller, but not less than either 4% of least horizontal dimension or 3 ft (1 m).
   h: Mean roof height, in feet (meters), except that eave height shall be used for $\theta \leq 10°$.
   θ: Angle of plane of roof from horizontal, in degrees.

Figure 2.5 External pressure coefficients, GC_{pf}, for loads on main wind-force resisting systems for enclosed or partially enclosed low-rise buildings with mean roof height h less than or equal to 60 ft (18 m).

TABLE 2.9 INTERNAL PRESSURE COEFFICIENT FOR BUILDINGS, GC_{pi}

Condition	GC_{pi}
Open buildings	0.00
Partially enclosed buildings	+0.80
	−0.30
Buildings satisfying the following conditions:	+0.80
	−0.30
(1) sited in hurricane-prone regions having a basic wind speed greater than or equal to 110 mph (49 m/s) or in Hawaii, and	
(2) having glazed openings in the lower 60 ft (18 m) which are not designed to resist wind-borne debris or are not specifically protected from wind-borne debris impact	
All buildings except those listed above	+0.18
	−0.18

NOTES:

1. Plus and minus signs signify pressures acting toward and away from the internal surfaces.
2. Values of GC_{pi} shall be used with q_z or q_h as specified in Table 6.1.
3. Two cases shall be considered to determine the critical load requirements for the appropriate condition: a positive value of GC_{pi} applied to all internal surfaces, and a negative value of GC_{pi} applied to all internal surfaces.
4. For buildings with mean roof height $h \leq 60$ ft (18 m) and sited within Exposure B, calculated internal pressures shall be multiplied by 0.85.
5. Hurricane-prone regions include areas vulnerable to hurricanes, such as the U.S. Atlantic and Gulf Coasts, Hawaii, Puerto Rico, Guam, Virgin Islands, and American Samoa.
6. If a building by definition complies with both the "Open" and "Partially Enclosed" definitions, it shall be treated as an "Open" building.

On low-rise buildings, the gust response factor and the pressure coefficient are combined into a single term (GC_{pf}) when computing the external pressure acting. Values of this combined term are presented in Figure 2.5 (ASCE 7-95, Figure 6.4). The gust response factor and the pressure coefficient also are combined when computing the internal pressure for either type of building. The values of this combined coefficient, GC_{pi}, are given in Table 2.9 (ASCE 7-95, Table 6.4).

EXAMPLE 2.2 _____

Determine the wind loads acting on the building shown in Figure 2.6. This building is located along the beach at Galveston on the Texas Gulf Coast. It is oriented so that the long side of the building is parallel with the beach. The primary use of the building will be hotel rooms; there are no areas where more than 300 people can congregate. Determine the design pressure at a point in the main wind-force resisting frame 70 feet above the ground.

Solution. From Figure 2.2, the basic wind speed for Galveston is 120 mph. Because this building is located on the coastline, it is an Exposure Category D structure. As such, the velocity pressure coefficient, K_z, is found to be 1.34 from

Table 2.8. The Importance Category is II because of the building use, so I is equal to 1.0. The topographic factor is taken to be 1.0 because there are no isolated topographic features along the beach on the windward side of the building. The external velocity pressure on the windward side is then

$$q_z = 0.00256(1.34)(1.0)(1.0)(120)^2 = 49.4 \text{ psf}$$

On the windward side the pressure coefficient is 0.8 from Figure 2.3 and the gust factor is 0.85. The external design pressure on the windward side is:

$$p = 49.4(0.8)(0.85) = 33.59 \text{ psf}$$

On the leeward and side walls of the building, the pressure is calculated based on the midheight of the roof or the height of the eaves in the case of low-pitch roofs. For this building the eave height is 100 feet, so the velocity pressure is

$$q_h = 0.00256(1.43)(1.0)(1.0)(120)^2 = 52.72 \text{ psf}$$

On the leeward side the gust factor still is 0.85, but the pressure coefficient is -0.5 from Figure 2.3. The external design pressure on the leeward side is:

$$p = 52.72(0.85)(-0.5) = -22.41 \text{ psf}$$

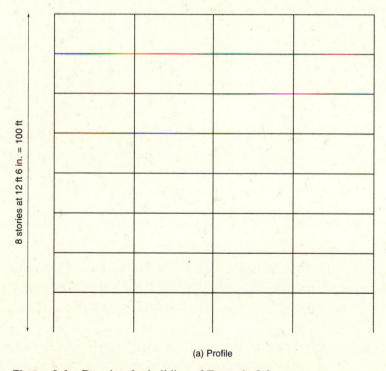

(a) Profile

Figure 2.6 Drawing for building of Example 2.2.

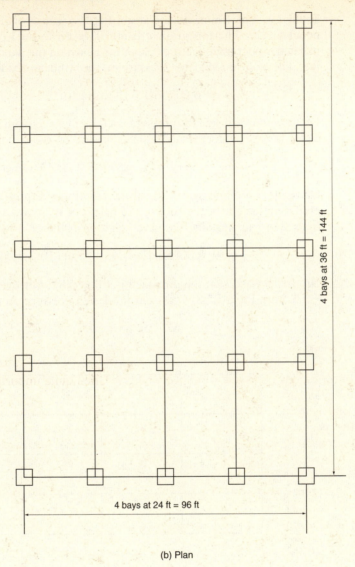

4 bays at 36 ft = 144 ft

4 bays at 24 ft = 96 ft

(b) Plan

Figure 2.6   (*cont.*)

On the side walls, the gust factor is 0.85, but the pressure coefficient is now -0.7 from Figure 2.4. The resulting external design pressure on the side walls is then

$$p = 52.72(0.85)(-0.7) = -31.37 \text{ psf}$$

The negative sign on the design pressure on the leeward and side walls indicates that the pressure is acting outward, away from the wall.

The design pressure on the windward, leeward, and side walls is modified by the internal pressure in the building. The internal pressure is either pressure acting against all walls simultaneously or a vacuum acting on all walls simultaneously. The internal pressure is based on the velocity pressure q_h. The combined gust and shape factor (GC_{pi}) is obtained from Table 2.9. It is found to be 0.8 and −0.3. Both values need to be used in the analysis or system response to determine the most critical case. The internal pressure to be used, then, is:

$$p_i = \begin{cases} 52.72(0.8) = 42.18 \text{ psf} \\ 52.72(-0.3) = -15.82 \text{ psf} \end{cases}$$

The final resulting wall pressures, then, are:

$$p = \begin{cases} \begin{cases} 33.59 - 42.18 = -8.59 \text{ psf} \\ 33.59 + 15.82 = 49.41 \text{ psf} \end{cases} \text{ on the windward wall at 70 ft} \\ \begin{cases} -22.41 - 42.18 = -63.27 \text{ psf} \\ -22.41 + 15.82 = -6.59 \text{ psf} \end{cases} \text{ on the leeward wall} \\ \begin{cases} -31.37 - 42.18 = -73.55 \text{ psf} \\ -31.37 + 15.82 = -15.55 \text{ psf} \end{cases} \text{ on the side walls} \quad \blacksquare \end{cases}$$

2.10 SNOW LOADS

In the colder states snow and ice loads are often quite important. One inch of snow is equivalent to approximately 0.5 psf, but it may be higher at lower elevations where snow is denser. For roof designs, snow loads of from 10 to 40 psf are usually specified, the magnitude depending primarily on the slope of the roof and to a lesser degree on the character of the roof surface. The larger values are used for flat roofs and the smaller ones for sloped roofs. Snow tends to slide off sloped roofs, particularly those with metal or slate surfaces. A load of approximately 10 psf might be used for 45° slopes and a 40-psf load for flat roofs. Studies of snowfall records in areas with severe winters may indicate the occurrence of snow loads much greater than 40 psf, with values as high as 100 psf in northern Maine.

Snow is a variable load that may cover an entire roof or only part of it. There may be drifts against walls or buildup in valleys or between parapets. Snow may slide off one roof onto a lower one. The wind may blow it off one side of a sloping roof or the snow may crust over and remain in position even during very heavy winds.

The snow loads that are applied to a structure are dependent upon many factors, including geographic location, the pitch of the roof, sheltering, and the shape of the roof. The discussion that follows is intended to provide only an introduction to the determination of snow loads on buildings. When estimating these loads the reader is urged to consult ASCE 7-95 for more complete information.

According to Section 7.3 of the aforementioned specification, the basic snow load to be applied to structures in the contiguous United States can be obtained from the expression:

$$p_f = 0.7 C_e C_t I p_g \tag{2.12}$$

This expression is for unobstructed flat roofs with slopes equal to or less than 5° (1 in./ft = 4.76°). In the equation C_e is the exposure index that is intended to account for the snow that can be blown from the roof because of the surrounding locality. It is lowest for highly exposed areas and is highest when there is considerable sheltering. Values of C_e are presented in Table 2.10 (Table 7.2 in ASCE 7-95).

The term C_t in Equation 2.12 is the thermal index. As shown in Table 2.11 (Table 7.3 in ASCE 7-95) it equals 1.0 for heated structures, 1.1 for structures that are minimally heated to keep them from freezing, and 1.2 for unheated structures.

The values of the importance factors I for snow loads are shown in Table 2.12 (Table 7.4 in ASCE 7-95). The categories of building use are the same as those used for the computation of wind loads (I through IV).

TABLE 2.10 EXPOSURE FACTOR, C_e

Terrain category	Exposure of roof*		
	Fully exposed	Partially exposed	Sheltered
A (see ASCE Section 6.5.3)	N/A	1.1	1.3
B (see ASCE Section 6.5.3)	0.9	1.0	1.2
C (see ASCE Section 6.5.3)	0.9	1.0	1.1
D (see ASCE Section 6.5.3)	0.8	0.9	1.0
Above the treeline in windswept mountainous areas.	0.7	0.8	N/A
In Alaska, in areas where trees do not exist within a 2-mile (3 km) radius of the site.	0.7	0.8	N/A

The terrain category and roof exposure condition chosen shall be representative of the anticipated conditions during the life of the structure.

* Definitions

Partially exposed: All roofs except as indicated below.
Fully exposed: Roofs exposed on all sides with no shelter** afforded by terrain, higher structures, or trees. Roofs that contain several large pieces of mechanical equipment or other obstructions are not in this category.
Sheltered: Roofs located tight in among conifers that qualify as obstructions.

** Obstructions within a distance of $10h_o$ provide "shelter," where h_o is the height of the obstruction above the roof level. If the only obstructions are a few deciduous trees that are leafless in winter, the "fully exposed" category shall be used except for terrain category "A." Note that these are heights above the roof. Heights used to establish the Terrain category in Section 6.5.3 are heights above the ground.

TABLE 2.11 THERMAL FACTOR, C_t

Thermal condition*	C_t
All structures except as indicated below	1.0
Structures kept just above freezing and others with cold, ventilated roofs having a thermal resistance (R-value) greater than 25°F · h · sq ft/Btu (4.4 $K \cdot m^2/W$)	1.1
Unheated structures	1.2

* These conditions shall be representative of the anticipated conditions during winters for the life of the structure.

TABLE 2.12 IMPORTANCE FACTOR, *I* (SNOW LOADS)

Category	*I*
I	0.8
II	1.0
III	1.1
IV	1.2

The last term in Equation 2.12, p_g, is the ground snow load in psf. Typical ground snow loads for the United States are shown in Figure 2.7 on pages 46–47. These values are dependent upon the climatic conditions at each site. If data are available that show that local conditions are more severe than the values given in the figure, the local conditions should always be used.

The minimum value of p_f is $p_g I$ in areas where the ground snow load is less than or equal to 20 psf. In other areas the minimum value of p_f is $20I$ psf.

Example 2.3 which follows illustrates the calculation of the design snow load for a building in Chicago according to the ASCE 7-95 specification.

EXAMPLE 2.3 _____

A shopping center is being designed for a location in Chicago. The building will be located in a residential area with minimal obstructions from surrounding buildings and from the terrain. It will contain large department stores and enclosed public areas in which more that 300 people can congregate. The roof will be flat, but to provide for proper drainage it will have a slope equal to 0.5 in./ft. What is the roof snow load that should be used for design?

Solution. Because the slope of the roof is less than 5°, it can be designed as a flat roof. From Figure 2.7 the ground snow load, p_g, for Chicago is 25 psf. The exposure factor, C_e, can be taken to equal 0.9 because there are minimal obstructions, though not necessarily an absence of all obstructions. Furthermore, because the building will be located in a residential area, it is unlikely there will be any obstructions to wind blowing across the roof. The thermal factor C_t is 1.0 because this will have to be a heated structure. Lastly, the importance factor *I* to be used is 1.1 because more than 300 people can congregate in one area. The design snow load to be used, then, is:

$$p_f = 0.7 C_e C_t I p_g = (0.7)(0.9)(1.0)(1.1)(25) = 17.3 \text{ psf}$$ ∎

Another expression is given in Section 7.4 of ASCE 7-95 for estimating the snow load on sloping roofs. It involves the multiplication of the flat roof snow load by a roof slope factor C_s. Values of C_s and illustrations are given for warm roofs, cold roofs, and for other situations in Sections 7.4.1 thru 7.4.4 of ASCE 7-95 and in their Figure 7.2.

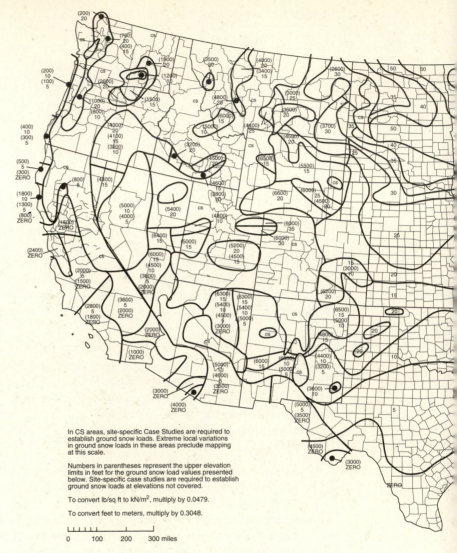

Figure 2.7   Ground snow loads, p_g, for the United States (lb/sq ft).

2.11 OTHER LOADS

There are quite a few other kinds of loads that the designer may occasionally face. These include the following:

Traffic Loads on Bridges   Bridge structures are subject to series of concentrated loads of varying magnitude caused by groups of truck or train wheels. Such loads are discussed in detail in Sections 9.20 and 9.22 of this text.

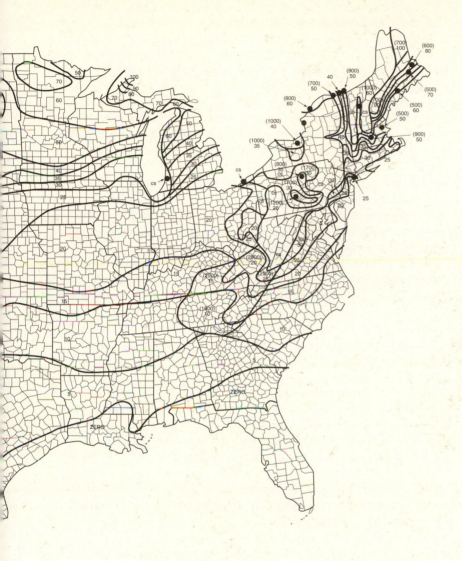

Seismic Loads Many areas of the world fall in "earthquake territory," and in those areas it is necessary to consider seismic forces in design for all types of structures. Through the centuries there have been catastrophic failures of buildings, bridges, and other structures during earthquakes. It has been estimated that as many as 50,000 people lost their lives in the 1988 earthquake in Armenia.[11] The 1989 Loma Prieta and 1994 Northridge earthquakes in California caused many billions of dollars of property damage as well as considerable loss of life.

[11] V. Fairweather, "The Next Earthquake," *Civil Engineering* (New York: ASCE, March 1990), 54–57.

Figure 2.8 shows seismic risk zones for various locations in the United States. These zones were established on the basis of past earthquakes. The blackened areas on the map are those with the most severe risks. As the shading becomes lighter the risk decreases, being least in the white areas. Much more detailed maps such as those in ASCE 7-95 are used today to estimate potential earthquake forces.[12]

Recent earthquakes have clearly shown that the average building or bridge that has not been designed for earthquake forces can be destroyed by an earthquake that is not particularly severe. Most structures can be economically designed and constructed to withstand the forces caused during most earthquakes. On the other hand, the cost of providing seismic resistance to existing structures (called *retrofitting*) can be extremely high.

Some engineers seem to think that the seismic loads to be used in design are merely percentage increases of the wind loads. This surmise is incorrect, however, as seismic loads are different in their action and are not proportional to the exposed area of the building, but rather are proportional to the distribution of the mass of the building above the particular level being considered.

Another factor to be considered in seismic design is the soil condition. Almost all of the structural damage and loss of life in the Loma Prieta earthquake occurred in areas that have soft clay soils. Apparently these soils amplified the motions of the underlying rock.[13]

It is well to understand that earthquakes load structures in an indirect fashion. The ground is displaced, and because the structures are connected to the ground, they also are displaced and vibrated. As a result, various deformations and stresses are caused throughout the structures.

From the above information you can understand that no external forces are applied above ground by earthquakes to structures. Procedures for estimating seismic forces such as the ones presented in Section 9 of ASCE 7-95 are very complicated. As a result, they usually are addressed in advanced structural analysis courses such as structural dynamics or earthquake resistance design courses.

Ice Loads   Ice has the potential of causing the application of extraordinarily large forces to structural members. Ice can emanate from two sources: (1) surface ice on frozen lakes, rivers, and other bodies of water; and (2) atmospheric ice (freezing rain and sleet). The latter can form even in warmer climates.

In colder climates ice loads often will greatly affect the design of marine structures. One such situation occurs when ice loads are applied to bridge piers. For such situations it is necessary to consider dynamic pressures caused by moving sheets of ice, pressures caused by ice jams, and uplift or vertical loads in water of varying levels causing adherence of ice.

The breaking up of ice and its movement during spring floods can significantly affect the design of bridge piers. During the breakup tremendous chunks of ice may be heaved upward, and when the jam breaks up the chunks may rush downstream,

[12] *American Society of Civil Engineers Minimum Design Loads for Buildings and Other Structures,* ASCE 7-95 (New York: ASCE), 46–104.

[13] V. Fairweather, "The Next Earthquake," *Civil Engineering* (New York: ASCE, March 1990), 54–57.

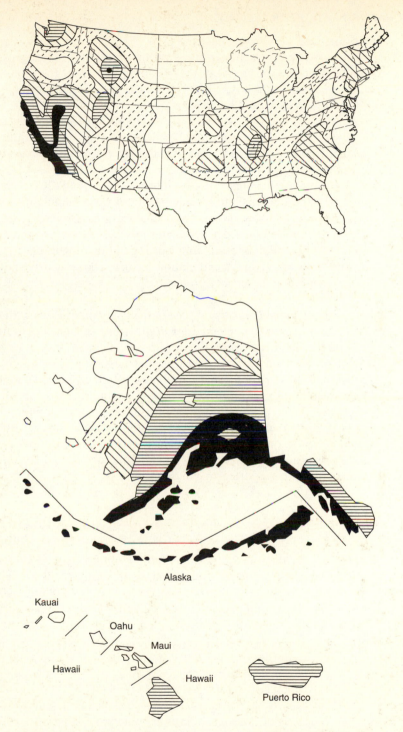

Kauai

Oahu

Maui

Hawaii

Hawaii

Alaska

Puerto Rico

Figure 2.8 Risk of earthquakes in various locations in the United States based upon past earthquakes. Blackened areas have the more severe risks. As the shading becomes lighter risk diminishes, with the white areas having very low risk.

striking and grinding against the piers. Furthermore, the wedging of pieces of ice between two piers can be extremely serious. It is thus necessary to either keep the piers out of the dangerous areas of the stream or to protect them in some way.

Section 3 of the AASHTO specifications provides formulas for estimating the dynamic forces caused by moving ice.

Bridges and towers—and any structure, for that matter—are sometimes covered with layers of ice from 1 to 2 inches thick. The weight of the ice runs up to about 10 psf. Another factor to be considered is the increased surface area of the ice-coated members as it pertains to wind loads.

Atmospheric icing is discussed in Section 10 of ASCE 7-95. Detailed information for estimating thicknesses and weight of ice accumulations is provided. This ice typically accumulates on structural members as shown on their Figure 10.1. The thickness to which the ice accumulates must be determined from historic data for the location in question or from a meteorological investigation of the conditions at that location. Extent of ice accumulation is such a localized phenomenon that general tables cannot be reasonably prepared.

Miscellaneous Loads   Some of the many other loads with which the structural designer will have to contend are *soil pressures* (such as the exertion of lateral earth pressures on walls or upward pressures on foundations); *hydrostatic pressures* (as water pressure on dams, inertia forces of large bodies of water during earthquakes, and uplift pressures on tanks and basement structures); *flooding conditions; blast loads* (caused by explosions, sonic booms, and military weapons); *thermal forces* (due to changes in temperature resulting in structural deformations and resulting structural forces); *centrifugal forces* (as those caused on curved bridges by trucks and trains or similar effects on roller coasters, etc.); and *longitudinal loads* (caused by stopping trucks or trains on bridges, ships running into docks, and the movement of traveling cranes that are supported by building frames).

PROBLEMS

Problems 2.1 through 2.5. Use the basic building layout shown in the figure for the solution of these problems. Determine the requested loads for your area of residence, or for an area specified by your instructor. Use the building code specified by your instructor. If no building code is specified, use ASCE 7-95. You do not need to determine the weight of the beams, girders, and columns for these problems.

2.1   The roof of the building is flat. It is composed of 4-ply felt and gravel on 3 in. of reinforced concrete. The ceiling beneath the roof is a suspended steel channel system.
a. The roof dead load
b. The live load in psf to be applied to Column B2

2.2   The roof of the building is flat. It is composed of 2 in. of reinforced concrete on 18-gauge metal decking. A single-ply waterproof sheet will be used. The ceiling beneath the roof is unfinished, but allowance for mechanical ducts should be provided.
a. The roof dead load (*Ans*. 32.7 psf)
b. The live load in psf to be applied to Column B1 (*Ans*. 12 psf)

2.3   The dead load and live load for the second floor of a library in which any area can be used for stacks. Assume that there will be a steel channel ceiling system and asphalt tile on the floors. The floors are 6-in. reinforced concrete. Allowance should be provided for mechanical ducts.

2.4   The dead load and live load for a floor in a light manufacturing warehouse/office complex in which any area can be used for storage. Assume that there will be no ceiling or floor finish and that the floors are 4-in. reinforced concrete. Allowance for mechanical ducts should be provided. (*Ans. D* = 54 psf, *L* = 125 psf)

2.5   The dead load and live load for a typical upper floor in an office building with movable partition walls. The ceiling is a steel channel system and the floors have a linoleum finish. The floors are 3-in. reinforced concrete. Allowance should be provided for mechanical ducts.

2.6   An upper floor in a school with steel stud walls (1/2-in. gypsum on each side), a steel channel ceiling system, and a 3-in. reinforced concrete floor with asphalt tile covering. Allowance should be made for mechanical ducts. (*Ans. D* = 52.5 psf, *L* = 40 psf classroom and 80 psf hallway)

Problems 2.7 through 2.9. Considering the basic building layout shown, determine the requested wind loads acting on the main wind-force resisting system of the building

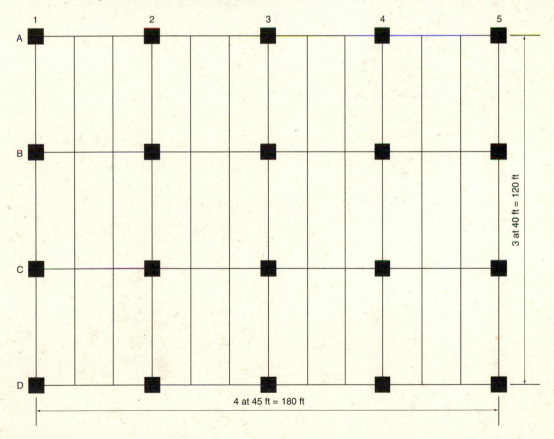

Basic building layout.

using ASCE 7-95. Assume that the long side of the building is perpendicular to the direction of the wind and there are no local changes in terrain.

2.7   The windward wall of a 70-ft tall building that is an essential facility and is located in Chicago, Illinois, on the shore of Lake Michigan.

2.8   The leeward wall if the building is a hospital, is 80 feet tall, and is located in a wooded residential suburb of Dallas, Texas. (*Ans.* total wall pressure = 26.6 psf)

2.9   The side walls of a 100-ft tall hotel located along the coast in Miami Beach, Florida.

2.10 Determine the design rain load on the roof of a building that is 250 feet wide and 500 feet long. The architect has decided to use 6-in diameter drains located uniformly around the perimeter of the building at 50-ft intervals for the secondary drainage system. The drains are located 1.5 inches above the roof surface. The rainfall intensity at the location of this building is 2.5 inches per hour. (*Ans.* 15.9 psf)

Problems 2.11 through 2.12. Use the basic building layout shown on the previous page for the solution of these problems. Determine the snow load for your area of residence or for an area specified by your instructor. Use ASCE 7-95 as the basis for computation.

2.11 The roof is flat and the building is an unheated warehouse. Assume this is a Category I structure and is sheltered by conifers.

2.12 The roof is flat and the building is an office building. This building is taller than the surrounding trees, buildings, and terrain. (*Ans.* 20.2 psf)

Chapter 3

System Loading and Behavior

3.1 INTRODUCTION

Chapter 2 discussed different types of loads that might be applied to structural systems, and methods for estimating their individual magnitudes were presented. In that discussion, however, we did not consider whether the loads acted at the same or at different times, nor did we address how and where to place them on the structure to cause maximum system response.

System response is a catchall phrase that really refers to a particular quantity of structural behavior. The response could be the negative bending moment in a floor beam, the displacement at a particular location in the structure, or the force at one of the structural supports. Although at this time we probably know very little about how to calculate these aspects of response, we will use the computer program SABLE, which is enclosed at the back of the book, to help us begin to develop an understanding of the interaction of system loading and response.

After the magnitudes of the loads have been computed, the next step in the analysis of a particular structure includes the placing of the loads on the structure and the calculation of its response to those loads. When placing the loads there are two distinct tasks that must be performed:

1. One must decide which loads can reasonably be expected to act concurrently in time. Because different loads act on the structure at different times, there will be several different loading conditions that must be evaluated. Each of these loading conditions will cause the structural system to respond in a different manner.
2. We need to determine how to place those loads on the structure. Loads are placed and the response is computed. If the same loads are placed on the structure in different positions, the response of the system will be

different. We need to determine how to place the loads so as to obtain maximum response. For example, would the bending moments in the floor beams be greater if we placed the floor live loads on every span or on every other span?

The placement of the live loads to cause the maximum or worst effects on any member of a structure is the responsibility of the structural engineer. Theoretically his or her calculations are subject to the review of the appropriate building officials, but seldom do such individuals have the time and/or the ability to make significant reviews. As a result, these calculations remain the responsibility of the designer.

3.2 TRIBUTARY AREAS

In Section 2.7 the term *tributary area* was briefly defined. In this section this term is discussed at greater length, and in the next section a related term, *influence area,* is introduced.

As previously stated, the tributary area is the loaded area of a particular structure that directly contributes to the load applied to a particular member in the structure. It is best defined as the area that is bounded by lines halfway to the next beam or to the next column. Tributary areas are shown for several columns in Figure 3.1.

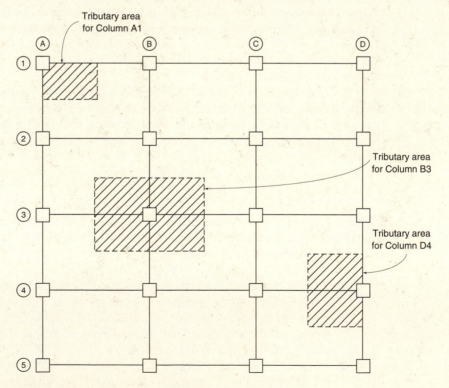

Figure 3.1   Column tributary areas.

Martin Towers, Bethlehem, Pennsylvania. (Courtesy Bethlehem Steel Corporation.)

In Figure 3.2 on page 56 the theoretical tributary areas for a beam and a girder are shown. (The term *girder* is used rather loosely, usually indicating a large beam and perhaps one into which smaller beams are framed.) If we consider the beam on the upper left part of the floor system of Figure 3.2 we see that in the middle of the beam the tributary area runs halfway to the next beam in each direction. At the ends of the beam, however, the floor is supported partly by the beams and partly by the girders that are perpendicular to the beams. The top surfaces of the beams and girders usually are assumed to be located at the same elevations. As a result, the boundary of the tributary area will fall halfway between the two—that is, at a 45° angle.

In a similar fashion the tributary areas for an interior floor girder are drawn in the same figure. The girder must support the loads on the tributary areas shown as well as the end reactions of the beams.

Let us look at the resulting loads on the beams and girders a little more closely. Using the tributary load as defined above, the load applied to a floor beam would be as shown in part (a) of Figure 3.3 and the load applied to a floor girder would be as shown in part (b) of the same figure. These loads are obtained by multiplying the tributary width (the width of the tributary area) at any point along the beam by the distributed load. The concentrated forces on the girder are the reactions from the floor beams. The distributed loads acting on girders are determined in the same manner as those for the floor beams. Depending on the spacing of the floor beams, the distributed loads on the members can be trapezoidal or triangular in shape.

Examples 3.1 and 3.2 illustrate the computation of loads for the beam and girder shown in Figures 3.2 and 3.3.

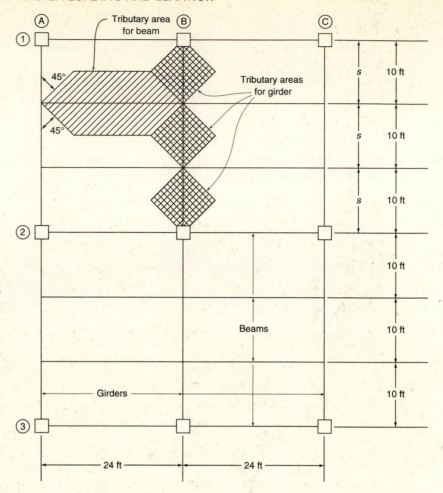

Figure 3.2 Tributary areas for beams and girders.

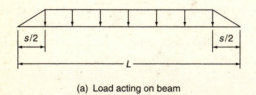

(a) Load acting on beam

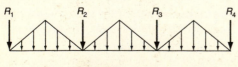

(b) Girder loads

Figure 3.3 Resulting loads acting on beams and girders.

EXAMPLE 3.1

The building floor of Figure 3.2 is to be designed to support a uniformly distributed load of 50 psf over its entire area. Using the dimensions given in the figure, determine the loads to be supported by (a) Column B1, (b) one of the interior floor beams not connected directly to the columns, and (c) the girder that runs between Columns B2 and B3.

Assume that the beams are pin connected to the girders and the girders are pin connected to the columns. (Note that when such connections are used, lateral bracing will be required to provide stability.) Further, assume that the floor beams that frame into the columns transfer their loads directly into the columns.

Solution.

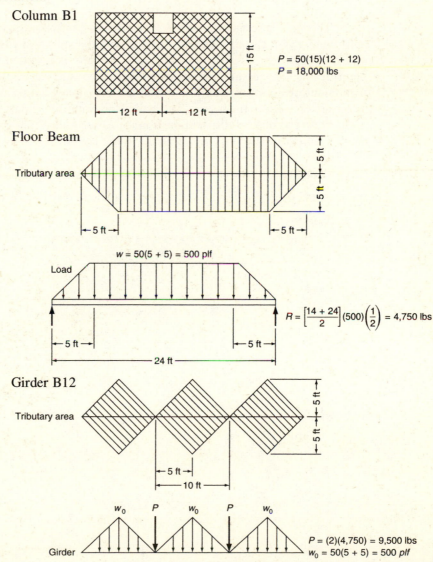

Column B1

$P = 50(15)(12 + 12)$
$P = 18,000$ lbs

15 ft
12 ft   12 ft

Floor Beam

Tributary area

5 ft
5 ft
5 ft
5 ft

$w = 50(5 + 5) = 500$ plf

Load

$R = \left[\dfrac{14 + 24}{2}\right](500)\left(\dfrac{1}{2}\right) = 4{,}750$ lbs

5 ft   5 ft
24 ft

Girder B12

Tributary area

5 ft
5 ft
5 ft
10 ft

w_0 P w_0 P w_0

Girder

$P = (2)(4{,}750) = 9{,}500$ lbs
$w_0 = 50(5 + 5) = 500$ *plf*   ∎

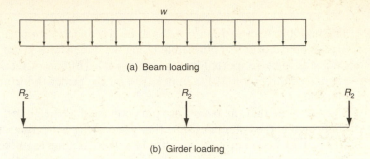

(a) Beam loading

(b) Girder loading

Figure 3.4   Usual assumed distribution of loads to beams and girders.

The load patterns of Figure 3.3 usually are considered to be too tedious to deal with in analysis. Usually the simpler load patterns shown in Figure 3.4 are used for beams and girders. The distributed load on the floor is assumed to act entirely on the floor beams. This assumption requires that the beams have a uniformly distributed load along their entire length and that the only loads acting on the girder are the reaction forces from the beams. The load in the columns is unaffected by this assumption. This type of load distribution is much easier to deal with and yields nearly the same results as the more complicated loading pattern.

EXAMPLE 3.2

Repeat Example 3.1, but use the approximate load distribution shown in Figure 3.4.

Solution.

Floor Beam

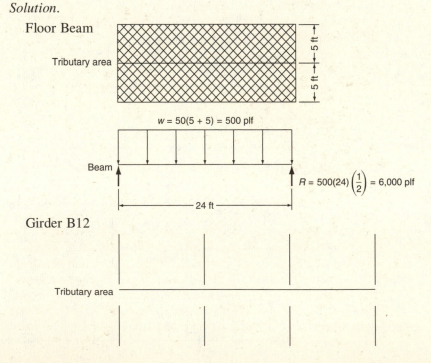

Girder B12

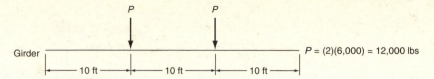

Girder $P = (2)(6,000) = 12,000$ lbs

|← 10 ft →|← 10 ft →|← 10 ft →|

3.3 LIVE LOAD REDUCTION

Under some circumstances the code-specified live loads for a building can be reduced. Before considering such reductions, however, the area that directly influences the forces in a particular member needs to be defined. That area is referred to as the *influence area*. Influence areas for several different beams and columns in a building structure are shown in Figure 3.5 on the following page. In this figure you can see that the influence area for a column is four times as large as its tributary area, while for beams the influence areas are twice as large as their tributary areas.

As the area contributing to the load on a particular member increases, the possibility of having the full design live load on the entire area at the same time decreases. As a result, building codes usually permit some reduction in the specified values. In Section 4.8 of ASCE 7-95 the following reduction factor is given:

$$L = L_0\left(0.25 + \frac{15}{\sqrt{A_1}}\right) \tag{3.1}$$

In this equation L is the reduced live load, L_0 is the code-specified live load, and the term in parentheses is the reduction factor. In this part of the expression A_1 is the influence area for the particular member.

From this equation we can see that live loads will be reduced only when the influence area is greater than 400 square feet. There are limits, though, as to how much the live load can be reduced. If the structural member supports one floor only, the live load cannot be reduced by more than 50 percent. For members supporting more than one floor, the live load cannot be reduced by more than 60 percent. Usually, beams are used to support one floor. Columns often support more than one floor. In fact, they support all of the floors above them. On page 61, a plot of live-load reduction factors, as obtained from Equation 3.1, is presented in Figure 3.6 for different influence areas.

However, live load reduction is not permitted in all cases by ASCE 7-95. If the unit live load is 100 psf or less and the loaded area is used for a place of public assembly, the reduction cannot be taken. Also, the reduction cannot be taken if the loaded area is a roof or a one-way slab (a slab that is primarily supported on two opposite edges only) and the unit live load is less than or equal to 100 psf.

When the unit live load exceeds 100 psf, Equation 3.1 does not apply. In these cases, a reduction of 20 percent can be taken for structural members supporting more than one floor. There is no permitted reduction for members that support only one floor when the unit live load is greater than 100 psf. The basis for this 20 percent reduction is that higher unit live loads tend to occur in buildings used for storage and

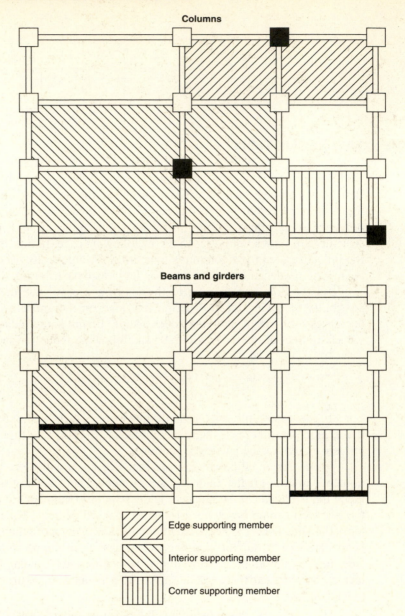

Figure 3.5   Typical influence areas.

warehousing. In this type of building, several adjacent spans may be loaded concurrently, but studies have indicated that rarely is an entire floor loaded to more than 80 percent of its rated design load.

The provision for live load reduction and the related limitations has two significant implications on structural analysis. First, the loads used to obtain column design forces and the loads used to obtain floor beam design forces may be different.

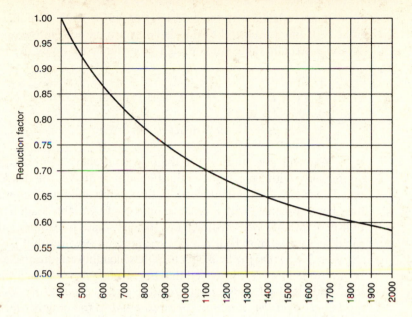

Figure 3.6   Influence area (sf).

This situation occurs because the live load reduction factors for each are likely to be different. Second, because roofs and floors are treated differently, it does not appear that the live load reduction is always permitted for those cases in which columns support a floor and the roof (such as the columns supporting the top story in a building). When a column supports a floor and the roof, that column should be considered to support a single floor for purposes of determining the permissible live load reduction.

3.4 LOADING CONDITIONS FOR ALLOWABLE STRESS DESIGN

Now that information has been presented concerning the calculation of the magnitudes of various loads, it is necessary to consider which of those loads can act concurrently. For this discussion the following nomenclature is used:

D = dead load W = wind load

E = earthquake load F = flood load

L = live load H = load due to the weight and

L_r = roof live load   lateral pressure of soil and

R = rain load   water in soil

S = snow load T = self-straining force

In accordance with the guidelines of ASCE 7-95, the following different conditions of loading need to be considered as a minimum in the analysis of a structural system when allowable stress design procedures are being used:

1. D
2. $D + L + F + H + T + (L_r$ or S or $R)$
3. $D + (W$ or $E)$
4. $D + L + (L_r$ or S or $R) + (W$ or $E)$

You will note that all of these loads (other than the dead loads) will vary appreciably with time. For the third condition above, the dead load has been combined with only one variable load. In the second and fourth conditions the dead loads have been combined with more than one type of variable load. It seems rather unlikely that two variable loads will achieve their absolute maximum values simultaneously. Load surveys seem to bear out this assumption. As a result it is stated in ASCE 7-95 (Section 2.4.3) that the load effects in these loading conditions, except dead load, may be multiplied by 0.75, provided that the result is not less than that produced by dead load and the load causing the greatest effect. *Remember, the code is listing the minimum conditions that must be considered.* If the designer feels that the maximum values of two variable loads (as, say, W and R) may occur at the same time in his or her area, it is not required that the 0.75 value be used.

These load combinations are only the recommended minimum load combinations that need to be considered. As with the determination of the loads themselves, the engineer must evaluate the structure being analyzed and determine whether these load combinations comprise all of the possible combinations for a particular structure. Under some conditions, other loads and load combinations may be appropriate.

EXAMPLE 3.3 _____

An observation deck at an airport has girders configured as shown in in Figure 3.7. These girders are spaced 15 feet on center. Assume that all of the loads are uniformly distributed over the deck and have been found to be as follows:

Dead load—32 psf

Live load—100 psf

Snow load—24 psf

Rain load—10 psf

What are the combined loads that can reasonably be expected to act on this girder?

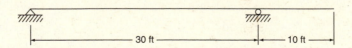

Figure 3.7

Solution.   Because the girders are 15 feet on center, the tributary area for this beam has a width of 15 feet. The applicable loading combinations for this beam are the following:

Combination 1 = 15[32] = 480 plf   Dead
Combination 2 = 15[32 + 100 + 10] = 2130 plf   Dead + Live + Rain
Combination 3 = 15[32 + 100 + 24] = 2340 plf   Dead + Live + Snow

The other load combinations either are not appropriate given the loads on the structure or are redundant. The beam should then be analyzed using these three conditions of load to determine the complete range of response. In this particular case, though, only Combination 3 will have significance from a design viewpoint because it is the largest.   ■

3.5 LOADING CONDITIONS FOR STRENGTH DESIGN

A design philosophy that is becoming quite common is the load and resistance factor (LRFD) procedure. It also is often referred to as strength design. With this method the estimated loads are multiplied by certain load or safety factors almost always larger than 1.0 and the resulting "ultimate" or "factored" loads are then used for designing the structure. The structure is proportioned to have a design ultimate or nominal strength sufficient to support the factored or ultimate loads.

The purpose of load factors is to increase the loads to account for the uncertainties involved in estimating the magnitudes of dead or live loads. For instance, how close in percent could you the reader estimate the largest wind or snow loads that ever will be applied to the building that you are now occupying?

The load factors used for dead loads are smaller than those used for live loads because designers can estimate so much more accurately the magnitudes of dead loads than they can the magnitudes of live loads. In this regard the reader will notice that loads that remain in place for long periods will be less variable in magnitude, while those applied for brief periods, such as wind loads, will have larger variations.

When a structure is to be designed using strength procedures, other load combinations and load factors may apply. Although we are not directly concerned about design while performing analysis, we must be cognizant of the design method that will be used so that the analysis results will have meaning for the design engineer.

The recommended load combinations and load factors for LRFD design usually are presented in the specifications of the different code-writing bodies, such as the International Council of Building Officials, the American Institute of Steel Construction, or the American Petroleum Institute. The load factors for use with strength design are determined statistically, and consideration is given to the type of structure upon which the loads are acting. As such, the load combinations and factors for a building will be different from those for a bridge. Both of these will likely be different than those for an offshore oil production platform. A structural analyst always should refer to the design guide or recommendation appropriate for the system being analyzed.

Following are the recommended load combinations for building structures as recommended by ASCE 7-95:

$$1.4D$$
$$1.2(D + F + T) + 1.6(L + H) + 0.5(L_r \text{ or } S \text{ or } R)$$
$$1.2D + 1.6[L_r \text{ or } S \text{ or } R] + [0.5L \text{ or } 0.8W]$$
$$1.2D + 1.3W + 0.5L + 0.5[L_r \text{ or } S \text{ or } R]$$
$$1.2D + 1.0E + 0.5L + 0.2S$$
$$0.9D + [1.3W \text{ or } 1.0E]$$

It may appear peculiar at first that only 90 percent of the dead load is used in the last load combination. After all, the dead load always is acting and is defined quite well. But if the nature of the other loads in the load combination is considered, using only 90 percent of the dead load may be a more severe condition than if the entire dead load had been applied. Wind and seismic loads cause overturning moments to occur in the structure; they try to tip the structure over. The dead load causes a righting moment— a resisting moment. It tries to keep the structure from tipping over. As such, there is less moment resisting the overturning effects of the wind and seismic loads.

As with the other load combinations previously discussed, these load combinations are only the recommended minimum load combinations that need to be considered. The engineer must decide whether these load combinations represent all of the possible combinations for the building being analyzed.

EXAMPLE 3.4

Repeat Example 3.2, but this time determine the load combinations for a structure to be designed using strength design procedures.

Solution.   The applicable load combinations are as follows:

Combination 1 = $15[1.4(32)] = 672$ plf   Dead
Combination 2 = $15[1.2(32) + 1.6(100) + 0.5(24)] = 3,156$ plf   Dead + Live + Snow
Combination 3 = $15[1.2(32) + 1.6(100) + 0.5(10)] = 3,051$ plf   Dead + Live + Rain
Combination 4 = $15[1.2(32) + 1.6(24) + 0.5(100)] = 1,902$ plf   Dead + Live + Snow
Combination 5 = $15[1.2(32) + 1.6(10) + 0.5(100)] = 1,566$ plf   Dead + Live + Rain

The other load combinations either are not appropriate or cause a lower total load to be placed on the structure.   ∎

3.6 PLACING LOADS ON THE STRUCTURE

After the magnitudes of the various loads that may be applied to a structure have been computed, it is necessary to determine where to place the loads on the structure so as to cause the most severe situations. There usually is little question about the position of the environmental loads, but this is not the case for the gravity live loads. The placement of these latter loads is discussed at length in Chapter 9 and is introduced only briefly in this section.

At some point in time the gravity live loads may not be acting at their maximum values everywhere on the structure. Then, at some other point in time, they may be acting at their full magnitudes everywhere. As a result, we need to learn where to place them for analysis so the worst possible effects can be found. The problem is complicated further because the pattern of loading that causes the maximum value of one response of a structure, such as a support reaction, may not be the same pattern that causes the maximum value of some other response, such as the bending moment or the deflection. An easy way to demonstrate this concept is through an example.

EXAMPLE 3.5 _____

The beam and loads of Example 3.3 are to be used for this example. With the enclosed computer program SABLE, calculate the reaction at the left support and the maximum negative moment at the right-hand support for each of the loading conditions shown in Figure 3.8. The term *bending moment,* with which you are already familiar, will be discussed at length in Chapter 5.

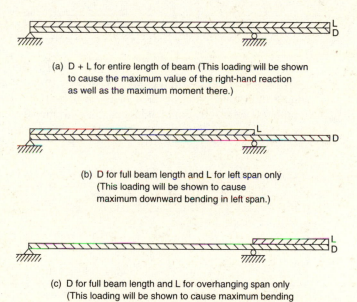

(a) D + L for entire length of beam (This loading will be shown
to cause the maximum value of the right-hand reaction
as well as the maximum moment there.)

(b) D for full beam length and L for left span only
(This loading will be shown to cause
maximum downward bending in left span.)

(c) D for full beam length and L for overhanging span only
(This loading will be shown to cause maximum bending
over right-hand support and the smallest downward bending
and deflection in the left span.)

Figure 3.8   Various loading conditions for a beam.

Solution.   Units of inches and pounds are used for this example. To begin the analysis, information about the joints and members in the structure must be entered into the computer. The nodes (or joints) and the members are numbered for identification by SABLE as shown in Figure 3.9.

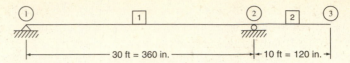

Figure 3.9 Joint and member numbers.

Detailed information regarding the joints, members, and loading data is entered into the computer and is shown in Tables 3.1 and 3.2. (By the way, the datafile for this example is included on the disk with the program SABLE.) Table 3.1 lists the locations of the joints and the types of restraint at each joint. You will note that the left hinge support has restraint in the x and y directions, but no moment restraint is provided.

TABLE 3.1 JOINT LOCATION AND RESTRAINT DATA

	Location		Restraints		
Joint	X	Y	Force X	Force Y	Mmnt Z
1	0	0	Y	Y	N
2	360	0	N	Y	N
3	480	0	N	N	N

In Table 3.2 detailed data for each of the members of the structure are provided. First, the connectivity of the members is shown, that is, the joint numbers at both ends of each member are given. Then the properties of the members (areas, moments of inertia, and moduli of elasticity) are entered. Next, the loads applied to each member are specified as described in the next paragraph.

TABLE 3.2 BEAM LOCATION, PROPERTY, AND LOAD DATA

	Connectivity		Properties			Load cases		
Beam	i joint	j joint	A	I	E	w_1	w_2	w_3
1	1	2	45	500	29×10^6	-40	-125	0
2	2	3	45	500	29×10^6	-40	0	-125

From Example 3.3 the dead load on the beam is 480 plf or 40 pounds per inch. The live load is 1500 plf or 125 pounds per inch. For this example we will consider three load cases. The first load case is the dead load acting along the entire beam. The second load case is the live load acting on the left span. The third load case is the live load acting on the cantilever. These loads are shown in Table 3.2 (and in Figure 3.10) together with other data for the beam. The negative sign indicates that the loads are acting downward. Using the load combination feature of SABLE, we will combine these individual load cases into the possible loading conditions acting on the structure.

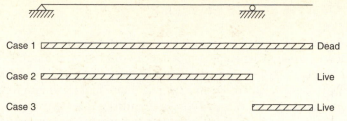

Figure 3.10   Load cases for Example 3.5.

In this example we will consider three loading combinations. These are: 1) the dead load on the entire beam and the live load on the left span; 2) the dead load on the entire beam and the live load on the cantilever span; and 3) the dead load and the live load on the entire span. These are the loading combinations that were shown previously in Figure 3.8.

We wish to find the maximum vertical reaction at the left support and the maximum negative moment at the right support. The forces of reaction are obtained by displaying joint forces after a static analysis has been performed. The maximum moment in the beam is obtained by plotting the shearing force and bending moment diagrams for the beam. The results obtained from the analysis are presented in Table 3.3 below.

Notice the range of forces in Table 3.3 for each of the quantities of response considered. The maximum vertical upward force at the left support varies from 3900 lbs to 28,900 lbs. The largest value occurs when load cases 1 and 2 are combined. The entire load is not acting on the beam.

The maximum negative moment at the right-hand support ranges from 288,000 to 1,188,000 ft lbs. Its maximum value occurs when the total load is acting on the beam (also when we have dead load on whole beam plus live load on overhanging end). From this simple example you can clearly see that the placement of live loads can cause very large variations in the magnitudes of the forces in a structural system.   ■

Other types of gravity loads that can vary with time are moving loads such as those that occur on bridges. Chapter 9 discusses more fully where and how to place loads that can change with time. For now, however, we should recognize that gravity live loads may not be acting at their full magnitude on all areas of the structure concurrently and that the maximum response does not necessarily occur in the fully loaded structure.

TABLE 3.3   RESULTS FOR EXAMPLE 3.5

	Basic load cases			Combinations of basic loads		
	Load 1	Load 2	Load 3	1 + 2	1 + 3	1 + 2 + 3
Force Y Joint 1 (1bs)	6,400	22,500	−2,500	28,900	3,900	26,400
Moment @ right support (ft 1bs)	−288,000	0	−900,000	−288,000	−1,188,000	−1,188,000

3.7 CONCEPT OF THE FORCE ENVELOPE

When loads are applied to a structure, the structure responds in reaction to those loads. The forces in any particular component in the system are a function of the loads on the structure and where those loads are located. From analysis of the response to these various forces, we can determine the maximum and minimum forces that can exist in any component. This range of forces is called a force envelope.

We saw a force envelope in the last example. The reaction in the left support had a minimum value of 3,900 pounds and maximum value of 28,900 pounds. These values were determined by placing loads on the system in locations to cause the maximum effect. For this reaction, the force envelope is 3,900 to 28,900 pounds. No matter where these loads are placed on the structure, the force at that reaction will always fall within this range. Force envelopes are very important to design engineers because they define the full range of response for which the structure must be designed. We will investigate force envelopes more fully in Chapter 17.

To further illustrate the idea of force envelopes, Example 3.6 is presented. A force envelope is prepared for the bending moment in a simple beam using the procedures learned in earlier courses.

EXAMPLE 3.6 ———

Using the principles of mechanics, draw diagrams showing the variation of bending moment throughout the beam of Figure 3.11 for each of the two loads shown. The loads do not act simultaneously.

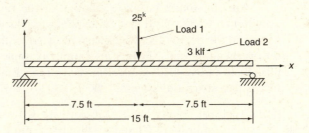

Figure 3.11

Solution.   Expressions for moment are written with the origin placed at the left end of the beam. The first two expressions given are for Load 1. You will note that due to symmetry the end reactions are each equal to one half the total load, or $1/2 \times 25 = 12.5^k$.

$M = 12.5x$ for x varying from 0 to 7.5 ft

$M = 12.5x - 25(x - 7.5)$ for x varying from 7.5 ft to 15 ft.

In a similar manner the moment caused by Load 2, where each reaction is 22.5^k, can be written as

$$M = 22.5x - 1.5x^2.$$

The preceding moment expressions are plotted in Figure 3.12. The shaded areas shown there represent the moment envelope.   ■

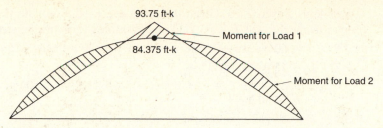

Figure 3.12   Moment envelope.

PROBLEMS

Problems 3.1 to 3.5. Use the basic floor framing plan shown in the adjacent figure when solving these problems. Consider live load reduction following the provisions of ASCE 7-95 as appropriate. For purposes of these problems assume that this is an upper story in a multi-story office building, but it is not the top story.

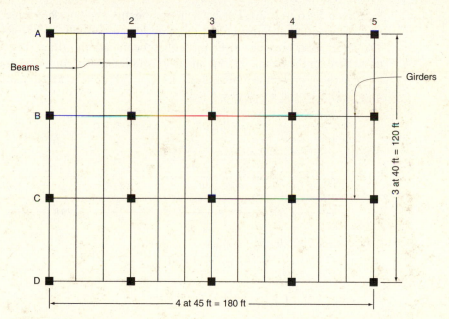

Basic building layout.

3.1   Load on Column B3 contributed by this floor if the live load is 150 psf

3.2   Load on Column A3 contributed by this floor if the live load is 75 psf (*Ans.* 33,750 pounds)

3.3   Load on Column A1 contributed by this floor if the live load is 60 psf

3.4   Load on an interior floor beam if the live load is 75 psf (*Ans.* 768.4 plf)

3.5   Load on Girder B2–B3 if the live load is 40 psf

Problems 3.6 to 3.10. Given the loads specified, compute the maximum combined load using the ASCE 7-95 load combinations for working stress design.

3.6 $D = 50$ psf, $L_r = 75$ psf, $R = 8$ psf, $S = 20$ psf (*Ans.* 125 psf)

3.7 $D = 45$ psf, $L = 60$ psf

3.8 $D = 2750$ lbs, $L = 4500$ lbs, $L_r = 1500$ lbs, $R = 1250$ lbs, $S = 1000$ lbs
(*Ans.* 8750 lbs)

3.9 $D = 87$ psf, $L = 150$ psf

3.10 $D = 75$ psf, $L_r = 35$ psf, $R = 12$ psf (*Ans.* 110 psf)

Problems 3.11 to 3.15. Repeat Problems 3.6 to 3.10 using the ASCE 7-95 LRFD load combinations.

3.11   Repeat Problem 3.6

3.12   Repeat Problem 3.7 (*Ans.* 150 psf)

3.13   Repeat Problem 3.8

3.14   Repeat Problem 3.9 (*Ans.* 344.4 psf)

3.15   Repeat Problem 3.10

Problems 3.16 to 3.20. For each of the beams shown, determine the range of the vertical reaction at Support B if the dead load is 540 plf and the live load is 960 plf. The dead load must act over the entire beam all the time. The live load can act anywhere on the beam, but does need to act on entire spans when it acts. Use SABLE to solve these problems. For analysis, let $A = 50$ in^2, $I = 550$ in^4, and $E = 29,000$ ksi.

3.16   (*Ans.* 13,500 lbs to 37,500 lbs)

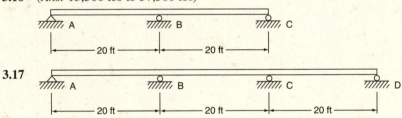

3.17

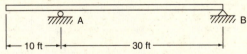

3.18   (*Ans.* 5,600 lbs to 21,600 lbs)

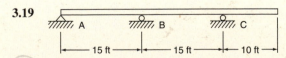

3.19

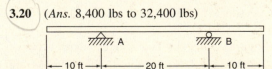

3.20   (*Ans.* 8,400 lbs to 32,400 lbs)

Chapter 4

Reactions

4.1 EQUILIBRIUM

A body at rest is said to be in *equilibrium*. The resultant of the external loads on the body and the supporting forces or reactions is zero. Not only must the sum of all forces (or their components) acting in any possible direction be zero, but also the sum of the moments of all forces about any axis must be zero.

If a structure or part thereof is to be in equilibrium under the action of a system of loads, it must satisfy the six statics equilibrium equations. Using the Cartesian x, y, and z system the equations can be written as

$$\Sigma F_x = 0 \qquad \Sigma F_y = 0 \qquad \Sigma F_z = 0$$
$$\Sigma M_x = 0 \qquad \Sigma M_y = 0 \qquad \Sigma M_z = 0$$

For purposes of analysis and design the large majority of structures can be considered as being plane structures without loss of accuracy. For these cases the sum of the forces in the x and y directions and the sum of the moments about an axis perpendicular to the plane must be zero.

$$\Sigma F_x = 0 \qquad \Sigma F_y = 0 \qquad \Sigma M_z = 0$$

These equations are commonly written as

$$\Sigma H = 0 \qquad \Sigma V = 0 \qquad \Sigma M = 0$$

These equations cannot be proved algebraically; they are merely statements of Sir Isaac Newton's observation that for every action on a body at rest there is an equal and opposite reaction. Whether the structure under consideration is a beam, a truss, a rigid frame, or some other type of assembly supported by various reactions, the equations of statics must apply if the body is to remain in equilibrium.

The structures presented in the first five chapters in this text are considered to be coplanar, whereas three-dimensional or space trusses are addressed in Chapter 8.

4.2 MOVING BODIES

The statement was made in the preceding section that a body at rest is in equilibrium. It is possible, however, for an entire structure to move and yet be in equilibrium. An airplane or ship moves, but its individual parts do not move with respect to each other.

For an accelerating body, additional forces must be included for the equations of statics to be applicable. These are the inertia forces, and with them included the body may be considered to be acted upon by a set of forces in equilibrium for purposes of structural analysis.

4.3 CALCULATION OF UNKNOWNS

To identify a force completely, there are three unknowns that must be determined: the magnitude, direction, and line of action of the force. All these values are known for external loads, but for a reaction only the point of application and perhaps the direction are known.

The total number of unknowns that can be determined by the equations of statics is controlled by the number of equations available. It does not make any difference how many reactions a two-dimensional structure has or how many unknowns each reaction has; there are three equations of statics and they can be used to determine but three unknowns for each structure. The determination of more than three unknowns requires additional equations or methods to use in conjunction with the statics equations. We will see that in some instances, owing to special construction features, equations of condition are available in addition to the usual equations.

4.4 TYPES OF SUPPORT

Structural frames may be supported by hinges, rollers, fixed ends, or links. These supports are discussed in the following paragraphs.

A *hinge* or pin-type support (represented herein by the symbol ⌗⌗⌗) is assumed to be connected to the structure with a frictionless pin. This type of support prevents movement in a horizontal or vertical direction, but does not prevent slight rotation about the hinge. There are two unknown forces at a hinge: the magnitude of the force required to prevent horizontal movement and the magnitude of the force required to prevent vertical movement. (The support supplied at a hinge may also be referred to as an inclined force, which is the resultant of the horizontal force and the vertical force at the support. Two unknowns remain: the magnitude and direction of the inclined resultant.)

City Park Lake Bridge, Baton Rouge, Louisiana. (Courtesy of the American Institute of Steel Construction, Inc.)

A *roller* type of support (represented herein by the symbol) is assumed to offer resistance to movement only in a direction perpendicular to the supporting surface beneath the roller. There is no resistance to slight rotation about the roller or to movement parallel to the supporting surface. The magnitude of the force required to prevent movement perpendicular to the supporting surface is the one unknown. *Rollers may be installed in such a manner that they can resist movement either toward or away from the supporting surface.*

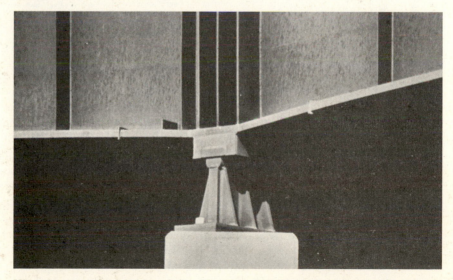

Hinged support for a bridge girder. (Courtesy Bethlehem Steel Corporation.)

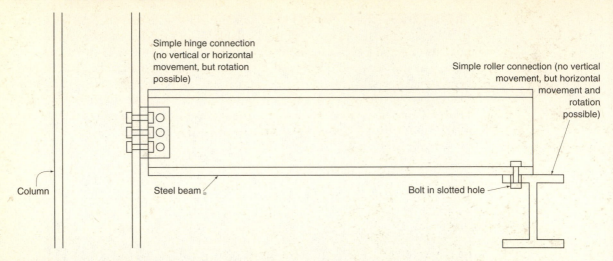

Figure 4.1   Simple connections for a steel beam.

A *fixed-end* support (represented herein by the symbol ⊬⊒) is assumed to offer resistance to rotation about the support and to movement vertically and horizontally. There are three unknowns: the magnitude of the force to prevent horizontal movement, the magnitude of the force to prevent vertical movement, and the magnitude of the force to prevent rotation.

A *link* type of support (represented herein by the symbol ⊬⊸) is similar to the roller in its action because the pins at each end are assumed to be frictionless. The line of action of the supporting force must be in the direction of the link and through the two pins. One unknown is present: the magnitude of the force in the direction of the link.

Figure 4.1 shows hinge and expansion (or roller) type connections as they might be used for a steel beam.

4.5 STABILITY, DETERMINACY, AND INDETERMINACY

The discussion of supports showed there are three unknown reaction components at a fixed end, two at a hinge, and one at a roller or link. If, for a particular structure, the total number of reaction components equals the number of statics equations available, the unknowns may be calculated and the structure is then said to be statically determinate externally. Should the number of unknowns be greater than the number of equations available, the structure is statically indeterminate externally; if less, it is unstable externally.

The internal arrangement of some structures is such that one or more equations of condition is available. The arch of Figure 4.13 has an internal pin (or hinge) at *C*. The internal moment at this "frictionless" pin is zero because no rotation can be transferred between the adjacent parts of the structure. A special condition exists because the internal moment at the pin must be zero regardless of the loading. A similar statement cannot be made for any continuous section of the beam.

By definition, a hinge transmits no rotation, and the three equations of statics plus a $\Sigma M = 0$ equation at hinge C are available to find the four unknown reaction components at A and B. The omission of a member in the truss of Figure 7.18 will be shown to give another condition equation. If the number of condition equations plus the three equations of statics equals the number of unknowns, the structure is statically determinate; if more, it is unstable; and if less, it is statically indeterminate.

Several structures are classified in Table 4.1 as to their statical condition. The beam labeled c in the table supported on its ends with rollers is stable under vertical

TABLE 4.1   STATICAL CLASSIFICATION OF STRUCTURES

Structure	Number of unknowns	Number of equations	Statical condition
a	3	3	Statically determinate (referred to as a simple beam)
b	5	3	Statically indeterminate to second degree (a continuous beam)
c	2	3	Unstable
d	3	3	Statically determinate (a cantilever beam)
e	6	3	Statically indeterminate to third degree (a fixed-ended beam)
f	4	3	Statically indeterminate to first degree (a propped beam)
g	3	3	Statically determinate
h	4	4	Statically determinate
i	7	3	Statically indeterminate to fourth degree
j	5	5	Statically determinate

loads but is unstable under inclined loads. A structure may be stable under one arrangement of loads, but if it is not stable under any other conceivable set of loads, it is unstable. This condition is sometimes referred to as *unstable equilibrium*. The structure labeled *j* has two internal hinges and thus two equations of condition. There are five equations available and five unknown reaction components: the structure is statically determinate. If one of the supporting hinges was changed to a roller, the structure would become unstable.

4.6 GEOMETRIC INSTABILITY

The ability of a structure to support adequately the loads applied to it is dependent not only on the number of reaction components but also on the arrangement of those components. It is possible for a structure to have as many or more reaction components as there are equations available and yet be unstable. This condition is referred to as *geometric instability*.

The frame of Figure 4.2(a) has three reaction components and three equations available for their solution: however, a study of the moment situation at *B* shows the structure to be unstable. The line of action of the reaction at *A* passes through *B*, and unless the line of action of the force *P* passes through the same point, the sum of the moments about *B* cannot equal zero. There is no resistance to rotation about *B*, and the frame will immediately begin to rotate. It may not collapse, but it will rotate until a stable situation is developed when the line of action of the reaction at *A* passes some little distance from *B*. Of prime importance to the engineer is for a structure to hold its position under load. One that does not do so is unstable.

Another geometrically unstable structure is shown in Figure 4.2(b). Four equations are available to compute the four unknown reaction components, but rotation

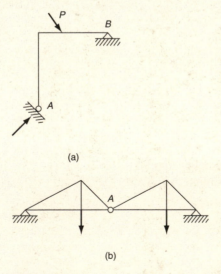

(a)

(b)

Figure 4.2

Mississippi River Bridge, St. Paul, Minnesota. (Courtesy of Kenneth M. Wright Studios, St. Paul, Minnesota.)

will instantaneously occur about the hinge at *A*. After a slight vertical deflection at *A*, the structure probably will become stable.

4.7 SIGN CONVENTION

The particular sign convention used for tension, compression, and so forth is of little consequence as long as a consistent system is used. The authors use the following signs in their computations.

1. For *tension* a positive sign is used, the thought being that pieces in tension become longer or have plus lengths.
2. A negative sign is used for pieces in *compression* because they are compressed or shortened and therefore have minus lengths.
3. For clockwise *moments* a positive sign usually is used; for counterclockwise moments, a negative sign is used. This system is of even less importance than the tension and compression system, the important thing being to use the same sign convention in taking moments throughout each complete problem to avoid confusion.
4. On many occasions it is possible to determine the direction of a *reaction* by inspection, but where it is not possible a direction is assumed and the appropriate statics equation is written. If on solution of the equation the numerical value for the reaction is positive, the assumed direction was correct; if negative, the correct direction is opposite that assumed.

4.8 HORIZONTAL AND VERTICAL COMPONENTS

It is good practice to compute the horizontal and vertical components of inclined forces for use in making calculations. If this practice is not followed, the perpendicular distances from the lines of action of inclined forces to the point where moment is being taken will have to be found. The calculation of these distances is often difficult, and the possibility of making mistakes in setting up the equations is greatly increased.

4.9 FREE-BODY DIAGRAMS

For a structure to be in equilibrium, each and every part of the structure must be in equilibrium. If the statics equations are applicable to an entire structure, they must also be applicable to any part of the structure, no matter how large or how small.

It is therefore possible to draw a diagram of any part of a structure and apply the statics equations to that part. The result, called a *free-body diagram,* must include all of the forces applied to that portion of the structure. These forces are the external reactions and loads and the internal forces applied from the adjoining parts of the structure.

A simple beam is cut into two free bodies in Figure 4.3. The effects of the right side of the beam on the left side are shown on the left free body and vice versa. For instance, the right-hand part of the beam tends to push the left-hand free body up while the left-hand part is trying to push the right-hand free body down.

Isolating certain sections of structures and considering the forces applied to those sections is the basis of all structural analysis. It is doubtful that this procedure can be overemphasized to the reader, who will, it is hoped, discover over and over that free-body diagrams open the way to the solution of structural problems.

4.10 REACTIONS BY PROPORTIONS

The calculation of reactions is fundamentally a matter of proportions. To illustrate this point, reference is made to Figure 4.4(a). The load P is three-fourths of the distance from the left-hand support A to the right-hand support B. By proportions, the right-hand support will carry three-fourths of the load and the left-hand support will carry the remaining one-fourth of the load.

Similarly, for the beam of Figure 4.4(b), the 10-kip (k or kip is the abbreviation for kilopound) load is one-half of the distance from A to B and each support will carry

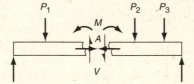

Figure 4.3

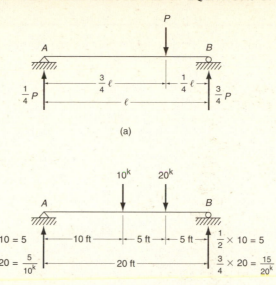

Figure 4.4

half of it, or 5 kips. The 20-kip load is three-fourths of the distance from *A* to *B*. The *B* support will carry three-fourths of it, or 15 kips, and the *A* support will carry one-fourth or 5 kips. In this manner the total reaction at the *A* support was found to be 10 kips and the total reaction at the *B* support 20 kips.

4.11 REACTIONS CALCULATED BY EQUATIONS OF STATICS

Reaction calculations by the equations of statics are illustrated by Examples 4.1 to 4.3. In applying the $\Sigma M = 0$ equation, a point may usually be selected as the center of moments so that the lines of action of all but one of the unknowns pass through the point. The unknown is determined from the moment equation, and the other reaction components are found by applying the $\Sigma H = 0$ and $\Sigma V = 0$ equations.

The beam of Example 4.1 has three unknown reaction components: vertical and horizontal ones at *A* and a vertical one at *B*. Moments are taken about *A* to find the value of the vertical component at *B*. All of the vertical forces are equated to zero, and the vertical reaction component at *A* is found. A similar equation is written for the horizontal forces applied to the structure, and the horizontal reaction component at *A* is found to be zero.

The solutions of reaction problems may be checked by taking moments about the other support, as illustrated on page 80 in Example 4.1. For future examples space is not taken to show the checking calculations. *A problem, however, should be considered incomplete until a mathematical check of this nature is made.*

EXAMPLE 4.1

Compute the reaction components for the beam shown in Figure 4.5.

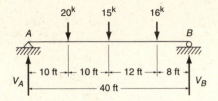

Figure 4.5

Solution.

$$\Sigma M_A = 0$$
$$(20)(10) + (15)(20) + (16)(32) - 40V_B = 0$$
$$V_B = 25.3^k \uparrow$$

$$\Sigma V = 0$$
$$20 + 15 + 16 - 25.3 - V_A = 0$$
$$V_A = 25.7^k \uparrow$$

Checking: $\Sigma M_B = 0$

$$(V_A)(40) - (20)(30) - (15)(20) - (16)(8) = 0$$
$$V_A = 25.7^k \uparrow \quad \blacksquare$$

The beam of Example 4.2, which is shown in Figure 4.6, is subjected to an inclined 50 k load. For convenience in taking moments this load is replaced with its vertical and horizontal components. The structure must also support for a short distance a uniform load of 3 klf (kips or kilopounds per linear foot).

EXAMPLE 4.2

Find all reaction components in the structure shown in Figure 4.6.

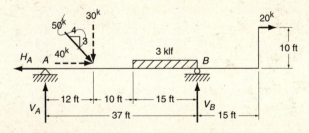

Figure 4.6

Solution.

$$\Sigma M_A = 0$$
$$(30)(12) + (3 \times 15)(29.5) + (20)(10) - (V_B)(37) = 0$$
$$V_B = 51^k \uparrow$$

$$\Sigma V = 0$$
$$30 + 45 - 51 - V_A = 0$$
$$V_A = 24^k \uparrow$$
$$\Sigma H = 0$$
$$40 + 20 - H_A = 0$$
$$H_A = 60^k \leftarrow \quad \blacksquare$$

The roller of the frame of Example 4.3 is supported by an inclined surface. The statics equations still are applicable, because the direction of the reaction at B is known (perpendicular to the supporting surface). If the direction of the reaction is known, the relationship between the vertical component, the horizontal component, and the reaction itself is known. Here the reaction has a slope of four vertically to three horizontally $(4:3)$, which is perpendicular to the slope of the supporting surface of three to four $(3:4)$. Moments are taken about the left support, which gives an equation including the horizontal and vertical components of the reaction at the inclined roller. But both components are in terms of that reaction; therefore only one unknown, R_B, is present in the equation and its value is easily obtained.

EXAMPLE 4.3

Compute the reactions for the frame shown in Figure 4.7.

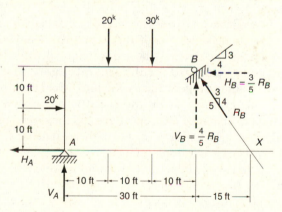

Figure 4.7

Solution.

$$\Sigma M_A = 0$$
$$(20)(10) + (20)(10) + (30)(20) - (H_B)(20) - (V_B)(30) = 0$$
$$200 + 200 + 600 - (\tfrac{3}{5}R_B)(20) - (\tfrac{4}{5}R_B)(30) = 0$$
$$R_B = 27.8^k \nwarrow$$

$$V_B = \tfrac{4}{5}R_B = 22.2^k \uparrow$$
$$H_B = \tfrac{3}{5}R_B = 16.7^k \leftarrow$$

$$\Sigma V = 0$$
$$20 + 30 - 22.2 - V_A = 0$$
$$V_A = 27.8^k \uparrow$$

$$\Sigma H = 0$$
$$20 - 16.7 - H_A = 0$$
$$H_A = 3.3^k \leftarrow$$

Alternate Solution.   Should the line of action of R_B be extended until it intersects a horizontal line through A at point X, another convenient location for taking moments will be available. Moments are taken at point X, and only one unknown (V_A) appears in the equation. This method may be simpler than the previous solution.

$$\Sigma M_x = 0$$
$$(V_A)(45) + (20)(10) - (20)(35) - (30)(25) = 0$$
$$V_A = 27.8^k \uparrow \quad \blacksquare$$

4.12 PRINCIPLE OF SUPERPOSITION

As we proceed with our study of structural analysis we will encounter structures subject to large numbers of forces and to different kinds of forces (concentrated, uniform, triangular, dead, live, impactive, etc.). To assist in handling such situations there is available an extremely useful tool called the *principle of superposition.*

The principle follows: **if the structural behavior is linearly elastic the forces acting on a structure may be separated or divided in any convenient fashion and the structure analyzed for the separate cases. The final results can then be obtained by adding algebraically the individual results.** The authors previously made use of this principle in Figure 4.4(b) when the reactions for the two loads were determined separately by proportions and then were added together to obtain the final values. The principle applies not only to reactions, but also to shears, moments, stresses, strains, and displacements.

There are two important situations in which the principle of superposition is not valid. The first occurs where the geometry of the structure is appreciably changed under the action of the loads. The second occurs where the structure consists of a material for which stresses are not directly proportional to strains. This latter situation can occur when the material is stressed beyond its elastic limit or when the material does not follow Hooke's law for any part of its stress–strain curve.

4.13 THE SIMPLE CANTILEVER

The simple cantilever pictured in Figure 4.8 has three unknown reaction components supporting it at the fixed end; they are the forces required to resist horizontal movement, vertical movement, and rotation. They may be determined with the equations of statics as illustrated in Example 4.4.

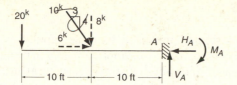

Figure 4.8

EXAMPLE 4.4 _____

Find all reaction components for the cantilever beam shown in Figure 4.8.

Solution.

$$\Sigma V = 0$$
$$20 + 8 - V_A = 0$$
$$V_A = 28^k \uparrow$$
$$\Sigma H = 0$$
$$6 - H_A = 0$$
$$H_A = 6^k \leftarrow$$
$$\Sigma M_A = 0$$
$$(20)(20) + (8)(10) - M_A = 0$$
$$M_A = 480 \text{ ft k} \curvearrowleft \quad \blacksquare$$

United Airlines hangar, San Francisco, California. (Courtesy of the Lincoln Electric Company.)

4.14 CANTILEVERED STRUCTURES

Moments in structures that are simply supported increase rapidly as their spans become longer. We will see that bending increases approximately in proportion to the square of the span length. Stronger and more expensive structures are required to resist the greater moments. For very long spans, moments are so large that it becomes economical to introduce special types of structures that will reduce the moments. One of these types is cantilevered construction, as illustrated in Figure 4.9(b).

A cantilevered type of structure is substituted for the three simple beams of Figure 4.9(a) by making the beam continuous over the interior supports B and C and introducing hinges in the center span as indicated in (b). An equation of condition ($\Sigma M_{hinge} = 0$) is available at each of the hinges introduced, giving a total of five equations and five unknowns. The structure is statically determinate.

The moment advantage of cantilevered construction is illustrated in Figure 4.10. The diagrams shown give the variation of moment in each of the structures of Figure 4.9 due to a uniform load of 3 klf for the entire spans. The maximum moment for the cantilevered type is seen to be considerably less than that for the simple spans, and this permits lighter and cheaper construction. The plotting of moment diagrams is fully explained in Chapter 5. For the cantilevered beam of Figure 4.9(b) it is possible to reduce the maximum moment even further by moving the unsupported hinges closer to supports B and C.

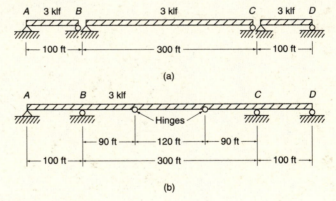

Figure 4.9

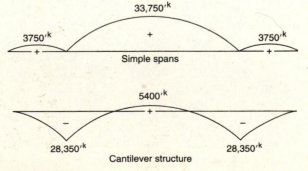

Figure 4.10

4.15 REACTION CALCULATIONS FOR CANTILEVERED STRUCTURES

Cantilevered construction consists essentially of two simple beams, each with an overhanging or cantilevered end as follows:

with another simple beam in between supported by the cantilevered ends:

The first step in determining the reactions for a structure of this type is to isolate the center simple beam and compute the forces necessary to support it at each end. Second, these forces are applied as downward loads on the respective cantilevers, and as a final step the end-beam reactions are determined individually. Example 4.5 illustrates the entire process.

EXAMPLE 4.5 _____

Calculate all reactions for the cantilevered structure of Figure 4.11.

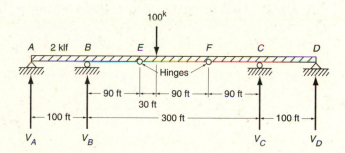

Figure 4.11

Solution.   By isolating the center section from the two end sections:

Computing reactions for center beam:

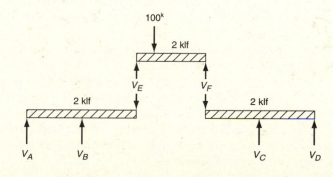

$$\Sigma M_E = 0$$
$$(100)(30) + (2 \times 120)(60) - 120V_F = 0$$
$$V_F = 145^k$$
$$\Sigma V = 0$$
$$100 + 240 - 145 - V_E = 0$$
$$V_E = 195^k$$

By computing reactions for left-end beam:

$$\Sigma M_A = 0$$
$$(2 \times 190)(95) + (195)(190) - 100V_B = 0$$
$$V_B = 731.5^k \uparrow$$
$$\Sigma V = 0$$
$$380 + 195 - 731.5 - V_A = 0$$
$$V_A = 156.5^k \downarrow$$

Similarly, the reactions for the right-end beam are found to equal

$$V_C = 636.5^k \uparrow$$
$$V_D = 111.5^k \downarrow$$

An examination of the reactions obtained for the left-end and right-end sections of this beam show why cantilevered bridges often are called "see-saw" bridges. These structures are primarily supported by the first interior supports on each end, where the reactions are quite large. The end supports may very well have to provide downward reaction components. Thus, an end section of a cantilevered structure seems to act as a seesaw over the first interior support with downward loads on both sides. ■

4.16 ARCHES

Historically, arches were the only feasible form that could be used for large structures made up of materials with negligible tensile strength such as bricks and stones. Masonry arches of such materials have been used for thousands of years.

In effect, an arch takes vertical loads and turns them into lateral thrusts that run around the arch and put the elements of the arch in compression. This is shown in Figure 4.12. The parts of a stone arch are called *voussoirs*. As you can see in the figure these are stones in the shape of truncated wedges. They are pushed together in compression.

Arches are very rigid stable structures that are not appreciably affected by movements of their foundations. It is rather interesting to note that excavations of ancient ruins show that they are the structures that generally have survived the best.

Theoretically, an arch can be designed for a single set of gravity loads so that only axial compression stresses are developed in the arch. Unfortunately, however, in practical structures the loads change and move so that bending stresses are developed. Nevertheless, arches generally are designed so that their predominant loading causes primarily compression stresses.

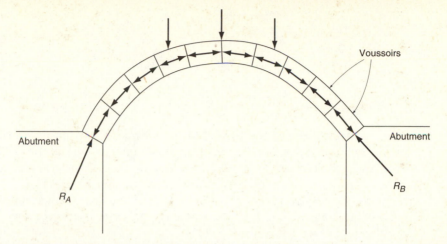

Figure 4.12   Stone arch.

Structural arch construction: U.S.A.F. hangar, Edwards Air Force Base, California. (Courtesy Bethlehem Steel Corporation.)

4.17 THREE-HINGED ARCHES

The reactions for structures discussed previously, which had horizontal supports, were vertical and parallel to each other under vertical loading. Arches are structures that produce horizontal converging reactions under vertical load. They tend to flatten out under load and must be fixed against horizontal movement at their supports. (There is an old saying to the effect the "an arch never sleeps," the thought being that the horizontal reactions are always present even when no live loads are applied, due to the structure's own weight.)

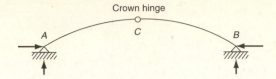

Figure 4.13

Arches may be constructed with three hinges, two hinges, one hinge (very rare), or no hinges (quite common in reinforced concrete construction). The three-hinged arch is discussed here because it is the only statically determinate arch.

Examination of the three-hinged arch pictured in Figure 4.13 reveals two reaction components at each support, or a total of four. Three equations of statics and one condition equation ($\Sigma M_C = 0$, where the subscript C means crown hinge) are available to find the unknowns.

The arch of Example 4.6 is handled by taking moments at one of the supports to obtain the vertical reaction component at the second support. Because the supports are on the same level, the horizontal reaction component at the second support passes through the point where moments are being taken. When one vertical reaction com-

Laminated timber beams, Maumee, Ohio. (Courtesy of Unit Structures, Inc.)

ponent has been found, the other may be obtained with the $\Sigma V = 0$ equation. The horizontal reaction components are obtained by taking moments at the crown hinge of the forces either to the left or to the right. The only unknown appearing in either equation is the horizontal reaction component on that side, and the equation is solved for its value. The other horizontal component is found by writing the $\Sigma H = 0$ equation for the entire structure. Once the reactions are determined it is easy to compute the moment and axial force in the arch at any point by statics.

EXAMPLE 4.6

Find all reaction components for the three-hinged arch shown in Figure 4.14.

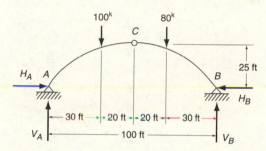

Figure 4.14

Solution.

$$\Sigma M_A = 0$$
$$(100)(30) + (80)(70) - 100V_B = 0$$
$$V_B = 86^k \uparrow$$

$$\Sigma V = 0$$
$$100 + 80 - 86 - V_A = 0$$
$$V_A = 94^k \uparrow$$

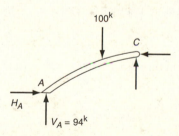

By using the free-body sketch shown,

$$\Sigma M_C \text{ to left} = 0$$
$$(94)(50) - (100)(20) - 25H_A = 0$$
$$H_A = 108^k \rightarrow$$

$$\Sigma H = 0$$
$$108 - H_B = 0$$
$$H_B = 108^k \leftarrow \quad \blacksquare$$

North Dakota State Teachers College fieldhouse, Valley City, North Dakota.
(Courtesy of the American Institute of Timber Construction.)

The computation of reactions for the arch of Example 4.7 is slightly more complicated because the supports are not on the same level. Taking moments at one support results in an equation involving both the horizontal and vertical components of reaction at the other support. Moments may be taken about the crown hinge of the forces on the same side as those two unknowns. The resulting equation contains the same two unknowns. Solving the equations simultaneously gives the values of the components, and the application of the $\Sigma H = 0$ and $\Sigma V = 0$ equations results in the values of the remaining components.

EXAMPLE 4.7 ────────────────────────────────────

For the structure diagrammed in Figure 4.15 determine reaction components at both supports.

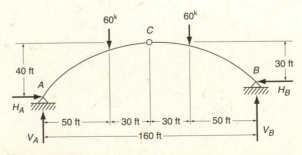

Figure 4.15

Solution.

$$\Sigma M_A = 0$$

$$(60)(50) + (60)(110) - 10H_B - 160V_B = 0$$

$$10H_B + 160V_B = 9600 \qquad (1)$$

$$\Sigma M_C \text{ to right} = 0$$

$$(60)(30) + 30H_B - 80V_B = 0$$

$$30H_B - 80V_B = -1800 \qquad (2)$$

Solving Equations (1) and (2) simultaneously gives

$$H_B = 85.7^k \leftarrow$$

$$V_B = 54.6k \uparrow$$

$$\Sigma H = 0$$

$$85.7 - H_A = 0$$

$$H_A = 85.7^k \rightarrow$$

$$\Sigma V = 0$$

$$60 + 60 - 54.6 - V_A = 0$$

$$V_A = 65.4^k \uparrow \quad \blacksquare$$

The reactions for the arch of Example 4.7 may be computed without using simultaneous equations. The horizontal and vertical axes (on the basis of which the $\Sigma H = 0$ and $\Sigma V = 0$ equations are written) are so rotated that the horizontal axis passes through the two supporting hinges (Figure 4.16). Components of forces can be computed parallel to the X' and Y' axes and the $\Sigma X' = 0$ and $\Sigma Y' = 0$ equations applied, but the computations are cumbersome.

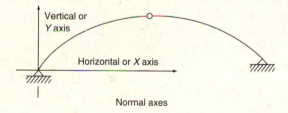

Normal axes

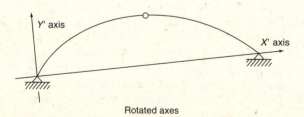

Rotated axes

Figure 4.16

Glued laminated three-hinged arches, The Jai Alai Fronton, Riveria Beach, Florida. (Courtesy of the Forest Products Association.)

4.18 USES OF ARCHES AND CANTILEVERED STRUCTURES

Three-hinged steel arches are used for short and medium-length bridge spans of up to approximately 600 ft. They are used for buildings where large clear spans are required underneath, as for hangars, field houses, and armories. Steel two-hinged arches generally are economical for bridges from 600 to 900 ft in length, with a few exceptional spans being over 1600 ft long. Reinforced-concrete hingeless arches are used for bridges of from 100- to 400-ft spans. Cantilever-type bridges are used for spans of from approximately 500 ft up to very long spans, such as the 1800-ft center span of the Quebec Bridge.

An arch is a structure that requires foundations capable of resisting the large horizontal reaction components or thrusts at the supports. In arches for buildings it is possible to carry the thrusts by tying the supports together with steel rods as illustrated in Figure 4.17, with steel sections, or even with specially designed floors. These are referred to as *tied arches*. For many locations the three-hinged arch is selected over the statically indeterminate arches because of poor foundation conditions with the possibility of settlement. It will become evident in later chapters that foundation settlement may cause severe stress changes in statically indeterminate structures. Ease of erection is another advantage of three-hinged arches. It often is convenient to assemble and ship the two halves of a precast-concrete, structural-steel, or laminated-timber arch separately and assemble them on the site.

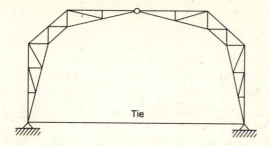

Figure 4.17

The fact that cantilever-type construction reduces bending moments for long spans has been previously demonstrated. Arch-type construction also reduces moments, because the reactions at the supports tend to cause bending in the arch in a direction opposite to that caused by the downward loads. Because of this characteristic of having small bending moments, arches were admirably suited for masonry construction as practiced by the ancient builders.

4.19 CABLES

Perhaps cables provide the simplest way of supporting loads. They are used for supporting bridge and roof systems; as guys for derricks, radio towers, and similar type structures; and for many other applications, as well. To the student their most common use may seem to be for the cable car systems used so successfully at hundreds of ski slopes around the world.

Steel cables are economically manufactured from high-strength steel wire, providing perhaps the lowest cost/strength ratio of any common structural members. Further, they are easily handled and positioned, even for very long spans.

The shape the cables take in resisting loads is called a funicular curve. (You may have noticed that the cable car systems in Europe often are called funiculars.) Cables are quite flexible and support their loads in pure tension as shown in Figure 4.18. It can be seen in this figure that the load P must be balanced by the vertical components of tension in the cable; thus, the cable must have a vertical projection in order to support the load. The greater the vertical projection, the smaller will be the cable tension. If the cable moves or if other loads are applied, the cable will change shape.

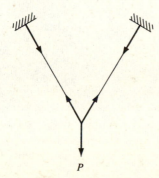

P

Figure 4.18

For the discussion to follow, the cable weight is neglected. When a cable of a given length is suspended between two supports, the shape it takes is determined by the applied loads. As a matter of fact, when vertical loads are supported a cable will take the shape of the bending-moment diagram (see Chapter 5) of a simple beam of the same span subjected to the same loads. For instance, if a cable is subjected to a set of concentrated loads such as those shown in Figure 4.19, the segments of the cable will fall along straight lines. (This shape actually is inverted from the shape of the moment diagrams drawn in later chapters of this book.)

A cable supporting a horizontal uniform load will assume a parabolic shape. Cables supporting the roadway of suspension bridges usually are assumed to fall into this class (though the loads are applied to the cables as closely spaced concentrated loads from the hangers). A cable supporting a load that is uniform along its length (such as a transmission line) will take the form of a catenary. Unless the sag of such a cable is quite large in proportion to its length, it may be assumed to be parabolic in shape, with the result that its analysis is appreciably simplified.

Cables are assumed to be so flexible that they cannot resist bending or compression. As a result they are in direct tension and a condition equation ($\Sigma M = 0$ at any point in the cable) is available for analysis. Should the position or sag of a cable at a particular point be known, the reactions at the cable ends and the sag at any other point in the cable can be determined with these equations. A numerical example of this type is presented in Example 4.8. The weight of the cable is assumed to be negligible in this case.

Cable-stayed Sitka Harbor Bridge, Sitka, Alaska. (Courtesy of the Alaska Department of Transportation.)

EXAMPLE 4.8

Determine the reactions for the cable in Figure 4.19 and the sag at the 40-kip load.

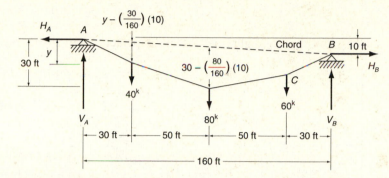

Figure 4.19

Solution.

$\Sigma M_B = 0$

$160V_A - 10H_A - (60)(30) - (80)(80) - (40)(130) = 0$

$$160V_A - 10H_A = 13,400 \qquad (1)$$

ΣM_{80} load to left $= 0$

$80V_A - 30H_A - (40)(50) = 0$

$$80V_A - 30H_A = 2000 \qquad (2)$$

Solving Equations (1) and (2) simultaneously gives

$$H_A = 188^k \leftarrow$$
$$V_A = 95.5^k \uparrow$$

By $\Sigma V = 0$ and $\Sigma H = 0$,

$$H_B = 188^k \rightarrow$$
$$V_B = 84.5^k \uparrow$$

ΣM to left of 40^k load letting the sag there $= y$

$$30V_A - H_A y = 0$$
$$(30)(95.5) - (188)(y) = 0$$
$$y = 15.24 \text{ ft} \quad \blacksquare$$

The resultant tension at any point can be obtained from the following equation in which H and V are the horizontal and vertical components of tensile force in the cable at that point.

$$T = \sqrt{(H)^2 + (V)^2}$$

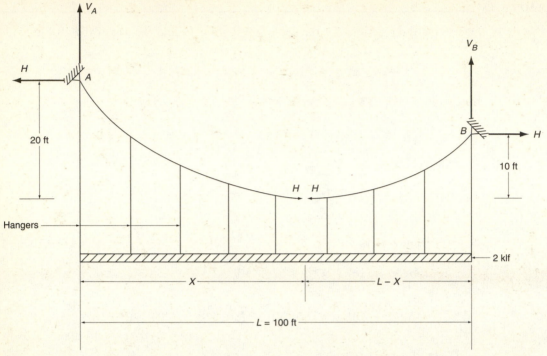

Figure 4.20   A cable loaded by hangers with an approximate uniform load along its horizontal projection.

From this equation it can be seen that the tension varies as one moves along the cable. If only vertical loads are present the value of H will be constant throughout and the maximum tensile force occurring can be determined by substituting the maximum value of the vertical force into the equation.

The distortion or deflection of most structures is assumed to be negligible when computing the forces produced in those structures. Such an assumption is not correct, however, for many cable structures, particularly the flat ones where a little sag can drastically affect cable tensions. This topic is not considered herein, but is described very well in a book by Firmage.[1] Flat cables cause very large horizontal reaction components and thus have very high tensile forces.

On many occasions concentrated loads are applied to cables by hangers. If they are closely spaced as shown in Figure 4.20, the loading situation will approach that of a uniform load along the horizontal projection of the cable. If this is the case, the cable will take a parabolic shape. If the load is uniform along the length of the cable, the shape taken will be that of a catenary. This is the situation that occurs when a cable is loaded only by its own weight.

The slope of the cable at its lowest point will be zero, and as a result, the vertical component of force there will be zero. Referring to Figure 4.20, an expression can be written for the moment about the left support of the forces from there down to the low

[1] D. A. Firmage, *Fundamental Theory of Structures* (New York: Wiley, 1963), 258–265.

point. (As before, there is assumed to be no moment resistance at any point in the cable.) A similar moment expression can be written about the right-hand support for the forces from there to the cable low point.

These two equations each will contain two unknowns: x (the horizontal distance from A to the cable low point) and H (the horizontal component of force in the cable). These values may be determined by solving the equations simultaneously, as illustrated in Example 4.9.

After H is obtained, the values of V_A and V_B can be calculated by summing the vertical forces on the free bodies to the left and right of the cable low point. Because the horizontal component of cable tension H is constant all along the cable, the tension at any point will equal $\sqrt{H^2 + V^2}$.

Its maximum value will occur at the steepest point in the cable, where V will be the largest. This will be at the highest support.

EXAMPLE 4.9

Determine the maximum tension in the cable shown in Figure 4.20.

Solution.

$$\underline{\Sigma M_A = 0}$$

$$(2)(x)\left(\frac{x}{2}\right) - 20H = 0$$

$$20H - x^2 = 0 \tag{1}$$

$$\underline{\Sigma M_B = 0}$$

$$-(2)(100 - x)\left(\frac{100 - x}{2}\right) + 10H = 0$$

$$10H - x^2 + 200x - 10,000 = 0 \tag{2}$$

Solving Equations (1) and (2) simultaneously

$$x = 58.58 \text{ ft}$$

$$H = 171.58 \text{ k}$$

Summing up the vertical forces to the left of the cable low point.

$$V_A - (2)(58.58) = 0$$

$$V_A = 117.16 \text{ k}$$

Then the maximum tension in the cable occurs at its steepest point, which is at the left support.

$$T = \sqrt{(117.16)^2 + (171.58)^2} = \underline{\underline{207.8^k}} \quad \blacksquare$$

The maximum tension occurring in a cable uniformly loaded along its length is addressed in Appendix A of this book.[2]

PROBLEMS

For Problem 4.1, determine which of the structures shown in the accompanying illustration are statically determinate, statically indeterminate (including the degree of indeterminacy), and unstable as regards outer forces.

4.1

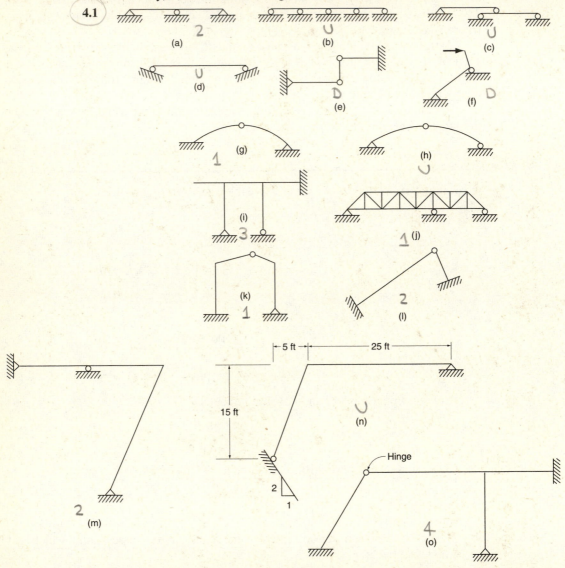

Problems 4.2 to 4.46 compute the reactions for the structures.

4.2   (*Ans.* $V_L = 65.71^k \uparrow$, $V_R = 94.29^k \uparrow$)

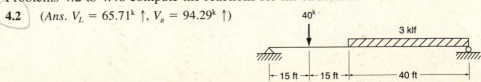

4.3

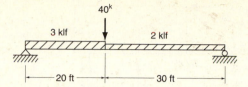

4.4   (*Ans.* $V_L = 28.33^k \uparrow$, $V_R = 21.67^k \uparrow$)

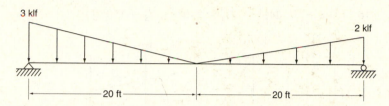

4.5

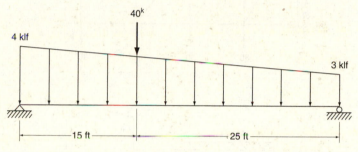

4.6   (*Ans.* $V_L = 99.29^k \uparrow$, $V_R = 80.71^k \uparrow$)

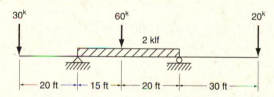

4.7

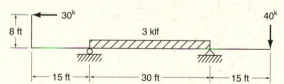

4.8   (*Ans.* $V_L = 115.82^k \uparrow$, $V_R = 5.10^k \uparrow$, $H_R = 66.97^k \rightarrow$)

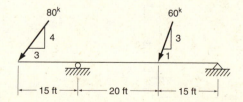

4.9

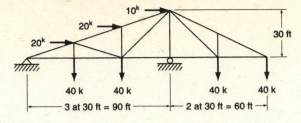

4.10   (Ans. $H_A = 240^k \rightarrow$, $H_B = 240^k \leftarrow$, $V_B = 150^k \uparrow$)

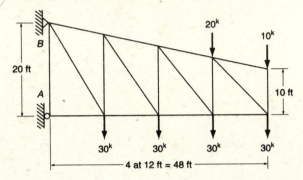

4.11

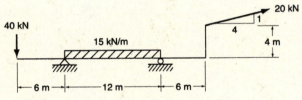

4.12   (Ans. $V_L = 41.11^k \uparrow$, $V_R = 15.81^k \uparrow$, $H_L = 18.97^k \leftarrow$)

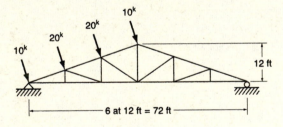

4.13

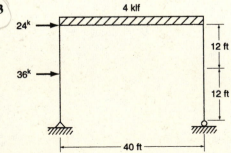

4.14 (*Ans.* $V_L = 78.75^k \uparrow$, $V_R = 71.25^k \uparrow$, $H_R = 45^k \leftarrow$)

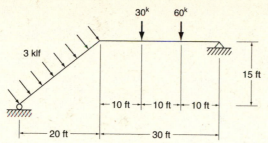

4.15

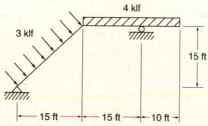

4.16 (*Ans.* $V_L = 43.64^k \uparrow$, $H_L = 12.27^k \rightarrow$, $V_R = 16.36^k \uparrow$, $H_R = 12.27^k \leftarrow$)

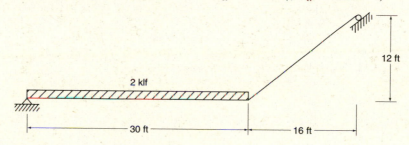

4.17

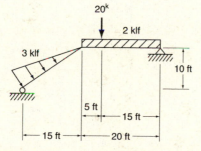

4.18 (*Ans.* $V_L = 127.09^k \uparrow$, $H_L = 19.07^k \leftarrow$, $V_R = 56.72^k \uparrow$, $H_R = 28.36^k \leftarrow$)

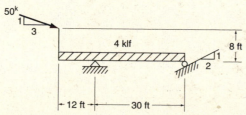

4.19

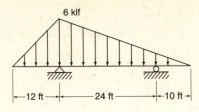

4.20   (*Ans.* $V_L = 58.32^k \uparrow$, $V_R = 16.67^k \uparrow$)

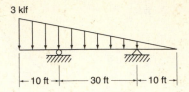

4.21

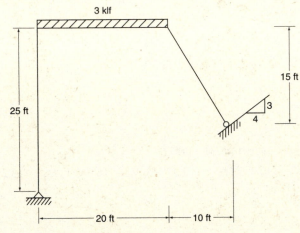

4.22   (*Ans.* $V_L = 0$, $H_L = 20^k \leftarrow$, $V_R = 40^k \uparrow$)

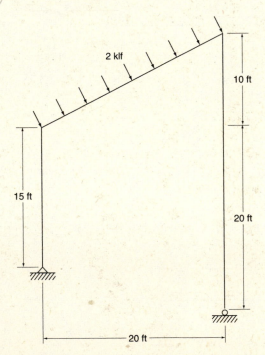

4.23

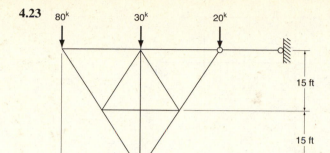

4.24 (*Ans.* $V_L = 159.6^k \uparrow$, $V_R = 170.4^k \uparrow$)

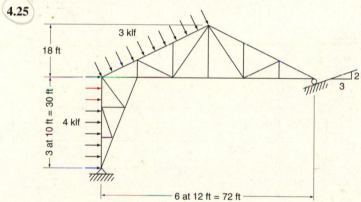

4.25

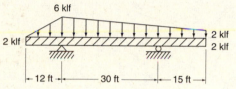

4.26 (*Ans.* $V_L = 66.96^k \uparrow$, $H_L = 22.30^k \rightarrow$, $V_R = 93.04^k$, $H_R = 5.82^k \rightarrow$)
Consider only 1 ft length of frame.

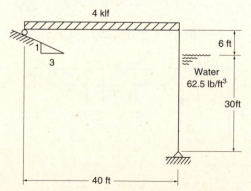

4.27

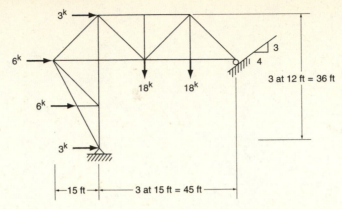

4.28 $(Ans.\ V_L = 75^k \uparrow,\ H_L = 37.5^k \leftarrow,\ V_R = 60^k \uparrow,\ H_R = 30^k \leftarrow)$

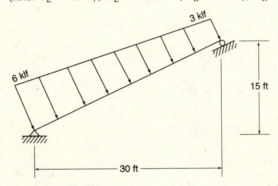

4.29

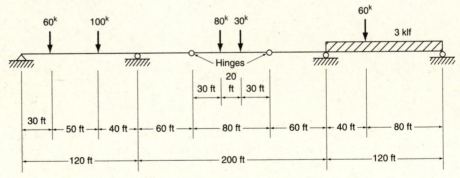

4.30 $(Ans.\ V_A = 60^k \downarrow,\ V_B = 280^k,\ V_C = 270^k \uparrow,\ V_D = 80^k \downarrow)$

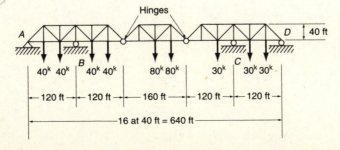

4.31

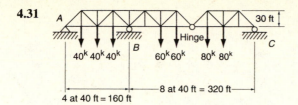

4.32   (*Ans.* $V_A = 198^k \uparrow$, $V_B = 753.2^k$, $V_C = 31.2^k$)

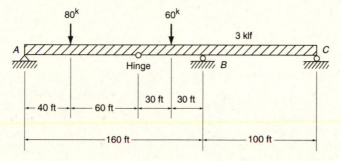

4.33

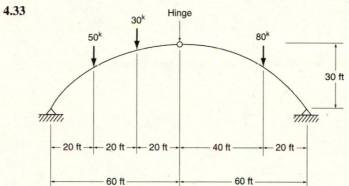

4.34   (*Ans.* $V_L = 94^k \uparrow$, $H_L = 108^k \rightarrow$, $V_R = 66^k \uparrow$, $H_R = 108^k \leftarrow$)

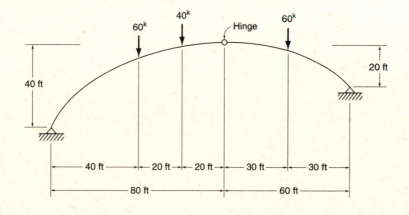

4.35

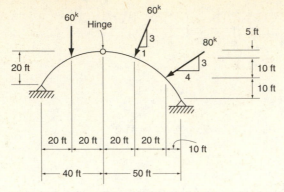

4.36   (*Ans.* $V_L = 73$ kN ↑, $H_L = 88.33$ kN →, $V_R = 137$ kN ↑, $H_R = 88.33$ kN ←)

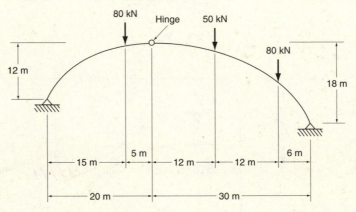

4.37

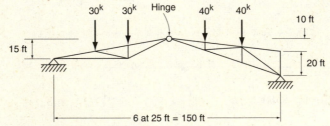

4.38   (*Ans.* $V_L = 5^k$ ↑, $H_L = 90^k$ →, $V_R = 5^k$ ↑, $H_R = 90^k$ →)

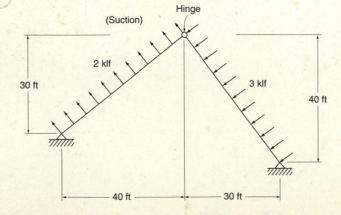

4.39 Repeat Problem 4.38 if the 2-klf load is increased to 3 klf and the 3-klf load is increased to 5 klf.

4.40 Repeat Problem 4.38 if the 3-klf load is removed.
(*Ans.* $V_L = 40^k \downarrow$, $H_L = 30^k \rightarrow$, $V_R = 40^k \downarrow$, $H_R = 30^k \rightarrow$)

4.41

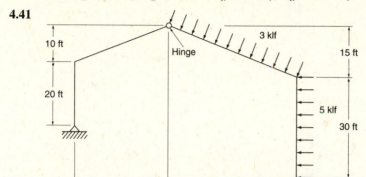

4.42 (*Ans.* $V_L = 64.5^k \downarrow$, $H_L = 101.9^k \leftarrow$, $V_R = 49.5^k \downarrow$, $H_R = 20.1^k \leftarrow$)

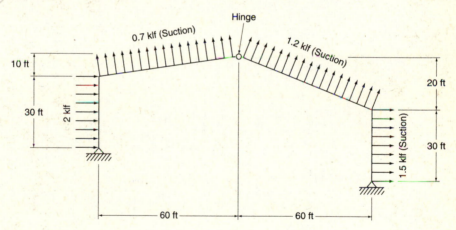

4.43

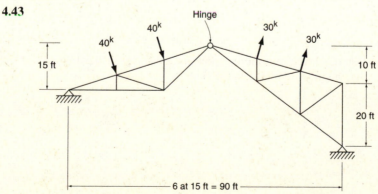

4.44   (*Ans.* $V_L = 65.8^k \uparrow$, $H_L = 48.8^k \rightarrow$, $V_R = 104.2^k \uparrow$, $H_R = 48.8^k \leftarrow$)

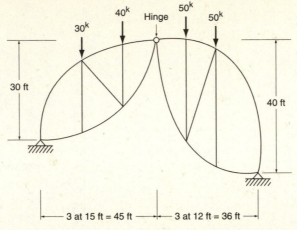

4.45

4.46   (*Ans.* $V_L = 96.3^k \uparrow$, $H_L = 97.2^k \rightarrow$, $V_R = 93.7^k \uparrow$, $H_R = 97.2^k \leftarrow$)

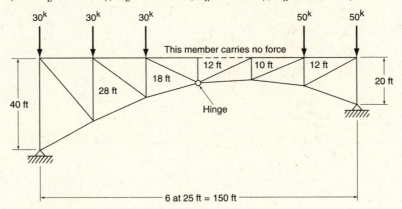

4.47   Determine the reaction components, cable sag at the 15- and 20-kip loads, and the maximum tensile force in the cable.

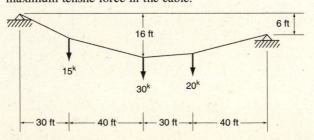

4.48 Determine the reaction components, cable sag at the 50-kN load, and the maximum tensile force in the cable.

(*Ans.* $V_L = 52.86$ kN ↑, $H_L = 142.8$ kN ←, $y = 5.70$ m)

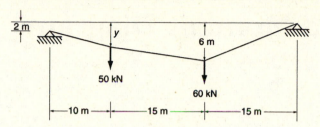

4.49 Determine the reaction components for the sag shown. Repeat the calculation if the center line sag is reduced to 4 ft.

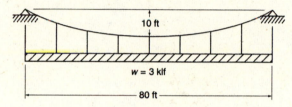

Chapter 5

Shear and Moment Diagrams

5.1 INTRODUCTION

An important phase of structural engineering is the understanding of shear and moment diagrams and their construction. It is doubtful that there is any other point at which careful study will give more reward in structural knowledge. These diagrams, from which values of shear and moment at any point in a beam are immediately available, are very convenient in design.

To examine the internal conditions of a structure, a free body must be taken out and studied to see what forces have to be present if the body is to be held in place, or in equilibrium. Shear and bending moments are two actions of the external loads on a structure that need to be understood to study properly the internal forces.

Shear is defined as the algebraic summation of the external forces to the left or to the right of a section that are perpendicular to the axis of the beam. In this book it is considered to be positive if the sum of the forces to the left is up or the sum of the forces to the right is down. The calculations for shear at two sections in a simple beam are given in Example 5.1. In each case the summations are made both to the left and to the right to prove that identical results are obtained. It is usually convenient to find a shear by considering the side of the section that has the least number of forces.

EXAMPLE 5.1

Find the shear at sections *a-a* and *b-b*, Figure 5.1.

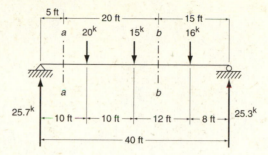

Figure 5.1

Solution.   Shear at section *a-a*:

$$V_{a-a} \text{ to left} = 25.7^k \uparrow, \text{ or } + 25.7^k$$
$$V_{a-a} \text{ to right} = 20 + 15 + 16 - 25.3 = 25.7^k \downarrow, \text{ or } + 25.7^k$$

Shear at section *b-b*:

$$V_{b-b} \text{ to left} = 25.7 - 20 - 15 = 9.3^k \downarrow = -9.3^k$$
$$V_{b-b} \text{ to right} = 16 - 25.3 = 9.3^k \uparrow = -9.3^k \quad \blacksquare$$

Bending moment is the algebraic sum of the moments of all of the external forces to the left or to the right of a particular section—the moments being taken about an axis through the centroid of the cross section. A positive sign herein indicates that the moment to the left is clockwise or the moment to the right is counterclockwise. A study of Figure 5.7 shows that positive moment at a section causes tension in the bottom fibers and compression in the top fibers of the beam at the section. Should the member under consideration be a vertical member, a fairly standard sign convention is to consider the right-hand side of the member to be the bottom side.

The bending moments at sections *a-a* and *b-b* in the beam of Example 5.1 are computed as follows:

Moment at section *a-a*:

$$M_{a-a} \text{ to left} = (25.7)(5) = 128.5'^k \curvearrowright, \text{ or } + 128.5'^k$$
$$M_{a-a} \text{ to right} = (25.3)(35) - (16)(27) - (15)(15) - (20)(5)$$
$$= 128.5'^k \curvearrowleft, \text{ or } + 128.5'^k$$

Moment at section *b-b*:

$$M_{b-b} \text{ to left} = (25.7)(25) - (20)(15) - (15)(5) = 267.5'^k \curvearrowright, \text{ or } + 267.5'^k$$
$$M_{b-b} \text{ to right} = (25.3)(15) - (16)(7) = 267.5'^k \curvearrowleft, \text{ or } + 267.5'^k$$

Later in this textbook when methods are introduced for the analysis of rather complicated statically indeterminate structures, a more detailed sign convention is

The Tridge, a triple-span pedestrian bridge using glued laminated timber, Midland, Michigan. (Courtesy of Unit Structures, Inc.)

presented for axial forces, shears, and bending moments. This is particularly necessary when analysis is made using matrix methods. (See Chaps. 20 through 21.)

5.2 SHEAR DIAGRAMS

Shear diagrams are quite simple to draw in most cases. The standard method is to start with the left end of the structure and work to the right. As each concentrated load or reaction is encountered, a vertical line is drawn to represent the quantity and direction of the force involved. Between the forces a horizontal line is drawn to indicate no change in shear.

Where uniform loads are encountered, the shear is changing at a constant rate per unit length and can be represented by a straight but inclined line on the diagram. When an ordinate on the shear diagram is above the line a positive shear is indicated because the sum of the forces to the left of that point is up. A shear diagram for a simple beam is drawn in Example 5.2.

EXAMPLE 5.2 ———————————————————————————————————————

Draw a shear diagram for the beam shown in Figure 5.2.

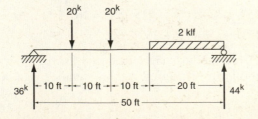

Figure 5.2

Solution.

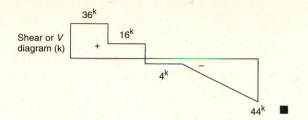

5.3 MOMENT DIAGRAMS

The moments at various points in a structure necessary for plotting a bending-moment diagram may be obtained algebraically by taking moments at those points, but the procedure is quite tedious if there are more than two or three loads applied to the structure. The method developed in the next section is much more practical.

5.4 RELATIONS AMONG LOADS, SHEARS, AND BENDING MOMENTS

There are significant mathematical relations among the loads, shears, and moments in a beam. These relations are discussed in the following paragraphs with reference to Figure 5.3.

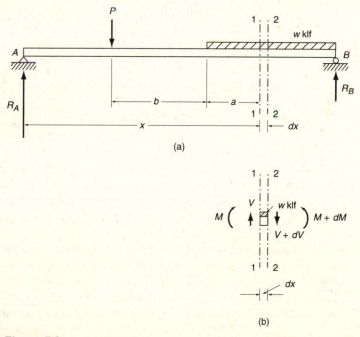

Figure 5.3

For this discussion an element of a beam of length dx [Figure 5.3(a)] is considered. This particular element is loaded with a uniform load of magnitude w klf (it doesn't have to be uniform). The shear and moment at the left end of this element at section 1-1 may be written as follows:

$$V_{1-1} = R_A - P - wa$$

$$M_{1-1} = R_A x - P(a + b) - \frac{wa^2}{2}$$

If we move a distance dx from section 1-1 to section 2-2 at the right end of the element, the new values of shear and moment can be written as at the end of this paragraph. The changes in these values may be expressed as dV and dM respectively.

$$V_{2-2} = V_{1-1} + dV = R_A - P - wa - w\,dx$$

$$M_{2-2} = M_{1-1} + dM = R_A x - P(a + b) - \frac{wa^2}{2} + V_{1-1}\,dx - \frac{w\,dx^2}{2}$$

From these expressions and with reference to part (b) of Figure 5.3, the changes in shear and moment in a dx distance are as follows:

$$\frac{dV}{dX} = -w$$

$$\frac{dM}{dX} = V \text{ neglecting the infinitesimal higher-order term involving } (dx)^2$$

From the preceding expressions the change in shear from one section to another can be written as follows (noting that some textbooks get rid of the minus sign by using a sign convention in which upward loads are given minus signs):

$$\Delta V = -\int_{1-1}^{2-2} w\,dx$$

And the change in moment in the same distance is

$$\Delta M = \int_{1-1}^{2-2} V\,dx$$

These two relationships are very useful to the structural designer. The first indicates that the rate of change of shear at any point equals the load per unit length at the point, meaning that the slope of the shear curve at any point is equal to the load at that point. The second equation indicates that the rate of change of moment at any point equals the shear. This relation means that the slope of the bending-moment curve at any point equals the shear.

The procedure for drawing shear and moment diagrams, to be described in Section 5.5, is based on the above equations and is applicable to all structures regardless of loads or spans. Before the process is described, it may be well to examine the equations more carefully. A particular value of dV/dx or dM/dx is good only for the portion of the structure at which the function is continuous. For instance, in the beam of Example 5.4 the rate of change of shear from A to B equals the uniform load, 4 klf. At the 30-kip load, which is assumed to act at a point, the rate of change of shear and the slope of the shear diagram are infinite, and a vertical line is drawn on the shear diagram to represent a concentrated load. The rate of change of moment from A to B has been constant, but at B the shear changes decidedly, as does the rate of change of moment. In other words, an expression for shear or moment from A to B is not the same as the expression from B to C beyond the concentrated load. The equations of the diagrams are not continuous beyond a point where the function is continuous.

5.5 MOMENT DIAGRAMS DRAWN FROM SHEAR DIAGRAMS

The change in moment between two points on a structure has been shown to equal the shear between those points times the distance between them ($dM = V\,dx$); therefore, the change in moment equals the area of the shear diagram between the points.

The relationship between shear and moment greatly simplifies the drawing of moment diagrams. To determine the moment at a particular section, it is necessary only to compute the total area beneath the shear curve, either to the left or to the right of the section, taking into account the algebraic signs of the various segments of the shear curve. Shear and moment diagrams are self-checking. If they are initiated at one end of a structure, usually the left, and check out to the proper value on the other end, the work probably is correct.

The authors usually find it convenient to compute the area of each part of a shear diagram and record that value on the part in question. This procedure is followed for the examples of this chapter where the values enclosed in the shear diagrams are areas. These recorded values appreciably simplify the construction of the moment diagrams.

The rate of change of moment at a point has been shown to equal the shear at that point ($dM/dx = V$). Whenever the shear passes through zero, the rate of change of moment must be zero ($dM/dx = 0$), and the moment is at a maximum or a minimum. If the moment diagram is being drawn from left to right and the shear diagram changes from positive to negative, the moment will reach a positive maximum at that point. Beyond that point it begins to diminish as the negative shear area is added. If the shear diagram changes from negative to positive, the moment reaches a negative maximum and then begins to taper off as the positive shear area is added.

This theory, which indicates that maximum moment occurs where the shear is zero, is not always applicable. In some cases (at the end of a beam or at a point of discontinuity) the maximum moment can occur where the shear is not zero. Should a cantilevered beam be loaded with gravity loads, the maximum shear and the maximum moment both will occur at the fixed end.

When shear and moment diagrams are drawn for inclined members, the components of loads and reactions perpendicular to the centroidal axes of the members are used and the diagrams are drawn parallel to the members. Examples 5.3 to 5.5 illustrate the procedure for drawing shear and moment diagrams for ordinary beams.

In studying these diagrams particular emphasis should be given to their shapes under uniform loads, between concentrated loads, and so on.

You will note that in the beam of Example 5.4 there is an axial tension force of 30 k for the entire length of the member. This force does not, however, affect the shear and moment in the beam.

EXAMPLE 5.3 _____

Draw shear and moment diagrams for the beam shown in Figure 5.4.

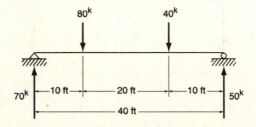

Figure 5.4

Solution.

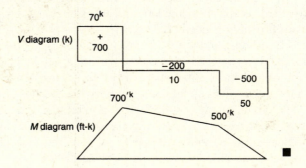

EXAMPLE 5.4 _____

Draw shear and moment diagrams for the structure shown in Figure 5.5.

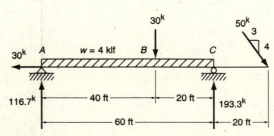

Figure 5.5

Solution.

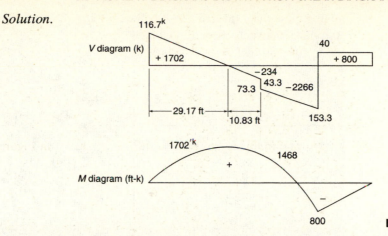

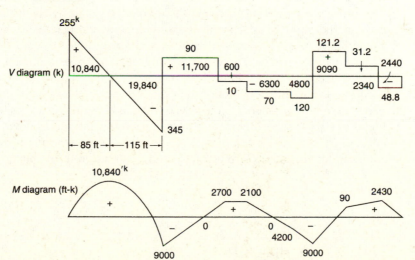

EXAMPLE 5.5 _____

Draw shear and moment diagrams for the cantilever-type structure shown in Figure 5.6.

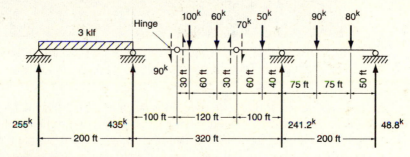

Figure 5.6

Solution.

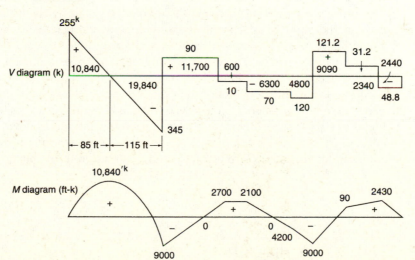

Pedestrian bridge in Pullen Park, Raleigh, North Carolina. (Courtesy of the American Institute of Timber Construction.)

Some structures have rigid arms (such as walls) fastened to them. If horizontal or inclined loads are applied to these arms, a twist or moment will suddenly be induced in the structure at the point of attachment. The fact that moment is taken about an axis through the centroid of the section becomes important because the lever arms of the forces applied must be measured to that centroid. To draw the moment diagram at the point of attachment, it is necessary to figure the moment at an infinitesimal distance to the left of the point and then add the moment applied by the arm. The moment exactly at the point of attachment is discontinuous and cannot be figured, but the moment immediately beyond that point is available.

The usual sign convention for positive and negative moments will apply in deciding whether to add or subtract the induced moment. It can be seen in Figure 5.7 that forces to the left of a section that tend to cause clockwise moments produce tension in the bottom fibers (+ moment), whereas those forces to the left that tend to cause counterclockwise moments produce tension in the upper fibers (− moment). Similarly, a counterclockwise moment to the right of the section produces tension in the bottom fibers; a clockwise moment produces tension in the upper fibers.

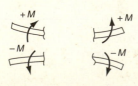

Figure 5.7

Shear and moment diagrams are shown in Example 5.6 for a beam that has a moment induced at a point by a rigid arm to which a couple is applied. The moment diagram is drawn from left to right. Considering the moment of the forces to the left of a section through the beam immediately after the rigid arm is reached, it can be seen that the couple causes a clockwise or positive moment, and its value is added to the moment obtained by summation of the shear-diagram areas up to the attached arm.

EXAMPLE 5.6

Draw shear and moment diagrams for the beam shown in Figure 5.8.

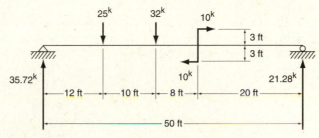

Figure 5.8

Solution.

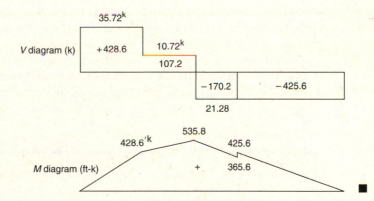

Students may at first have a little difficulty in drawing shear and moment diagrams for structures that are subjected to triangular loads. Example 5.7 is presented to demonstrate how to cope with them. The reactions are calculated for the beam shown in Figure 5.9 and the shear diagram is sketched. Note, however, that the point of zero shear is unknown and is shown in the figure as being located a distance x from the left support. The ordinate on the load diagram at this point is labeled y and can be expressed in terms of x by writing the following expression:

$$\frac{x}{30} = \frac{y}{2}$$

$$y = \frac{1}{15}x$$

At this point of zero shear the sum of the vertical forces to the left can be written as the upward reaction, 10 kips, minus the downward uniformly varying load to the left.

$$10 - \left(\frac{1}{2}\right)(x)\left(\frac{1}{15}x\right) = 0$$

$$x = 17.32 \text{ ft}$$

This value of x also can be determined by writing an expression for moment in the beam at a distance x from the support and then taking $dM/dx = 0$ of that expression and solving the result for x.

Finally, the moment at a particular point can be determined by taking (to the left or right of the point) the sum of the forces times their respective lever arms. For this example the moment at 17.32 ft is

$$M = (10)(17.32) - (10)\left(\frac{17.32}{3}\right) = 115.5'^k$$

The student may wonder why the authors (once x was determined) did not just sum up the area under the shear diagram from the left support to the point of zero shear. Such a procedure is correct, but the designer must be sure that he or she determines the area properly. When partial parabolas are involved it may be necessary to determine the areas by calculus instead of with the standard parabolic formulas. As a result, it may be simpler on some occasions simply to take moments of the loads and reactions about the points where moments are desired.

In Chapter 10, Figure 10.4 presents the properties (centers of gravity and areas) of several geometric figures. These values may be quite useful to the student preparing shear and moment diagrams for complicated loading situations.

EXAMPLE 5.7 ———————————————————————————————

Draw the shear and moment diagrams for the beam shown in Figure 5.9.

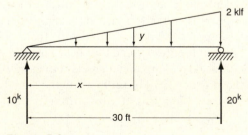

Figure 5.9

Solution.

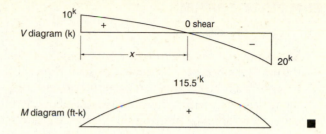

The reactions for the beam of Example 5.8 cannot be obtained using only the equations of statics. They have been computed by a method discussed in a later chapter, and the shear and moment diagrams have been drawn to show that the load, shear, and moment relationships are applicable to all structures.

EXAMPLE 5.8

Draw the shear and moment diagrams for the continuous beam shown in Figure 5.10 for which the reactions are given.

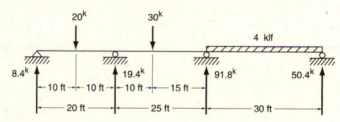

Figure 5.10

Solution.

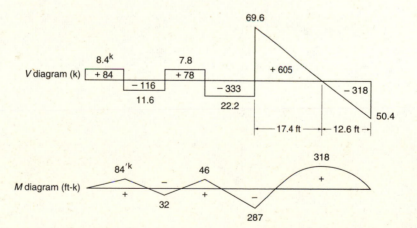

5.6 SHEAR AND MOMENT DIAGRAMS FOR STATICALLY DETERMINATE FRAMES

Shear and moment diagrams are very useful for rigid frames as well as for individual flexural members. The members of such frames cannot rotate with respect to each other at their connections. As a result, axial forces, shear forces, and bending moments are transferred between the members at the joints. These values must be accounted for in the preparation of the shear and moment diagrams.

For a first example, the frame of Example 5.9 shown in Figure 5.11 is considered. After the reactions are computed free-body diagrams are prepared for each of the members. The forces involved at the bottom of the column are obviously the reactions. Those at the top represent the effect of the rest of the frame on the top of the column and can be easily obtained by statics. Finally, the shear and moment diagrams are prepared.

EXAMPLE 5.9 _____

Draw shear and moment diagrams for the frame shown in Figure 5.11.

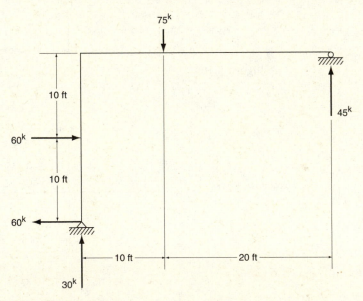

Figure 5.11

Solution.

Free Body Diagrams

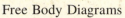

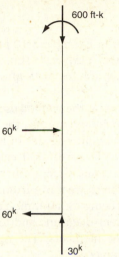

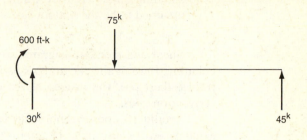

Shear Diagrams

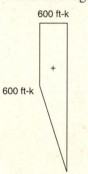

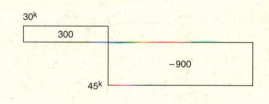

Moment Diagrams

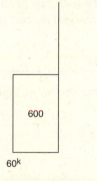

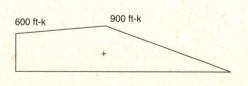

For a sign convention for this first example the authors assumed that they were standing underneath the frame. Thus, the right-hand side of the column was assumed to be the bottom side.   ■

Shear and moment diagrams are drawn in Example 5.10 for a frame with two columns. Here again, the authors assumed that they were standing underneath the frame; therefore, the insides of the columns are the bottom sides so far as sign convention goes.

Should a frame have more than two columns, the method used here for these two examples will cause us to be confused in deciding which is the bottom side of the columns. For such a situation we can simply assume one side of all the columns (say, the right-hand side) is the bottom side. This, however, will cause us to have sign convention problems at the tops of all the columns, except for the far left-hand one. That is, we will have to change sign convention as we move from the beam to the top of all the other columns. Perhaps a better procedure for multiple-column frames, then, is simply to leave the signs off of the moment diagrams and to draw the moment diagrams on the compression sides of all the members: The results for horizontal members will be the same as previously obtained.

Free body diagrams were not shown for the various members of this frame, as the authors felt that the reader would now be capable of drawing the shear and moment diagrams directly from the external loads and reactions applied to the frame.

EXAMPLE 5.10

Draw shear and moment diagrams for the frame shown in Figure 5.12.

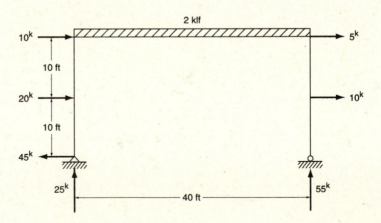

Figure 5.12

Solution.   (Assuming insides of columns are bottom sides)

Shear Diagrams

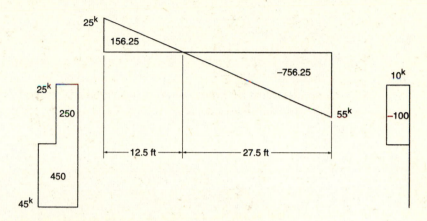

Moment Diagrams

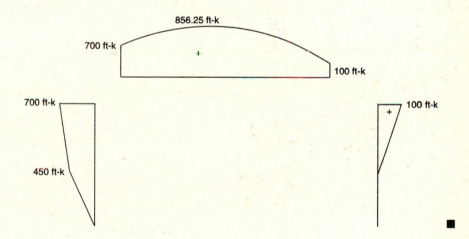

PROBLEMS

For Problems 5.1 to 5.36 draw shear and moment diagrams for the structures.

5.1

5.2   Repeat Problem 5.1 if the 40 k load is changed to 100 k.
(*Ans.* max $V = $ 104 k, max $M = 1120$ ft-k)

5.3

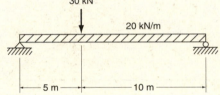

3 klf

40^k

|←——————— 30 ft ———————→|←— 10 ft —→|

5.4   Repeat Problem 5.3 if the 3 klf load is changed to 1.5 klf.
(*Ans.* max $V = 40$ k, max $M = -400$ ft-k)

5.5

30^k

2 klf

|←——— 12 ft ———→|←——— 12 ft ———→|←——— 12 ft ———→|

5.6   (*Ans.* max $V = 170$ kN, max $M = 640$ kN · m)

30 kN

20 kN/m

|←—— 5 m ——→|←——— 10 m ———→|

5.7

60^k

50^k

30^k

4

3

1

3

|←— 20 ft —→|← 15 ft →|←——— 30 ft ———→|←——— 30 ft ———→|

5.8   (*Ans.* max $V = 100.5$ k, max $M = 968$ ft-k)

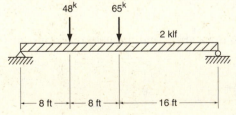

48^k

65^k

2 klf

|←— 8 ft —→|←— 8 ft —→|←——— 16 ft ———→|

5.9

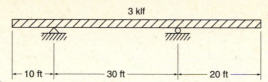

3 klf

|← 10 ft →|← 30 ft →|← 20 ft →|

5.10   (*Ans.* max V = 160 kN, max M = −640 kN · m)

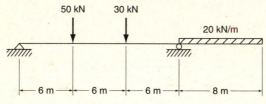

50 kN   30 kN   20 kN/m

|← 6 m →|← 6 m →|← 6 m →|← 8 m →|

5.11

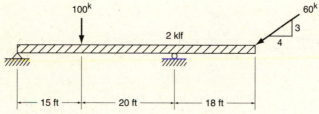

100^k 2 klf 60^k 3   4

|← 15 ft →|← 20 ft →|← 18 ft →|

5.12   (*Ans.* max V = 54 k, max M = −1098 ft-k)

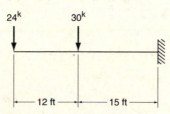

24^k 30^k

|← 12 ft →|← 15 ft →|

5.13

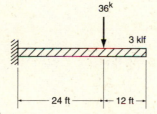

36^k 3 klf

|← 24 ft →|← 12 ft →|

5.14   (*Ans.* max V = 120 kN, max M = −960 kN · m)

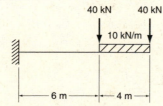

40 kN   40 kN   10 kN/m

|← 6 m →|← 4 m →|

5.15

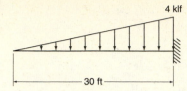

4 klf

30 ft

5.16   (*Ans.* max V = 50 k, max M = −900 ft-k)

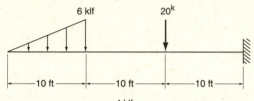

6 klf   20^k

10 ft   10 ft   10 ft

5.17

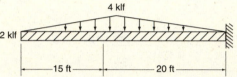

4 klf

2 klf

15 ft   20 ft

5.18   (*Ans.* max V = 120 k, max M = −2362.5 ft-k)

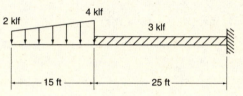

2 klf   4 klf

3 klf

15 ft   25 ft

5.19

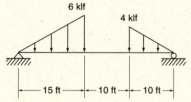

6 klf

4 klf

15 ft   10 ft   10 ft

5.20   (*Ans.* max V = 57.68 k, max M = 278.6 ft-k)

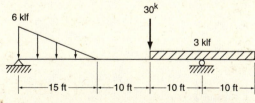

6 klf   30^k

3 klf

15 ft   10 ft   10 ft   10 ft

5.21

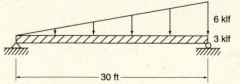

6 klf

3 klf

30 ft

5.22   (*Ans.* max V = 79.64 k, max M = 644.4 ft-k)

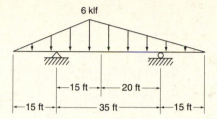

5.23

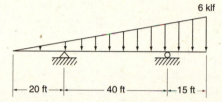

5.24   (*Ans.* max V = 317.9 kN, max M = 2526 kN · m)

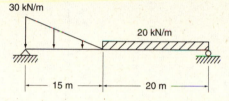

5.25

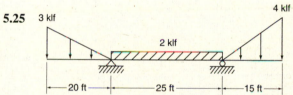

5.26   (*Ans.* max V = 115 k, max M = −1166.7 ft-k)

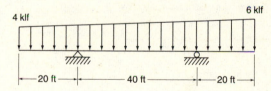

5.27

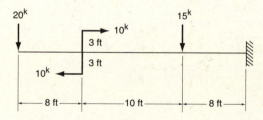

5.28 (*Ans.* max $V = 22.4$ k, max $M = 83.7$ ft-k)

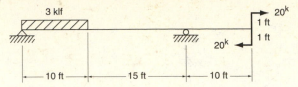

5.29

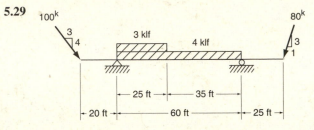

5.30 (*Ans.* max $V = 100$ k, max $M = -2417.8$ ft-k)

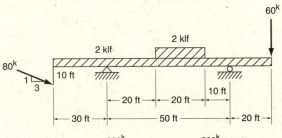

5.31

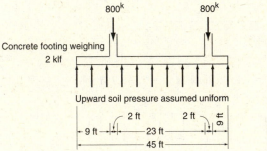

5.32 (*Ans.* max $V = 89.66$ k, max $M = 2933$ ft-k)

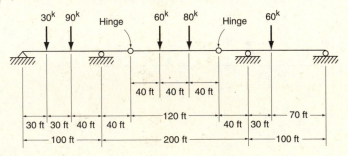

5.33

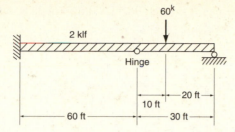

5.34   (*Ans.* max $V = 80^k$, max $M = -2600$ ft-k)

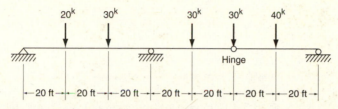

5.35   *Given:* Moment at interior support $= -1274$ ft-k

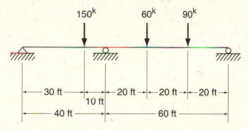

5.36   *Given:* Moment at fixed end $= -147.7$ ft-k; other reactions as shown.
(*Ans.* max $V = 25.48^k$, max $M = -164.5$ ft-k)

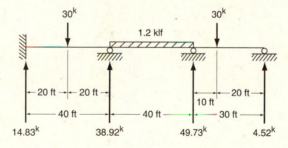

For Problems 5.37 to 5.39, for the moment diagrams and dimensions given draw the shear diagrams and load diagrams. Assume upward forces are reactions.

5.37

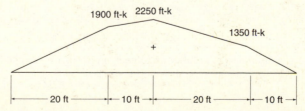

5.38 (*Ans.* Reactions and loads left to right: 96.5 kN, 80 kN, 40 kN, 50 kN, 73.5 kN)

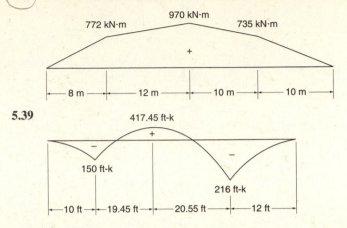

5.39

For Problems 5.40 to 5.49 draw shear and moment diagrams for the frames.

5.40 (*Ans.* max $V = 58^k$, max $M = 841$ ft-k)

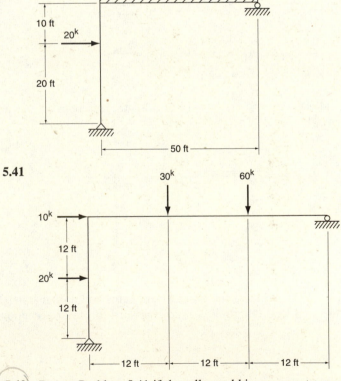

5.41

5.42 Repeat Problem 5.41 if the roller and hinge supports are swapped.
(*Ans.* max $V = 46.67$ k, max $M = 520$ ft-k)

5.43

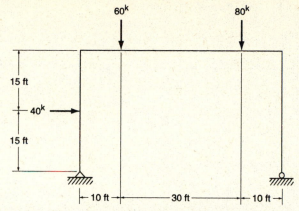

5.44   (*Ans.* max V = 113.33 k, max M = 1066.7 ft-k)

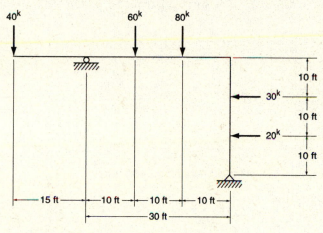

5.45

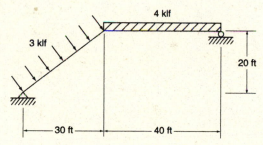

5.46   (*Ans.* max $V = 90^k$, max $M = 1350$ ft-k)

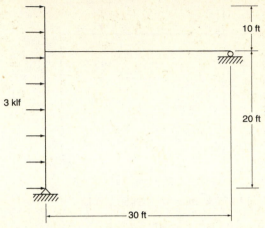

5.47

5.48   (*Ans.* max $V = 88.85$ k, max $M = 2400$ ft-k)

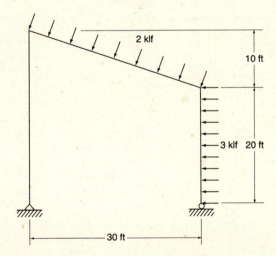

5.49

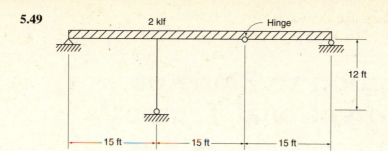

2 klf

Hinge

12 ft

15 ft   15 ft   15 ft

Chapter 6

Introduction to Plane or Two-Dimensional Trusses

6.1 GENERAL

Italian architect Andrea Palladio (1518–1580) is thought to have first used modern trusses, although his designs were not rational. He may have revived some old Roman designs and probably sized the members by some rules of thumb (perhaps also revivals of old Roman rules). Palladio's extensive writing in architecture included detailed descriptions and drawings of wooden trusses quite similar to those used today. After his time trusses were forgotten for 200 years, until they were reintroduced by Swiss designer Ulric Grubermann.

A truss is defined in Chapter 1 as a structure formed by a group of members arranged in the shape of one or more triangles. Because the members are assumed to be connected with frictionless pins, the triangle is the only stable shape. Study of the truss of Figure 6.1(a) shows that it is impossible for the triangle to change shape under load unless one or more of the sides is bent or broken. Figures of four or more sides are not stable and may collapse under load, as seen in Figure 6.1(b) and (c). These structures may be deformed without a change in length of any of their members. We will see, however, that there are many stable trusses that include one or more figures that are not triangles. A careful study will show they consist of separate groups of triangles that are connected together according to definite rules, forming nontriangular but stable figures in between.

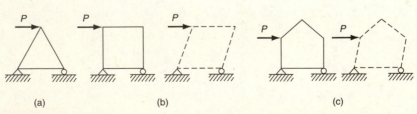

(a)   (b)   (c)

Figure 6.1

Truss for conveyor of West Virginia quarry of Pennsylvania Glass Sand Corporation. (Courtesy Bethlehem Steel Corporation.)

6.2 ASSUMPTIONS FOR TRUSS ANALYSIS

The following assumptions are made in order to simplify the analysis of trusses:

1. Truss members are connected with frictionless pins. (Pin connections are used for very few trusses erected today, and no pins are frictionless. A heavy bolted or welded joint is a far cry from a frictionless pin.)
2. Truss members are straight. (If they were not straight, the axial forces would cause them to have bending moments.)
3. The deformations of a truss under load, caused by the changes in lengths of the individual members, are not of sufficient magnitude to cause appreciable changes in the overall shape and dimensions of the truss. Special consideration may have to be given to some very long and flexible trusses.
4. Members are so arranged that the loads and reactions are applied only at the truss joints.

Examination of roof and bridge trusses will prove this last statement to be generally true. In buildings with roof trusses the beams, columns, and bracing frame directly into the truss joints. Roof loads are transferred to trusses by horizontal beams, called *purlins,* that span the distance between the trusses. The roof is supported directly by the purlins or is supported by rafters, or subpurlins, which run parallel to

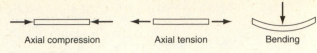

Axial compression   Axial tension   Bending

Figure 6.2

the trusses and are supported by the purlins. The purlins are placed at the truss joints unless the top-chord panel lengths become exceptionally long, in which case it is sometimes economical to place purlins between the joints, although some bending will develop in the top chords. (Some types of roofing, such as corrugated steel, gypsum slabs, and others may be laid directly on the purlins. The purlins then have to be spaced at intermediate points along the top chord so as to provide a proper span for the roofing they directly support.) The loads supported by a highway bridge are transferred to the trusses at the joints by beams running underneath the roadway as shown in Figure 9.16 and described in Section 9.13.

6.3 EFFECT OF ASSUMPTIONS

The effect of the foregoing assumptions is to produce an ideal truss, whose members have only axial forces. A member with axial force only is subject to a push or pull with no bending present, as illustrated in Figure 6.2. (Even if all the assumptions were perfectly true, there would be some bending in a member caused by its own weight.)

Forces obtained on the basis of these simplifying assumptions are very satisfactory in most cases and are referred to as *primary forces*. Structures are sometimes analyzed without using some or all of these assumptions. Forces caused by conditions not considered in the primary force analysis are said to be *secondary forces*.

6.4 TRUSS NOTATION

A common system of denoting the members of a truss is shown in Figure 6.3. The joints are numbered from left to right, the bottom labeled L (for lower) and the top joints labeled U (for upper). Should there be, in more complicated trusses, joints between the lower and upper joints, they may be labeled M (for middle).

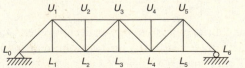

Figure 6.3

The various members of a truss often are referred to by the following names, references being made to Figure 6.3.

1. *Chords* are those members forming the outline of the truss, such as members U_1U_2 and L_4L_5.
2. *Verticals* are named on the basis of their direction in the truss, such as members U_1L_1 and U_3L_3.
3. *Diagonals* also are named on the basis of their direction in the truss, such as members U_1L_2 and L_4U_5.
4. *Web members* include the verticals and diagonals of a truss, and most engineers consider them to include the end diagonals, or *end posts,* such as L_0U_1 and U_5U_6.

There are other frequently used truss-notation systems. For instance, for computer-programming purposes it is convenient to assign a number to each joint and each member of a truss. Such a system is illustrated in Chapter 7 of this text.

6.5 ROOF TRUSSES

The purposes of roof trusses are to support the roofs that keep the elements out (rain, snow, wind) and to support the loads connected underneath (ducts, piping, ceiling), as well as their own weights.

The designer often is concerned with the problem of selecting a truss or a beam to span a given opening. Should no other factors be present, the decision probably would be based on consideration of economy. The smallest amount of material will nearly always be used if a truss is selected for spanning a certain opening; however, the cost of fabrication and erection of trusses probably will be appreciably higher than that required for beams. For shorter spans the overall cost of beams (material plus fabrication and erection) will definitely be less, but as the spans become greater, the higher fabrication and erection costs of trusses will be more than offset by their weight saving. A further advantage of trusses is that for the same amounts of material they have greater stiffnesses than do beams.

It is impossible to give a lower economical span for trusses. They may be used for spans as small as 30 to 40 ft and as large as 300 to 400 ft. Beams may be economical for some applications for spans much greater than the lower limits mentioned for trusses.

Roof trusses can be flat or peaked. In the past the peaked roof trusses have probably been used more for short-span buildings and the flatter trusses for the longer spans. The trend today for both long and short spans, however, seems to be away from the peaked trusses and toward the flatter ones, the change being due to the appearance desired and perhaps a more economical construction of roof decks.

On the following page, Figure 6.4 shows quite a few of the different types of roof trusses that have been used in the past.

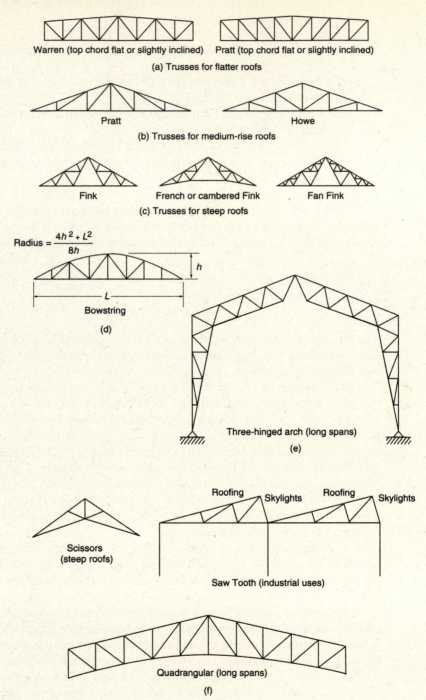

Figure 6.4   Various types of roof trusses.

Willard Bridge over Kansas River, north of Willard, Kansas. (Courtesy of the American Institute of Steel Construction, Inc.)

6.6 BRIDGE TRUSSES

As bridge spans become longer and loads heavier, trusses begin to be competitive economically. Early bridge trusses were constructed with wood, but they had several disadvantages. First, they were subject to deterioration from wind and water. As a result covered bridges were introduced, and such structures often would last for quite a few decades. Nevertheless, wooden truss bridges were subject to destruction by fire, particularly railroad bridges. In addition, there were some problems with the loosening of the fasteners under moving loads over time and as the loads became heavier.

As a result of the several preceding disadvantages, wooden truss bridges faded from use toward the end of the 19th century and structural steel bridges, which did not require extensive protection from the elements and whose joints had higher fatigue resistance, took over the market. (There were also some earlier iron truss bridges.)

Today existing steel truss bridges are being steadily replaced with steel, precast-concrete, or prestressed-concrete beam bridges. It seems that the age of steel truss bridges is over except for spans of more than several hundred feet (a very small percentage of the total). Even for these longer spans there is much competition from other types of structures such as cable-stayed bridges, prestressed-concrete box girder bridges, arches, and so on.

The reader has often seen highway bridges in which the trusses were on the sides. As you rode across the bridge you could see overhead lateral bracing between the trusses. This type of bridge is said to be a *through bridge*. The floor system is supported by floor beams that run under the roadway and between the bottom chord joints of the trusses.

In the *deck bridge* the roadway is placed on top of the trusses or girders. Deck construction has every advantage over through construction except for underclearance. There is unlimited overhead horizontal and vertical clearance, and future expansion is more feasible. Another very important advantage is that supporting trusses or girders can be moved close together, reducing lateral moments in the floor system. Other advantages of the deck truss are simplified floor systems and possible reduction in the sizes of piers and abutments due to reductions in their heights. Finally, the very pleasing appearance of deck structures is another reason for their popularity.

Today's bridge designer tries to prevent any sense of confinement for the users of his or her bridges. In trying to achieve this goal the designer attempts to eliminate any overhead bracing or truss members that will protrude above the roadway level. The result is that the deck structure is again desirable unless underclearance requirements prevent its use or the spans are so large as to make it impractical.

Sometimes short-span through bridge trusses were so shallow that adequate depth was not available to provide overhead bracing and at the same time leave sufficient vertical clearance above the roadway for the traffic. As a result, the bracing was placed underneath the roadway. Bridges of the through type without overhead

Brown's Bridge, Forsyth and Hall Counties, Gainesville, Georgia. (Courtesy of the American Institute of Steel Construction, Inc.)

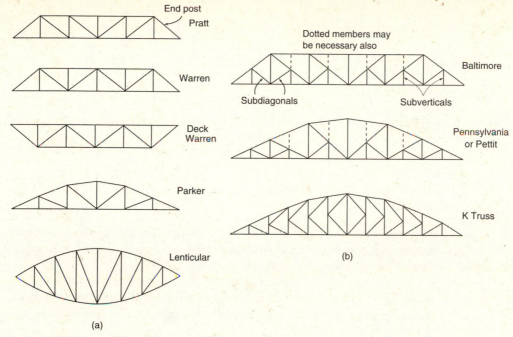

Figure 6.5

bracing are referred to as *half-through* or *pony bridges*. One major problem with pony truss bridges was the difficulty of providing adequate lateral bracing for the top-chord compression members of the trusses. It would be very unlikely for a pony truss to be economical today in an age where beams have captured the short-span bridge market.

Figure 6.5(a) shows some of the types of bridge trusses that have been used in the past for medium spans.

The deeper a truss is made for the same size chord members, the greater will be its resisting moment. If the depth of a truss is varied across its span in some proportion to its bending moments the result will be a lighter truss, but its fabrication cost per pound of steel will be higher than for a parallel-chord truss. As spans become longer, the weight saving achieved by varying truss depths with bending moments will outweigh the extra fabrication costs and the so-called curved-chord trusses become economical.

Figure 6.5(b) shows some types of trusses that have been frequently used for long spans. The Baltimore and Pennsylvania trusses shown are said to be subdivided trusses in that the unsupported lengths of some of the members have been reduced by introducing short members called subdiagonals and subverticals.

6.7 ARRANGEMENT OF TRUSS MEMBERS

The triangle has been shown to be the basic shape from which trusses are developed because it is the only stable shape. (For this discussion the reader needs to remember that the members are assumed to be connected at their joints with frictionless pins.)

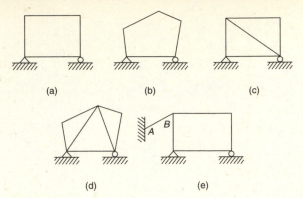

Figure 6.6

Other shapes, such as the ones shown in Figures 6.6(a) and (b) are obviously unstable and may possibly collapse under load. Structures such as these can, however, be made stable by one of the following methods:

1. Addition of members so that the shapes are made to consist of triangles. The structures of Figures 6.6(a) and (b) are stabilized in this manner in (c) and (d), respectively.
2. Using a member to tie the unstable structure to a stable support. Member *AB* performs this function in Figure 6.6(e).
3. Making some or all of the joints of an unstable structure rigid, so they become moment resisting. A figure with moment-resisting joints, however, does not coincide with the definition of a truss (that is, members connected with frictionless pins, and so on).

6.8 STATICAL DETERMINACY OF TRUSSES

The simplest form of truss, a single triangle, is illustrated in Figure 6.7(a). To determine the unknown forces and reaction components for this truss, it is possible to isolate the joints and write two equations, $\Sigma H = 0$ and $\Sigma V = 0$, for each.

The single-triangle truss may be expanded into a two-triangle one by the addition of two new members and one new joint. In Figure 6.7(b), triangle *ABD* is added by installing new members *AD* and *BD* and the new joint *D*. A further expansion with

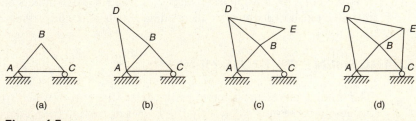

Figure 6.7

a third triangle is made in part (c) of the figure by the addition of members *BE* and *DE* and joint *E*. For each of the new joints *D* and *E*, a new pair of equations is available for calculating the two new-member forces. As long as this procedure of expanding the truss is followed, the truss will be statically determinate internally. Should new members be installed without adding new joints, such as member *CE* in Figure 6.7(d), the truss will become statically indeterminate because no new joint equations are made available to find the new-member forces.

From the information above an expression can be written for the relationship that must exist between the number of joints and the number of members and reaction components for a particular truss if it is to be statically determinate internally. (The identification of externally determinate structures has previously been discussed in Chapter 4.) In the following discussion, *m* is the number of members, *j* is the number of joints, and *r* is the number of reaction components.

If the number of equations available ($2j$) is sufficient to obtain the unknowns, the structure is statically determinate, from which the following relation may be written:

$$2j = m + r$$

or, as more commonly written,

$$m = 2j - r$$

Before an attempt is made to apply this equation, it is necessary to have a structure that is stable externally or the results are meaningless; therefore, *r* is the least number of reaction components required for external stability. Should the structure have more external reaction components than necessary for stability (and thus be statically indeterminate externally), the value of *r* remains the least number of reaction components required to make it stable externally. This statement means that *r* will equal 3 for the usual statics equations plus the number of any additional condition equations that may be available.

It is possible to build trusses that have too many members to be analyzed by statics, in which case they are statically indeterminate internally, and *m* will exceed $2j - r$ because there are more members present than are absolutely necessary for stability. The extra members are said to be redundant members. If *m* is 3 greater than $2j - r$, there are three redundant members, and the truss is internally statically indeterminate to the third degree. Should *m* be less than $2j - r$, there are not enough members present for stability.

A brief glance at a truss usually will show if it is statically indeterminate. Trusses having members that cross over each other or members that serve as the sides for more than two triangles may quite possibly be indeterminate. The $2j - r$ expression should be used, however, if there is any doubt about the determinacy of a truss, because it is not difficult to be mistaken. On the following page, Figure 6.8 shows several trusses and the application of the expression to each. The small circles on the trusses indicate the joints.

Little explanation is necessary for most of the structures shown, but some remarks may be helpful for a few. The truss of Figure 6.8(e) has five reaction components and is statically indeterminate externally to the second degree; however, two of the reaction components could be removed and leave a structure with sufficient

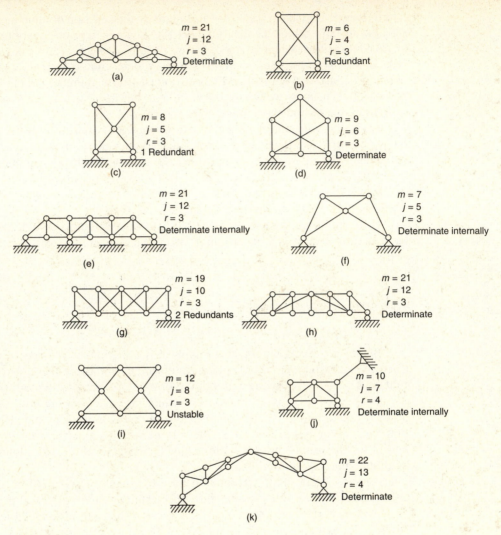

Figure 6.8

reactions for stability. The least number of reaction components for stability is 3, m is 21, and j is 12; applying the equation $m = 2j - r$ yields

$$21 = 24 - 3 = 21 \qquad \text{statically determinate internally}$$

The truss of Figure 6.8(j) is externally indeterminate because there are five reaction components and only four equations available. With $r = 4$ the structure is shown to be statically determinate internally. The three-hinged arch of Figure 6.8(k) has four reaction components, which is the least number of reaction components required for stability; so, $r = 4$. Application of the equation shows the arch to be statically determinate internally.

In Chapter 13, pertaining to the analysis of statically indeterminate structures, it will be seen that the values of the redundants may be obtained by applying certain simultaneous equations. The number of simultaneous equations equals the total number of redundants, whether internal, external, or both. It therefore may seem a little foolish to distinguish between internal and external determinacy. The separation is particularly questionable for some types of internally and externally redundant trusses where no solution of the reactions is possible independently of the member forces, and vice versa.

If a truss is externally determinate and internally indeterminate, the reactions may be obtained by statics. If the truss is externally indeterminate and internally determinate, the reactions are dependent on the internal-member forces and may not be determined by a method independent of those forces. If the truss is externally and internally indeterminate, the solution of the forces and reactions will be performed simultaneously. (For any of these situations, it may be possible to obtain a few forces here and there by statics without going through the indeterminate procedure necessary for complete analysis.) This entire subject is discussed in detail in later chapters.

6.9 USE OF SECTIONS

An indispensable part of truss analysis, as in beam analysis, is the separation of the truss into two parts with an imaginary section. The part of the truss on one side of the section is removed and studied independently. The loads applied to this free body include the axial forces of the members that have been cut by the section and any loads and reactions that may be applied externally.

Application of the equations of statics to isolated free bodies enables one to determine the forces in the cut members if the free bodies are carefully selected so that the sections do not pass through too many members whose forces are unknown. There are only three equations of statics, and no more than three unknowns may be determined from any one section.

After the student has analyzed a few trusses, he or she will have little difficulty in most cases in selecting satisfactory locations for the sections. The student is not encouraged to remember specific sections for specific trusses, although he or she will probably unconsciously fall into such a habit as time goes by. At this stage the student needs to consider each case individually without reference to other similar trusses.

6.10 HORIZONTAL AND VERTICAL COMPONENTS

It is convenient to work with horizontal and vertical components in the computation of forces in truss members, as in the computation of reactions. The $\Sigma H = 0$ and $\Sigma V = 0$ equations of statics generally are written on the basis of a pair of axes that are horizontal and vertical. Because the forces in truss members are determined successively across a truss, much time will be saved if the vertical and horizontal components of forces in inclined members are recorded for use in applying the equations to other members. The use of components is clearly illustrated in the example problems of the sections that follow.

6.11 ARROW CONVENTION

The sign convention for tensile and compressive forces ($+$ and $-$, respectively) has already been mentioned. Arrows also are used throughout the text to represent the character of forces. The arrows indicate what members are doing to resist the axial forces applied to them by the remainder of the truss. For example, if a truss is compressing a certain member from each end ($\rightarrow$ ———— $\leftarrow$), the member will push back against the compressive forces ($\leftarrow$ ———— $\rightarrow$). This arrow convention is used for members in compression. The arrow convention for a member in tension is just the opposite, because a member that is being pulled or stretched from the ends ($\leftarrow$ ———— $\rightarrow$) will resist by pulling back ($\rightarrow$ ———— $\leftarrow$).

After some practice in the analysis of trusses, it is possible to determine by examination the character of the forces in many of the members of a truss. The reader should try to picture whether a member is in tension or compression before making the actual calculations. In this way a better understanding of the action of trusses under load will be obtained. The following paragraphs will show that it is possible to determine entirely by mathematical means the character as well as the numerical value of the forces.

Ford Automobile Assembly Plant, Milpitas, California. (Courtesy of the American Institute of Steel Construction.)

6.12 METHOD OF JOINTS

An imaginary section may be completely passed around a joint in a truss, regardless of its location, completely isolating it from the remainder of the truss. The joint has become a free body in equilibrium under the forces applied to it. The equations $\Sigma H = 0$ and $\Sigma V = 0$ may be applied to the joint to determine the unknown forces in members meeting there. It should be evident that no more than two unknowns can be determined at a joint with these two equations.

A person learning the method of joints may initially find it necessary to draw a free-body sketch for every joint in a truss he or she is analyzing. After you have computed the forces in two or three trusses, it will be necessary to draw the diagrams for only a few joints, because you will be able to visualize with ease the free bodies involved. The most important thing for the beginner to remember is that you are interested in only one joint at a time. You must keep your attention away from the loads and forces at other joints. Your concern is only with the forces at the one joint on which you are working. Another very helpful suggestion for the reader is to *draw large sketches*. The method of joints is illustrated by Example 6.1.

EXAMPLE 6.1

By using the method of joints, find all forces in the truss shown in Figure 6.9.

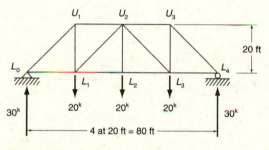

Figure 6.9

Solution. Considering joint L_0,

$$\Sigma V = 0$$

$$30 - V_{L_0 U_1} = 0$$

$$V_{L_0 U_1} = 30^k \text{ compression}$$

An examination of the joint shows a vertical reaction of 30^k acting upward. The equation $\Sigma V = 0$ indicates the members meeting there must supply 30^k downward. A member that is horizontal, such as $L_0 L_1$, can have no vertical component of force; therefore, $L_0 U_1$ must supply the entire amount and 30^k will be its vertical component. The arrow convention shows that $L_0 U_1$ is in compression. From its slope ($20:20$, or $1:1$) the horizontal component can be seen to be 30^k also.

$$\Sigma H = 0$$

$$-30 + F_{L_0 L_1} = 0$$

$$F_{L_0 L_1} = 30^k \text{ tension}$$

The application of the $\Sigma H = 0$ equation shows $L_0 U_1$ to be pushing horizontally to the left against the joint with a force of 30^k. For equilibrium, $L_0 L_1$ must pull to the right away from the joint with the same force. The arrow convention shows the force is tensile.

Considering joint U_1,

$$\Sigma V = 0$$

$$30 - F_{U_1 L_1} = 0$$

$$F_{U_1 L_1} = 30^k \text{ tension}$$

The force in $L_0 U_1$ has previously been found to be compressive with vertical and horizontal components of 30^k each. Since it is pushing upward at joint U_1 with a force of 30^k, $U_1 L_1$ (the only other member at the joint that has a vertical component) must pull down with a force of 30^k in order to satisfy the $\Sigma V = 0$ equation.

$$\Sigma H = 0$$

$$30 - F_{U_1 U_2} = 0$$

$$F_{U_1 U_2} = 30^k \text{ compression}$$

Member $L_0 U_1$ is pushing to the right horizontally with a force of 30^k. For equilibrium $U_1 U_2$ is pushing back to the left with 30^k.

Considering joint L_1,

$$\Sigma V = 0$$

$$30 - 20 - V_{L_1 U_2} = 0$$

$$V_{L_1 U_2} = 10^k \text{ compression}$$

$$\Sigma H = 0$$

$$-30 - 10 - \; + F_{L_1 L_2} = 0$$

$$F_{L_1 L_2} = 40^k \text{ tension}$$

The forces in all of the truss members may be calculated in a similar manner with the following results:

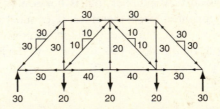

The resultant forces for inclined members may be determined from the square root of the sum of the squares of the vertical and horizontal components of force. An easier method is to write ratios comparing the resultant axial force of a member and its horizontal or vertical component with the true length of the member and its horizontal or vertical component. By letting F, H, and V represent the force and its components and l, h, and v the length and its components, the ratios of Figure 6.10 are developed.

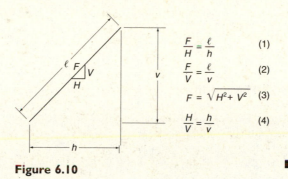

$$\frac{F}{H} = \frac{l}{h} \quad (1)$$

$$\frac{F}{V} = \frac{l}{v} \quad (2)$$

$$F = \sqrt{H^2 + V^2} \quad (3)$$

$$\frac{H}{V} = \frac{h}{v} \quad (4)$$

Figure 6.10

The method of joints may be used to compute the forces in all of the members of many trusses. The trusses of Examples 6.2 and 6.3 and all of the home problems at the end of this chapter fall into this category. There are, however, a large number of trusses that need to be analyzed by a combination of the method of joints and the methods discussed in Chapter 7. The authors like to calculate as many forces as possible in a truss by using the method of joints. At joints where they have a little difficulty they take moments to obtain one or two forces, as described in Chapter 7. They then continue the calculations as far as possible by joints until they reach another point of difficulty where they again take moments, and so on.

EXAMPLE 6.2

Find all the forces in the truss of Figure 6.11.

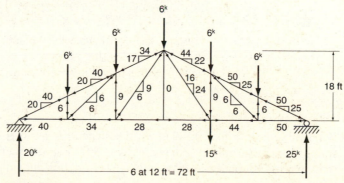

Figure 6.11

EXAMPLE 6.3

Find all forces in the truss of Figure 6.12.

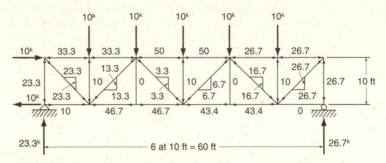

Figure 6.12   ■

Fairview-St. Mary's Skyway System, Minneapolis, Minnesota. (Courtesy of the American Institute of Steel Construction, Inc.)

PROBLEMS

For Problems 6.1 to 6.22 classify the structures as to their internal and external stability and determinacy. For statically indeterminate structures include the degree of redundancy internally or externally. (The small circles on the trusses indicate the joints.)

6.1

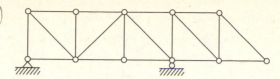

j=11   m=19
r=3

6.2   (*Ans.* Statically indeterminate internally to first degree)

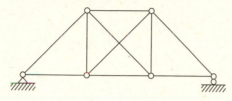

6.3

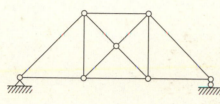

j=7   m=12
r=3
ind. 1st degree internally

6.4   (*Ans.* Statically indeterminate internally to first degree)

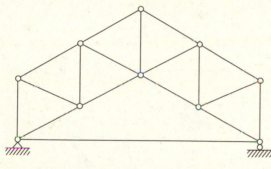

6.5

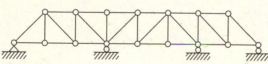

j=16   m=29
r=3
indet. externally 2nd

6.6   (*Ans.* Statically determinate externally and internally)

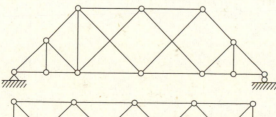

6.7

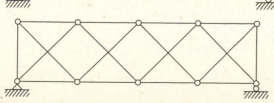

6.8   (*Ans.* Statically determinate externally and internally)

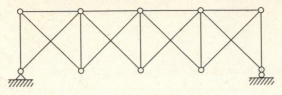

6.9

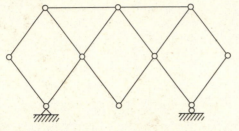

6.10   (*Ans.* Unstable)

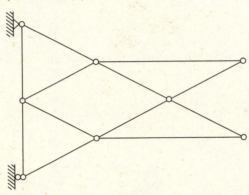

6.11

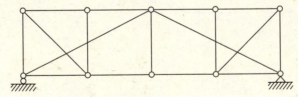

6.12   (*Ans.* Unstable externally)

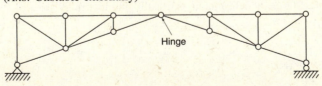

Hinge

6.13

Hinge

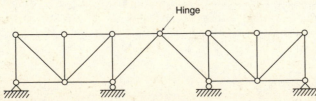

$j = 13$
$r = 4$
$m = 22$

ind. ext. 2nd

6.14   (*Ans.* Statically determinate externally and internally)

Hinge

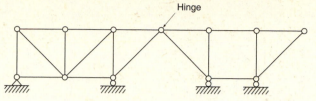

6.15

Hinges

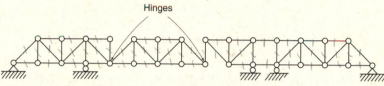

6.16   (*Ans.* Statically determinate externally and internally)

$j = 30$ $m = 54$ $r = 6$

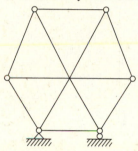

6.17

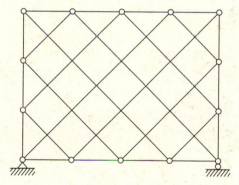

6.18   (*Ans.* Statically determinate externally and internally)

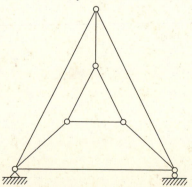

6.19

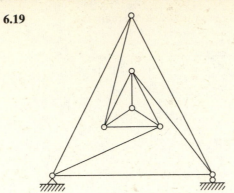

6.20   (*Ans.* Statically indeterminate internally to second degree)

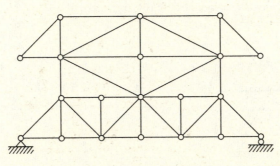

6.21

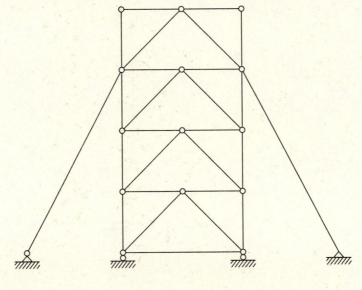

6.22   (*Ans.* Statically indeterminate internally to first degree)

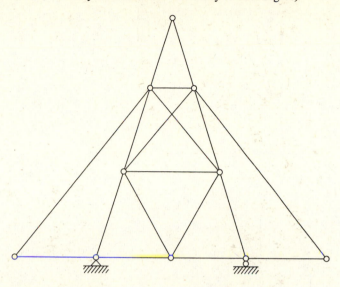

For Problems 6.23 to 6.37, compute the forces in all the members of the trusses using the method of joints.

6.23

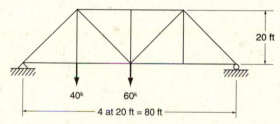

40ᵏ   60ᵏ

20 ft

4 at 20 ft = 80 ft

6.24   Rework Problem 6.23 if the truss depth is reduced to 15 ft and the loads doubled. (*Ans.* $L_0L_1 = +160$ k, $U_1U_2 = -213.3$ k, $L_2U_3 = +133.3$ k)

6.25

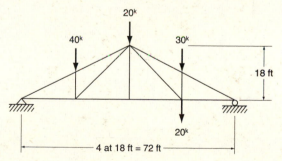

20ᵏ

40ᵏ   30ᵏ

18 ft

20ᵏ

4 at 18 ft = 72 ft

6.26 (*Ans.* $U_0U_1 = -110$ kN, $L_1L_2 = +140$ kN, $U_2L_3 = -62.5$ kN)

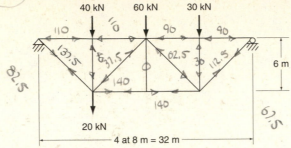

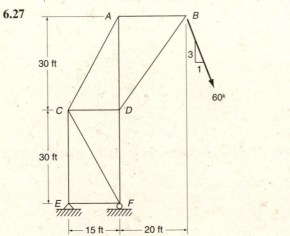

6.27

6.28 (*Ans.* $U_0L_0 = +55$ k, $U_0L_1 = +33.5$ k, $L_1L_2 = -40$ k)

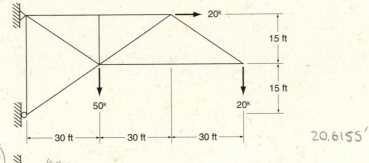

20.6155′

105 ◊

160

6.29

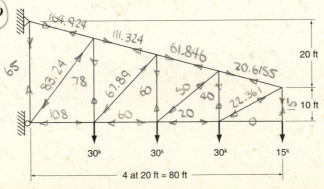

6.30   (*Ans.* $U_0 L_0 = +10$ k, $L_0 U_1 = -94.9$ k, $U_1 L_2 = -126.5$ k)

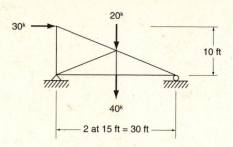

6.31   Repeat Problem 6.30 if the roller support (due to friction, corrosion, etc.) is assumed to supply one-third of the total horizontal force resistance needed, with the other two-thirds supplied by the pin support.

6.32   (*Ans.* $U_0 U_1 = -28$ k, $U_1 L_1 = -27.6$ k, $U_2 L_2 = -10.8$ k)

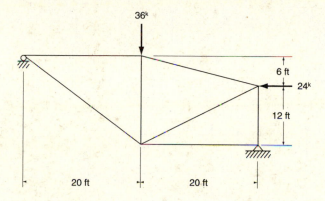

6.33

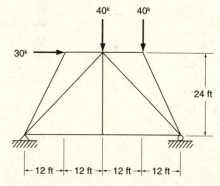

6.34   Rework Problem 6.33 if the supporting surface beneath the roller is changed as follows (*Ans.* $L_0 U_2 = -21.2$ k, $U_2 U_3 = -20$ k, $U_3 L_4 = -44.7$ k)

6.35

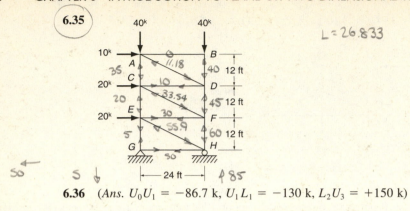

$L = 26.833$

6.36   (*Ans.* $U_0U_1 = -86.7$ k, $U_1L_1 = -130$ k, $L_2U_3 = +150$ k)

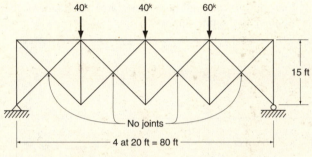

No joints

4 at 20 ft = 80 ft

6.37

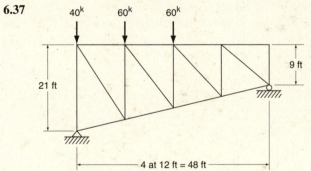

4 at 12 ft = 48 ft

Chapter 7

Plane Trusses, Continued

7.1 METHOD OF MOMENTS

The equilibrium of free bodies is the basis of force computation by the method of moments, as it is by the method of joints. To obtain the value of the force in a particular member, an imaginary section is passed completely through the truss to divide it into two free bodies. A section is placed to cut the member whose force is desired and as few other members as possible.

The moment of all the forces applied to the free body under consideration about any point in the plane of the truss is zero. If it is possible to take moments of the forces about a point so that only one unknown force appears in the equation, the value of that force can be obtained. This objective usually can be attained by selecting a point along the line of action of one or more of the forces of the other members. Some familiar trusses have special locations for placing sections that greatly simplify the work involved. These cases will be discussed in the pages to follow.

One advantage of the method of moments is that if the force in only one member of a truss is desired and the member is not near the end of the truss, it may be obtained directly in most cases without first determining the forces in other members. If the method of joints were used, it would be necessary to calculate the forces in the members joint by joint from the end of the truss until the member in question was reached.

7.2 FORCES IN MEMBERS CUT BY SECTIONS

Should tension and compression members actually be cut, the results would be as described in the following paragraphs and as pictured in Figure 7.1.

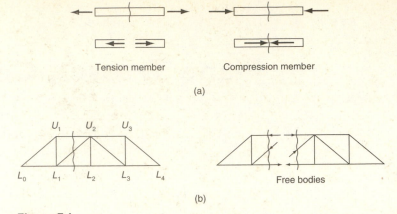

Figure 7.1

1. A tension member is being stretched, and should it be cut in half, it would tend to resume its original length, leaving a gap at the section. A tension member pulls away from the free body in Figure 7.1(a).
2. A compression member has been shortened, and should it be cut in half would tend to resume its original length, that is, try to expand. A compression member pushes against the free body from the outside in Figure 7.1(a).

Final truss slipped into place for Newport Bridge linking Jamestown and Newport, Rhode Island. (Courtesy Bethlehem Steel Corporation.)

A truss is divided into two free bodies in Figure 7.1(b). Members U_1U_2 and L_1U_2 are assumed to be in compression, and member L_1L_2 is assumed to be in tension. On the basis of these assumptions the directions of the forces on the two free bodies are shown.

7.3 APPLICATION OF THE METHOD OF MOMENTS

Examples 7.1 to 7.5 illustrate in detail the computation of forces with the $\Sigma M = 0$ equation. In writing the moment equation note that the unknown force may be assumed to be tension or compression. If the mathematical solution yields a negative number, the character of the force is opposite that which was assumed. The numerical answer is correct regardless of the sign.

It is probably simpler always to assume the unknown force to be in tension, that is, pulling away from the free body. If the solution yields a positive number, the force is tensile; if a negative number, the force is compressive. Therefore, the sign always agrees with the normal sign convention of $+$ for tension and $-$ for compression. This practice is followed in the illustrative problems throughout this book.

EXAMPLE 7.1 _____

Find the forces in members L_1L_2 and U_2U_3 of the truss shown in Figure 7.2 by moments.

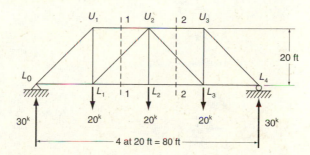

Figure 7.2

Solution. Member L_1L_2. Section 1-1 is passed through the truss, and the part of the truss to the left of the section is considered to be the free body. The forces acting on the free body are the 30-kip reaction, the 20-kip load at L_1, and the axial forces in the members cut by the section (U_1U_2, L_1U_2, and L_1L_2). Moments of these forces are taken about U_2, which is the point of intersection of L_1U_2 and U_1U_2. The moment equation contains one unknown force, L_1L_2, and its value may be found by solving the equation.

$$\Sigma M_{U_2} = 0$$
$$(30)(40) - (20)(20) - 20F_{L_1L_2} = 0$$
$$F_{L_1L_2} = +40^k \text{ tension}$$

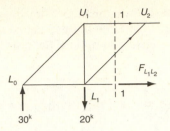

Member U_2U_3. Section 2-2 is passed through the truss, and the portion of the truss to the right of the section is considered to be the free body. Members U_2U_3, U_2L_3 and L_2L_3 are cut by the section. Taking moments at the intersection of L_2L_3 and U_2L_3 at L_3 eliminates those two members from the equation because

Timber trusses for tannery, South Paris, Maine. (Courtesy of the American Wood Preservers Institute.)

the lines of action of their forces pass through the center of moments. The force in U_2U_3 is the only unknown appearing in the equation and its value may be determined.

$$\Sigma M_{L_3} = 0$$
$$-(30)(20) - 20F_{U_2U_3} = 0$$
$$F_{U_2U_3} = -30^k \text{ compression}$$

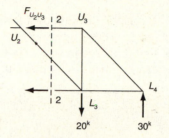

EXAMPLE 7.2

Find the forces in all of the members of the truss shown in Figure 7.3. Use both the method of joints and the method of moments as convenient.

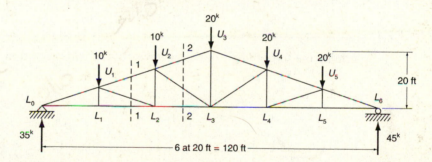

Figure 7.3

Solution.   The forces of members meeting at L_0 and L_1 are quickly determined by the method of joints. To calculate the force in U_1U_2, section 1-1 is passed and moments are taken at L_2. Because U_1U_2 is an inclined member, the force is resolved into its vertical and horizontal components. The components of a force may be assumed to act anywhere along its line of action. It is convenient in this case to break the force down into its components at joint U_2, because the vertical component will pass through the center of moments and the moment equation may be solved for the horizontal component of force.

$$\Sigma M_{L_2} = 0$$
$$(35)(40) - (10)(20) + 13.33H_{U_1U_2} = 0$$
$$H_{U_1U_2} = -90^k \text{ compression}$$

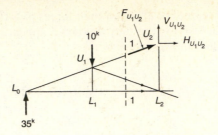

By joints, the unknown forces in members meeting at U_1 and L_2 may now be obtained. Section 2-2 is passed through the truss, and moments are taken at L_3 to find the force in U_2U_3. By knowing this force, the remaining forces in the truss can be found by joints, with the results shown.

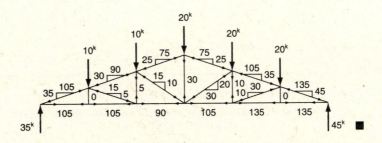

EXAMPLE 7.3 ──

Determine the forces in all of the members of the truss shown in Figure 7.4.

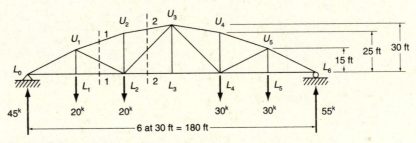

Figure 7.4

Solution.   The forces meeting at joints L_0 and L_1 are determined by joints. Then section 1-1 is passed through the truss, the free body to the left is considered, and moments are taken at L_2 to determine $H_{U_1U_2}$.

$$\Sigma M_{L_2} = 0; \text{ free body to left of section 1-1}$$
$$(45)(60) - (20)(30) + 25H_{U_1U_2} = 0$$
$$H_{U_1U_2} = -84^k \text{ compression}$$

The remaining forces at joints U_1 and U_2 are determined by joints. Section 2-2 is passed through the truss and with respect to the left free-body moments are taken at U_3 to determine L_2L_3.

$\Sigma M_{U_3} = 0$; free body to left of section 2-2

$$(45)(90) - (20)(30) - (20)(60) - 30F_{L_2L_3} = 0$$

$$F_{L_2L_3} = +75^k \text{ tension}$$

The same procedure is continued for the rest of the truss and the results are shown in the following sketch.

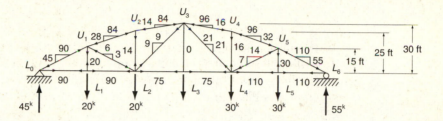

EXAMPLE 7.4

Determine the forces in all of the members of the Fink truss shown in Figure 7.5.

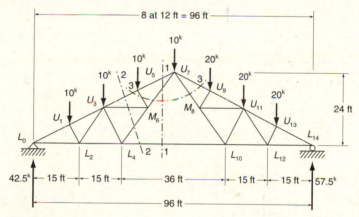

Figure 7.5

Solution. The forces in members meeting at joints L_0, L_2, and U_1 can be found by the methods of joints and moments without any difficulty. At each of the next two joints, U_3 and L_4, there are three unknown forces that cannot be determined directly with sections. It is convenient to compute the forces in some members further over in the truss and then work back to these joints. The sections numbered 1-1, 2-2, and 3-3 may be used to advantage. From the first of these sections the forces in any of the three members cut may be obtained by moments. By using section 2-2 and taking moments at U_3, the force in L_4M_6 may be found. It is important to note that four members have been cut by the section and only two of them pass through the point where moments are being taken; however, the force in one of these members, L_4L_{10}, was previously found with section 1-1, and only one unknown is left in the equation. The remaining forces in the truss may be calculated by the usual methods. These two sections are

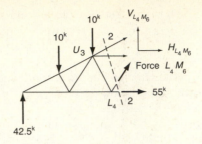

sufficient for analyzing the truss, but should another approach be desired, a section such as 3-3 may be considered. From this section the force in $U_5 M_6$ can be found by taking moments at U_7 because all of the other members cut by the section pass through U_7. The forces in all of the truss members are as shown.

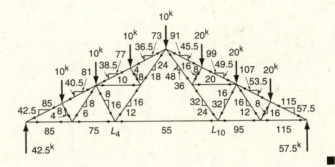

EXAMPLE 7.5

Calculate the force in member cg of the truss of Figure 7.6.

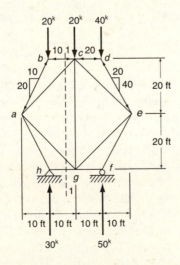

Figure 7.6

Solution.   The force in the member in question cannot be determined immediately by joints or moments. It is necessary to know the force values for several other members before the value for *cg* can be found. The forces in members *ba*, *bc*, *dc*, and *de* may be found by joints as shown, and the force in member *ac* can be found by moments. Considering section 1-1 and the free body to the left, moments may be taken about *g*. By noting that the force in *bc* is in compression and pushes against the free body from the outside, and by assuming member *ac* to be in tension, the following equation may be written. The unknown force is broken into its vertical and horizontal components at *c*.

$$\Sigma M_g = 0$$

$$(30)(10) - (20)(10) - (10)(40) + (H_{ac})(40) = 0$$

$$H_{ac} = +7.5^k \text{ tension}$$

Having the force in *ac*, the forces in *ce* and *cg* can be determined by joints as shown.

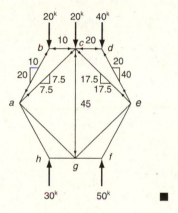

7.4 METHOD OF SHEARS

It should be obvious by this time that if a vertical section is passed through a truss and divides it into two separate free bodies, the sum of the vertical forces to the left of the section must be equal and opposite in direction to the sum of the vertical forces to the right of the section. The summation of these forces to the left or to the right of a section has been defined as the *shear*.

The inclined members cut by a section must have vertical components of force equal and opposite to the shear along the section, because the horizontal members can have no vertical components of force. For most parallel-chord trusses there is only one inclined member in each panel, and the vertical component of force in that inclined member must be equal and opposite to the shear in the panel. The vertical components of force are computed by shears for the diagonals of the parallel-chord truss of Example 7.6.

EXAMPLE 7.6

Determine the vertical components of force in the diagonals of the truss shown in Figure 7.7. Use the method of shears.

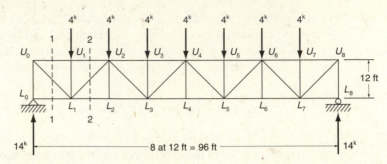

Figure 7.7

Solution.   By considering section 1-1 and free body to the left,

$$\text{shear to the left} = 14^k \uparrow$$

$$V_{U_0 L_1} = 14^k \downarrow \text{ tension (pulling away from free body)}$$

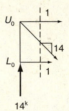

By considering section 2-2 and free body to the left,

$$\text{shear to left} = 10^k \uparrow$$

$$V_{L_1 U_2} = 10^k \downarrow \text{ compression (pushing against free body)}$$

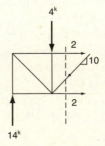

The vertical components of force in all the diagonals are as follows:

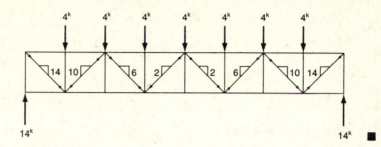

Nonparallel-chord trusses have two or more diagonals in each panel, and they all may have vertical components of force; however, their sum must be equal and opposite to the shear in the panel. If all but one of the diagonal forces in a panel are known, the remaining one may be determined by shears, as illustrated in Example 7.7.

EXAMPLE 7.7 _____

Referring to sections 1-1 and 2-2 of the truss of Example 7.3 and assuming that the forces in the chords U_1U_2 and U_2U_3 are known, find the vertical components of force in U_1L_2 and L_2U_3 by the method of shears.

Solution. By considering section 1-1 and free body to the left,

$$\text{shear to left} = 25^k \uparrow$$
$$V_{U_1U_2} = 28^k \downarrow$$
$$V_{U_1L_2} = 3^k \uparrow \text{ compression}$$

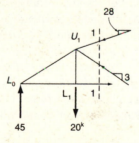

By considering section 2-2 and free body to the right,

$$\text{shear to right} = 5^k \downarrow$$
$$V_{U_2U_3} = 14^k \uparrow$$
$$V_{L_2U_3} = 9^k \downarrow \text{ tension}$$

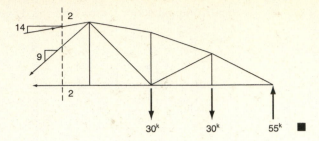

7.5 ZERO-FORCE MEMBERS

Frequently, some readily identifiable truss members have zero forces (assuming secondary forces due to member weights, load eccentricities, etc., are neglected). The ability to spot these members will on occasion appreciably expedite truss analysis. Zero-force members usually can be identified by making brief examinations of the truss joints. Several illustrations are presented here, with reference being made to the trusses of Figure 7.8.

1. *If only one member at a joint has a possible force in a particular direction and if there is no external load applied at the joint with a component in the direction of the member, the force in the member must be zero.* An examination of joint L_3 of the truss of part (a) of Figure 7.8 reveals that member $U_3 L_3$ has a zero force as long as no external load with a vertical component is applied at the joint. If the member had a force, the sum of the vertical forces at L_3 could not be zero. A similar examination of joint U_2 shows member $U_2 L_2$ has a zero force.
2. *Not only must the sum of the forces in the x and y directions at a particular joint be zero, but also the sum of the forces in any direction at the joint must be zero.* Therefore, the force in member $U_1 L_2$ of the truss of part (b) of the figure must be zero, as there are no other forces at the joint with components perpendicular to the members $L_0 U_1$ and $U_1 U_3$.
3. *Should two members be joined (the members not being in line with each other), they both will have zero forces unless an external load is applied at the joint in the plane of the members.* In the truss of part (c) of Figure 7.8 both members $M_0 U_1$ and $M_0 L_1$ are zero-force members. If one of the members had a force, it would be impossible to have a force in the other member such that both $\Sigma H = 0$ and $\Sigma V = 0$. See the following sketches.

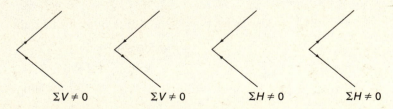

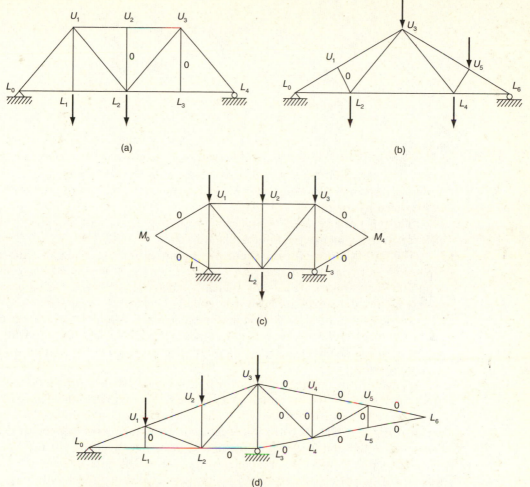

Figure 7.8 Zero-force members.

4. Based on the preceding illustrations it is easy to spot twelve zero-force members in the truss of part (d) of Figure 7.8. These members can be identified by examining the following joints in the order given: L_6, L_5, U_5, U_4, L_4, L_3, and L_1.

7.6 WHEN ASSUMPTIONS ARE NOT CORRECT

The engineer should realize that often his or her assumptions regarding the behavior of a structure (pinned joints, loads applied at joints only, frictionless rollers, or whatever) may not be entirely valid. As a result he or she should give consideration to what might happen to a structure if the assumptions made in analysis were

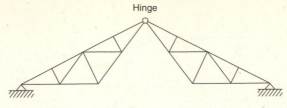

Figure 7.9

appreciably in error. Perhaps an expansion device or roller will resist (due to friction) a large proportion of any horizontal forces present. How would this affect the member forces of a particular truss? For this very reason the truss of Problem 6.31 was included at the end of Chapter 6, where it was assumed that half of the horizontal load was resisted by the roller.

The types of end supports used can have an appreciable effect on the magnitude of forces in truss members caused by lateral loads.

For fairly short roof trusses generally no provisions are made for temperature expansion and contraction and both ends of the trusses are bolted down to their supports. These trusses are actually statically indeterminate, but the usual practice is to assume that the horizontal loads split equally between the supports.

For longer roof trusses, provisions for expansion and contraction are considered necessary. Usually the bolts at one end are set in a slotted hole so as to provide space for the anticipated length changes. A base plate is provided at this end on which the truss can slide.

Actually it is impossible to provide a support that has no friction. From a practical standpoint the maximum value of the horizontal reaction at the expansion end equals the vertical reaction times the coefficient of friction (one-third being a reasonable guess). Should corrosion occur, thus preventing movement (a rather likely prospect), a half-and-half split of the lateral loads may again be the best estimate.

The authors once read of an interesting case in which the owners of a building with a roof supported by a series of Fink trusses decided that the middle lower chord member (L_4L_{10} in Figure 7.5) was in their way. They therefore removed the member from quite a few of the trusses and much to the designer's amazement the roof did not collapse. Apparently the roller or expansion device on one end of the trusses (perhaps bolts in a slot) permitted very little or no movement. As a result each of the trusses apparently behaved as a three-hinged arch, as shown in Figure 7.9. In many situations in which the assumptions do not prove to be correct, the results are more unpleasant than they were for these Fink trusses, however.

7.7 SIMPLE TRUSSES

The first step in forming a truss has been shown to be the connecting of three members at their ends to form a triangle. Subsequent figures are formed by adding two members and one joint; the new members meet at the new joint and each is pinned at its opposite ends into one of the existing joints. Trusses formed in this way are said to be *simple trusses*. (Some of these trusses, however, are not very "simple" to analyze.)

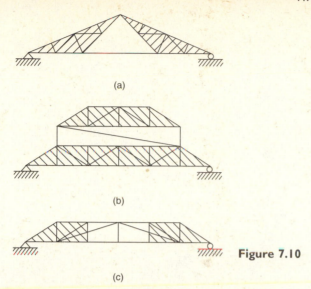

(a)

(b)

(c)

Figure 7.10

7.8 COMPOUND TRUSSES

A *compound truss* is a truss made by connecting two or more simple trusses. The simple trusses may be connected by three nonparallel nonconcurrent links, by one joint and one link, by a connecting truss, by two or more joints, and so on. An almost unlimited number of trusses may be formed in this way. The Fink truss shown in Figure 7.10(a), consisting of the two crosshatched trusses connected by one joint and one link, is one example. Other compound trusses are shown in Figures 7.10(b) and (c). The $2j - r$ equation applies equally well to compound trusses and simple trusses.

7.9 THE ZERO-LOAD TEST

In the next section of this chapter another set of trusses called complex trusses is discussed. Based on the information presented up to this point, the reader will find that it is extremely difficult to determine whether complex trusses are stable or unstable without conducting a complete analysis. If such an analysis is made and all the truss joints balance, the truss will be stable. If all the truss joints do not balance, the truss is unstable. The analysis procedure can be very time-consuming, however, as well as discouraging if calculations show that the structure is unstable. Indeed, we would like to know before performing an analysis. As a result, a simple method for checking stability for any type of truss is presented in this section before complex trusses are considered. This method is called the *zero-load test*.

A statically determinate truss has only one possible set of forces for a given loading and is said to have a *unique solution*. Therefore, if it is possible to show that more than one solution can be obtained for a structure for a given set of conditions, the structure is unstable.

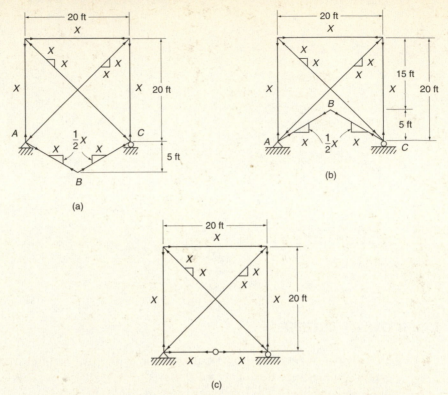

Figure 7.11 (a) Joints *A*, *B*, and *C* do not balance. Truss is stable. (b) Joints *A*, *B*, and *C* do not balance. Truss is stable. (c) All joints balance. Truss is unstable.

This discussion leads to the idea of the so-called *zero-load test*. If no external loads are applied to a truss, it is logical to assume that all of the members will have zero forces. Should an assumed force (not zero) be given to one of the members of a truss that has no external loads and the forces be computed in the other members, the results must be incompatible if the truss is stable. If the calculated forces are compatible, the truss is unstable.

To illustrate this procedure, the top horizontal member of each of the three trusses of Figure 7.11 is assumed to have tension of *X* and the other member forces are computed working down from the top by joints. It will be seen that the lower three joints of the trusses of parts (a) and (b) of the figure cannot be balanced, whereas they can for the truss of part (c). A structure that has zero external loads should also have zero internal forces. Truss (c) must therefore be unstable or have *critical form*. For the first two arrangements, (a) and (b), it is impossible to assume a set of member forces other than zero for which the joints will balance, and they must therefore be stable.[1]

For another illustration of the zero-load test the reader might work with the truss of Figure 7.12(a). He or she will find that it also has critical form.

[1] G. L. Rogers and M. L. Causey, *Mechanics of Engineering Structures* (New York: John Wiley and Sons, Inc., 1962), 19–20.

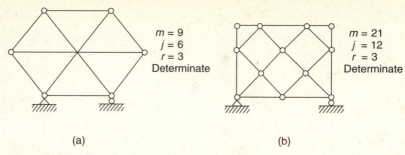

$$m = 9$$
$$j = 6$$
$$r = 3$$
Determinate

$$m = 21$$
$$j = 12$$
$$r = 3$$
Determinate

(a)   (b)

Figure 7.12   Complex trusses. Truss (a) can be proved to be unstable.

7.10 COMPLEX TRUSSES

There are a few trusses that are statically determinate that do not have the requirements necessary to fall within the classification of either simple or compound trusses. These are referred to as *complex trusses*. The members of simple and compound trusses usually are arranged so that sections may be passed through three members at a time, moments taken at the intersection of two of them, and the force found in the third.

Complex trusses may not be analyzed in this manner. Not only does the method of moments result in failure, but the methods of shears and joints are also of no avail. The difficulty lies in the fact that there are three members meeting at almost every joint, and therefore there are too many unknowns at every location in the truss to pass a section and obtain the force in any member directly by the equations of statics. Two complex trusses are shown in Figure 7.12. The number of joints and members is sufficient for them to be statically determinate.

One method of computing the forces in complex trusses is to write the equations $\Sigma H = 0$ and $\Sigma V = 0$ at each joint, giving a total of $2j$ simultaneous equations. These equations may be solved simultaneously for the member forces and external reactions. (It is often possible to calculate the external reactions initially, and their values may be used as a check against the results obtained from the solution of the simultaneous equations.) This method will work for any complex truss, but the solution of the equations is very tedious unless a digital computer is available. Usually, other methods are more desirable.

One procedure that works for some complex trusses involves the assumption of the force in one of the members.[2] A convenient member is selected and given a force of X. Forces in the surrounding members are then computed in terms of X. The process is continued until it is possible to pass a section completely through the truss and write one of the equations of statics so that the only member unknowns appearing in the equation have forces that have been calculated in terms of X. Solution of the resulting equation may give the value of X, but frequently will just reduce to $0 = 0$. If this happens we will have to try other sections and see if we can determine the value of X. This method, which is not easy to apply for many complex trusses, is illustrated in Example 7.8.

[2] C. H. Norris, J. B. Wilbur, and S. Utku, *Elementary Structural Analysis,* 3rd ed. (New York: John Wiley and Sons, Inc., 1976), 121.

EXAMPLE 7.8 _____

Check the truss of Figure 7.13 for stability with the zero-load test and (as it will prove to be stable) then determine the forces in all of its members.

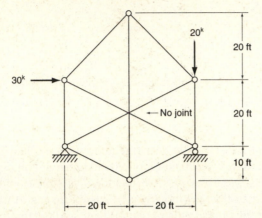

Figure 7.13

Solution.   The bottom left-hand member of the truss is assumed to be in tension with a horizontal component equal to $+X$. Then the forces in the other members are determined with the method of joints. As the top three joints cannot be balanced, the truss is stable.

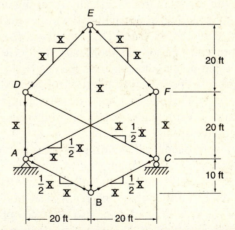

After the reactions for the loaded truss are determined, the bottom left-hand member is assumed to be in tension with a horizontal component equal to X. Then the forces in members *AB, BC, BE, DE,* and *EF* are computed in terms of X.

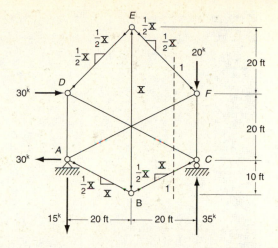

Section 1-1 is passed through the truss and the free body to the right of the section is considered. Then moments are taken about the intersection point of the diagonals and the value of X is determined.

$$-\left(\frac{X}{2}\right)(30) + (20)(20) - (35)(20) + (X)(20) = 0$$

$$X = +60^k \text{ tension}$$

Having the value of X, the forces in all of the members can easily be determined with the following results:

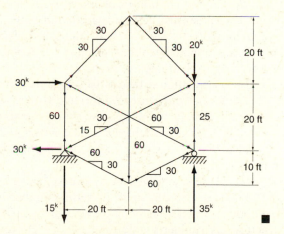

The equation $m = 2j - r$ may be used to determine whether a complex truss is statically determinate or indeterminate, but to determine whether it is stable or unstable may not be so simple. Complex trusses often consist not only of triangles, but also of other figure shapes, with the result that they are particularly susceptible to

College fieldhouse, Largo, Maryland. (Courtesy Bethlehem Steel Corporation.)

geometrical instability. They may be completely unstable and yet their instability may not be obvious until a solution is attempted. *If an analysis is attempted on an unstable truss, the results will always be inconsistent or noncompatible.*

Should the truss be geometrically unstable, it is said to have *critical form.* The critical form of a truss may or may not be obvious, with the result that an analysis may have to be attempted before its stability is known. One arrangement or configuration of a truss may be stable and a slightly varying configuration may be unstable. The $m = 2j - r$ equation will be satisfied whether the truss is stable or has critical form.

The trusses of Figures 7.14(a) and (b) are stable under the action of a vertical load at joint B. In between these two arrangements lies the truss of part (c) of the same figure. This truss is unstable because it will not hold its position under the action of the load at B. Joint B will begin to deflect downward without causing calculable forces in the members.

It is possible to determine by matrix analysis (a subject presented in Chapters 20 and 21) if a structure is stable by setting up the simultaneous equations at each joint and finding the determinant of the resulting matrix. If the determinant is not zero, the structure has a unique solution and is stable. However, should the determinant be zero, the structure is unstable, as there is an infinite set of answers that will satisfy the simultaneous equations.

The student will happily note that the enclosed computer program SABLE will easily handle complex trusses, just as it will other types. Should a truss be unstable, the computer will clearly indicate that fact and no analysis will be performed.

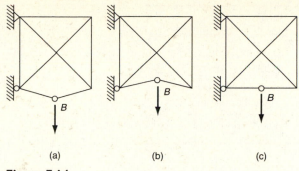

(a)   (b)   (c)

Figure 7.14

For a more comprehensive discussion of complex trusses the reader may refer to the method of substitute members described in *Theory of Structures* by S. Timoshenko and D. H. Young.[3]

Generally speaking, there is little need for complex trusses because it is possible to select simple of compound trusses that will serve the desired purpose equally well.

7.11 STABILITY

The following paragraphs discuss several situations that may cause a structure to be unstable.

Less Than $2j - r$ Members

A truss that has less than $2j - r$ members is obviously unstable internally, but a truss may have as many or more than $2j - r$ members and still be unstable. On the following page, the truss of Figure 7.15(a) satisfies the $2j - r$ relationship and is statically determinate and stable; however, if the diagonal in the second panel is removed and added to the first panel as shown in Figure 7.15(b), the truss is unstable even though the number of members remains equal to $2j - r$. The part of the truss to the left of panel 2 can move with respect to the part of the truss to the right of panel 2 because panel 2 is a rectangle. (As previously indicated, a rectangular shape is unstable unless restrained in some way.)

Similarly, the addition of diagonals to panels 3 and 4, as shown in Figure 7.15(c), will not prevent the truss from being unstable. There are two more than $2j - r$ members, and the truss is seemingly statically indeterminate to the second degree; but it is unstable because panel 2 is unstable.

[3] S. Timoshenko and D. H. Young, *Theory of Structures* (New York: McGraw-Hill, 1965), 92–103.

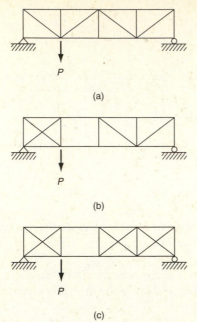

(a)

(b)

(c)   **Figure 7.15**

Trusses Consisting of Figures That Are Not All Triangles

As the reader becomes more familiar with trusses, he or she will be able in most cases to tell with a brief glance if a truss is stable or unstable. For the present, though, it may be a good idea for the reader to study trusses in detail if he or she thinks there is a possibility of instability. When a truss has some nontriangular figures in its makeup, we should be aware that instability is indeed a possibility. The trusses of Figures 7.15(b) and (c) fall in this category.

The fallacy of this idea is that the number of perfectly stable trusses that can be assembled not consisting entirely of triangles is endless. As an example, consider the truss of Figure 7.16(a). The basic triangle *ABC* has been extended by the addition of joint *D* and members *AD* and *CD*. A stable truss is maintained, even though figure *ABCD* is not a triangle. Joint *D* is firmly held in position and cannot move without changing the length of one or more members. Compound trusses such as the ones of Figures 6.8(h) and (i) and the subdivided types of Figure 6.5(b) often present a

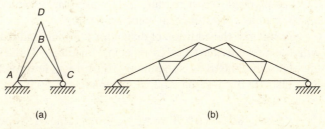

(a)   (b)

Figure 7.16

nontriangular but stable situation. Another example is presented in Figure 7.16(b). Structures of these types have the joints around the nontriangular figures tied to the rest of the truss so that they are not free to move. If a truss consists of figures that are not all triangles, it should be carefully examined to see if any of the joints can possibly move in any direction without causing changes in length of one or more of the truss members.

Analysis As Means of Finding Instability

The members of a truss must be arranged to support the external loads. What will support the external loads satisfactorily is a rather difficult question to answer with only a glance at the truss under consideration, but an analysis of the structure will always provide the answer. If the structure is stable, the analysis will yield reasonable results, but if it is unstable, the analysis will never balance. (The zero-load test, previously discussed in Section 7.9, will often enable the analyst to detect stability or instability.)

Unstable Supports

A structure cannot be stable if its supports are unstable. To be stable, it must be supported by at least three nonparallel, nonconcurrent forces. This subject was discussed in Chapter 4.

7.12 EQUATIONS OF CONDITION

On some occasions two or more separate structures are connected so that only one type of force can be transmitted through the connection. The three-hinged arch and cantilever types of structures of Chapter 4 have been shown to fall into this class because they are connected with interior hinges unable to transmit rotation.

Perhaps the simplest way to produce a hinge in a truss is by omitting a chord member in one of the panels, as shown in Figure 7.17(a) on the following page. It is obvious that the moment of all the external forces on the part of the structure to the left or the right of the pin connection at joint L_3 must be zero. The truss is statically determinate because there are three statics equations and one condition equation available for calculating the four reaction components.

The omission of members in some other situations may produce equations of condition. A diagonal of the truss of Figure 7.17(b) has been omitted between the two interior supports. With no members in the panel able to have a vertical component of force, no shear can be transmitted through the panel, and an equation of condition is available. The supports on each side of the usually unstable rectangular shape prevent it from collapsing.

Practically speaking, the bars mentioned as being omitted will probably not be omitted because their omission would detract from the appearance of the structure and might frighten the users. Furthermore, the presence of these members might be useful during erection. They are frequently assembled so they can be adjusted to be inactive in the completed truss.

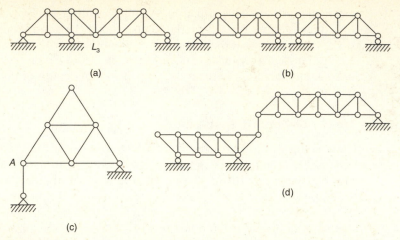

Figure 7.17

Figure 7.17(c) and (d) present two more situations in which equations of condition are produced. In the first of these there are four reaction components, and the structure may appear to be statically indeterminate externally: however, the joint at A is pin connected and cannot transmit rotation. This equation of condition makes the structure statically indeterminate externally. Figure 7.17(d) shows two separate trusses that are connected by a link. The link makes available two equations of condition, because rotation may not be transmitted at either end.

Mr. E. M. Wichert patented a type of statically determinate continuous truss in 1932. His truss falls into the omitted-member class, because the verticals over interior supports are left out. (Müller-Breslau, a professor in Berlin, had discussed this form of truss as early as 1887.[4]) The Wichert truss of Figure 7.18, which has 32 members, 18 joints, and 4 reaction components, is shown to be statically determinate as follows:

$$m = 2j - r$$
$$32 = 36 - 4$$
$$32 = 32$$

The reader is reminded that satisfying the $m = 2j - r$ criterion is a *necessary* but not a *sufficient* condition for statical determinancy. Unless the criterion is satisfied, the structure cannot be statically determinate. If it is satisfied, the structure may be statically determinate or it may be unstable. The structure will indeed be statically determinate and stable if the equations have a unique solution.

In Chapter 12 the advantages and disadvantages of statically determinate and statically indeterminate structures are discussed. The continuous but statically determinate Wichert truss has all of the advantages of statically determinate structures (ease of analysis and design) but none of the disadvantages of the continuous statically indeterminate ones (major stress variations caused by support settlements, fabrication

---

[4]H. Sutherland and H. L. Bowman, *Structural Theory* (New York: Wiley, 1954), Chap. 5.

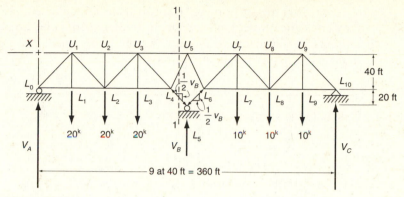

Figure 7.18

errors, and so on). The method of analysis is based on the conditions at the interior support. The support at L_5 in the truss of Figure 7.18 is a roller, and the reaction there is vertical with no horizontal components; therefore, the horizontal components of force in the members L_4L_5 and L_5L_6 must be equal and opposite for equilibrium (that is, both tension or both compression). Members having the same slope and the same horizontal force components must have the same forces and the same vertical components of force. Assuming the reaction V_B at the roller to be up, the two forces are in compression, and the sum of their vertical components is equal and opposite to the reaction. This relation may be stated as follows:

$$V_{L_4L_5} = V_{L_5L_6} = \tfrac{1}{2}V_B$$

The horizontal components of the two forces can be expressed in terms of the vertical components, which are in terms of V_B. For this particular truss the members have an inclination of 45°, and the vertical and horizontal components are equal, as shown on the members in the figure. Moments may be taken about joint U_5 of all the forces to the left of section 1-1. Since the components of force in L_4L_5 are expressed in terms of V_B, the equation will contain two unknowns, V_A and V_B. Moments may then be taken about L_{10} of all the external forces acting on the truss. The resulting equation has the same two unknowns, V_A and V_B, and their values may be determined by solving the two equations simultaneously. Once two of the reactions are obtained, the analysis may be completed by the usual method of statics. Example 7.9 presents a complete analysis of this truss.

Analysis of several Wichert trusses of different arrangements will show that instability is possible under some circumstances. This situation will occur when the lower-chord members meeting at the interior supports become very flat, and instability will be indicated by the tremendous values of those forces. There is a definite slope of the bottom-chord members at which the truss becomes unstable, and this slope can be seen for the truss in Figure 7.18. Should L_4L_5 be so flat that its line of action intersects the line of action of the top chord as far to the left as point X in the figure, the truss will be unstable.

Wichert trusses of more than two spans are quite tedious to analyze, although they are statically determinate. Dr. D. B. Steinman, in his book *The Wichert Truss*,[5] presents a detailed discussion of the various types, including methods of analysis for multispan trusses, design, and economy.

EXAMPLE 7.9 _____

Calculate the reactions and member forces of the Wichert truss of Figure 7.18.

Solution.

$\Sigma M_{U_5} = 0$ (to left of section 1-1)

$$180V_A - (20)(60 + 100 + 140) + 60H_{L_4L_5} = 0$$
$$180V_A - 6000 + (60)(\tfrac{1}{2}V_B) = 0$$
$$180V_A + 30V_B = 6000 \tag{1}$$

$\Sigma M_{L_{10}} = 0$ (entire structure)

$$-(10)(40 + 80 + 120) - (20)(240 + 280 + 320)$$
$$+ 360V_A + 180V_B = 0$$
$$360V_A + 180V_B = 19.200 \tag{2}$$

Solving Eqs. (1) and (2) simultaneously gives

$$V_A = 23.3^k$$
$$V_B = 60.0^k$$

by $\Sigma V = 0$

$$V_C = 6.7^k$$

By statics the following forces are obtained:

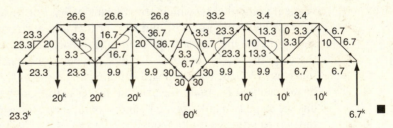

7.13 COMPUTER EXAMPLE

Example 7.10 illustrates the analysis of a statically determinate plane truss using the program SABLE. In the paragraphs to follow the fundamental bits of information needed to apply the program to such a truss are discussed.

Numbering and Coordinate System   It is first necessary to set up a numbering system for the members and joints of the truss. In Figure 7.19 the joint or node numbers selected are enclosed in circles while the member numbers are placed in

_____

[5] D. B. Steinman. *The Wichert Truss* (New York: D. Van Nostrand, 1932).

squares. The coordinates of the joints are specified with respect to a convenient origin. In Example 7.10 the lower-left joint is used as the origin (any point will do) and the dimensions are given in feet. The sign convention shown below is used both for specifying coordinates and for specifying the direction of loads.

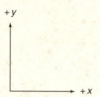

Member Data   It is necessary to provide the node or joint numbers at both ends of each member. Once this is done a drawing of the truss can be obtained on the screen, enabling the analyst to check the configuration of the structure.

Joint Restraints   The joints of a two-dimensional truss can be restrained against translation in the x and y directions as well as against rotation about the x axis perpendicular to the xy plane (z axis). Such information is input to the computer for each of the joints by selecting each of the possible restraints.

Areas, Moments of Inertia, and Moduli of Elasticity   The analyst may feel that all of these data are not needed for a statically determinate truss. The program, however, is a general one applicable to both statically determinate and statically indeterminate trusses alike. In addition, it is set up to compute the displacements of the truss joints. To make these latter calculations it is necessary to have A, I, and E for each member. These data are given in foot and kip units in Example 7.10.

EXAMPLE 7.10 ───

Using SABLE, determine the forces in the members of the truss of Figure 7.19. All member areas, moments of inertia, and moduli of elasticity are, respectively, 0.03 ft², 1 ft⁴, and 29,000 ksi.

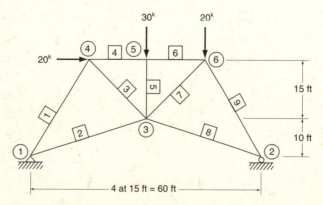

Figure 7.19

Solution.

Structural Data

Data: Nodal location and restraint data

	Coordinates			Restraints		
Node	X	Y	Z	X	Y	Rot
----	---------	----------	---------	---	---	---
1	0.000E+00	0.000E+00	0.000E+00	Y	Y	Y
2	6.000E+01	0.000E+00	0.000E+00	N	Y	Y
3	3.000E+01	1.000E+01	0.000E+00	N	N	Y
4	1.500E+01	2.500E+01	0.000E+00	N	N	Y
5	3.000E+01	2.500E+01	0.000E+00	N	N	Y
6	4.500E+01	2.500E+01	0.000E+00	N	N	Y

Data: Beam location and property data

				Beam properties		
Beam	i	j	Type	Area	Izz	E
----	---	---	----	---------	---------	---------
1	1	4	P-P	3.000E-02	1.000E+00	2.900E+04
2	1	3	P-P	3.000E-02	1.000E+00	2.900E+04
3	4	3	P-P	3.000E-02	1.000E+00	2.900E+04
4	4	5	P-P	3.000E-02	1.000E+00	2.900E+04
5	5	3	P-P	3.000E-02	1.000E+00	2.900E+04
6	5	6	P-P	3.000E-02	1.000E+00	2.900E+04
7	3	6	P-P	3.000E-02	1.000E+00	2.900E+04
8	3	2	P-P	3.000E-02	1.000E+00	2.900E+04
9	6	2	P-P	3.000E-02	1.000E+00	2.900E+04

Data: Applied joint loads

Node	Case	Force-X	Force-Y	Moment-Z
----	----	---------	---------	---------
1	1	0.000E+00	0.000E+00	0.000E+00
2	1	0.000E+00	0.000E+00	0.000E+00
3	1	0.000E+00	0.000E+00	0.000E+00
4	1	2.000E+01	0.000E+00	0.000E+00
5	1	0.000E+00	−3.000E+01	0.000E+00
6	1	0.000E+00	−2.000E+00	0.000E+00

Results

Data: Calculated beam end forces

Beam	Case	End	Axial	Shear-Y	Moment-Z
----	----	---	---------	---------	---------
1	1	i	2.673E+01	0.000E+00	0.000E+00
		j	−2.673E+01	0.000E+00	0.000E+00
2	1	i	−3.558E+01	0.000E+00	0.000E+00
		j	3.558E+01	0.000E+00	0.000E+00
3	1	i	−3.241E+01	0.000E+00	0.000E+00
		j	3.241E+01	0.000E+00	0.000E+00
4	1	i	5.667E+01	0.000E+00	0.000E+00
		j	−5.667E+01	0.000E+00	0.000E+00

Beam	Case	End	Axial	Shear-Y	Moment-Z
5	1	i	3.000E+01	0.000E+00	0.000E+00
		j	−3.000E+01	0.000E+00	0.000E+00
6	1	i	5.667E+01	0.000E+00	0.000E+00
		j	−5.667E+01	0.000E+00	0.000E+00
7	1	i	−3.948E+01	0.000E+00	0.000E+00
		j	3.948E+01	0.000E+00	0.000E+00
8	1	i	−3.031E+01	0.000E+00	0.000E+00
		j	3.031E+01	0.000E+00	0.000E+00
9	1	i	5.588E+01	0.000E+00	0.000E+00
		j	−5.588E+01	0.000E+00	0.000E+00

Note. Printouts of nodal displacements, nodal forces, and drawing of overall displaced shape of structure are not shown here due to space limitations. ■

PROBLEMS

Simple Trusses

For Problems 7.1 through 7.23, determine the forces in all the members of these trusses.

7.1

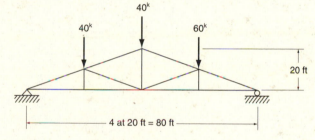

7.2 (*Ans.* $L_1 L_2 = +300^k$, $U_3 L_3 = +50^k$, $U_4 U_5 = -158.1^k$)

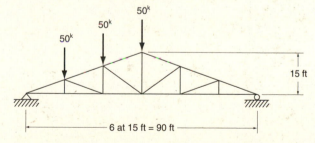

7.3 Rework Problem 7.1 if the panels are changed from 4 @ 20 ft to 4 @ 15 ft and the 40^k loads are doubled.

7.4 Rework Problem 7.1 if a uniform load of 2 klf is applied for the entire span in addition to the loads shown. This additional load is to be applied to the bottom-chord joints. (*Ans.* $L_0 U_1 = -279.5^k$, $U_2 L_2 = +130^k$, $L_2 L_3 = +270^k$)

7.5

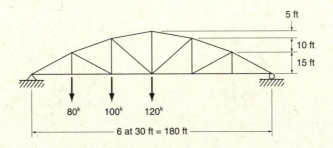

7.6   (*Ans. $U_2L_2 = +109.3^k$, $L_3U_4 = +83.2^k$, $L_4L_5 = +213.3^k$*)

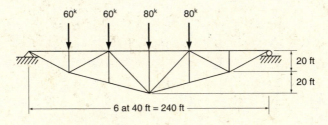

7.7

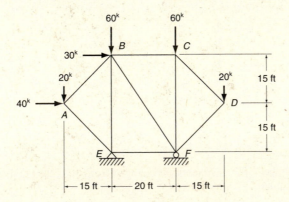

7.8   (*Ans. $AE = -42.4^k$, $BC = +10^k$, $DF = -14.1^k$*)

7.9

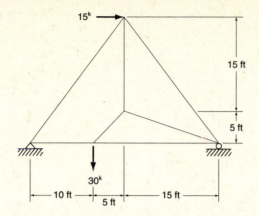

15^k 15 ft   5 ft   30^k 10 ft   5 ft   15 ft

7.10   (Ans. $AE = +20^k$, $AD = +58.3^k$, $DC = -51.0^k$)

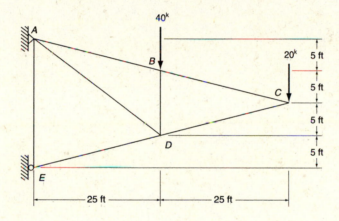

40^k 20^k 5 ft   5 ft   5 ft   5 ft   A   B   C   D   E   25 ft   25 ft

7.11

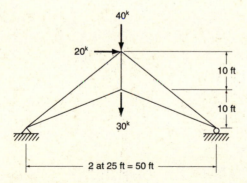

40^k 20^k 10 ft   10 ft   30^k 2 at 25 ft = 50 ft

7.12   (*Ans.* $U_0L_1 = +561.4$ kN, $U_2L_2 = -466.7$ kN, $L_2L_3 = +421.2$ kN)

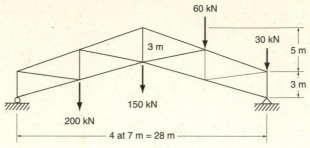

7.13

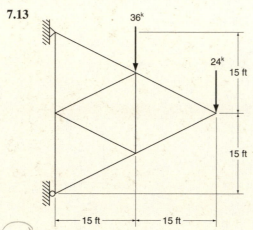

7.14   (*Ans.* $L_0U_1 = -164.4^k$, $U_2L_2 = +85^k$, $L_2L_4 = +146.3^k$)

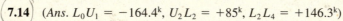
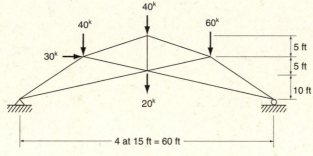

7.15

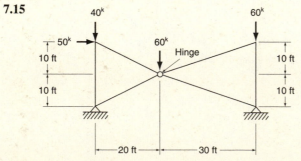

7.16 (*Ans. AC* $= -254.9^k$, *BC* $= -152^k$, *DE* $= +36.5^k$)

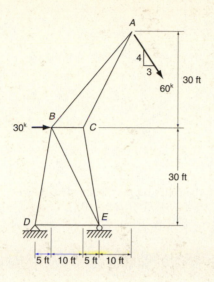

7.17

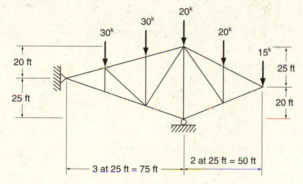

7.18 (*Ans.* $L_0U_1 = -82.6^k$, $U_1L_1 = -9.2^k$, $U_1U_2 = -42.2^k$)

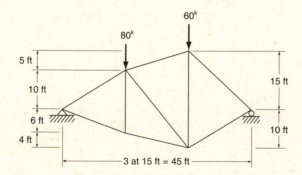

7.19

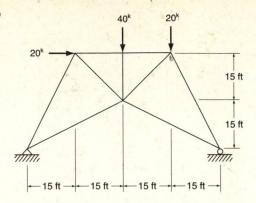

7.20 Repeat Problem 7.19 with the roller support inclined as shown.
(*Ans.* $L_0 U_1 = -20.5^k$, $U_2 U_3 = -47.5^k$, $L_2 L_4 = 0$)

7.21

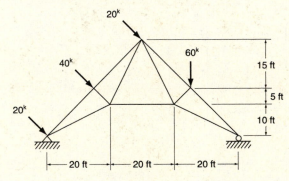

7.22 (*Ans.* $AD = -22.6^k$, $BE = -77.8^k$, $CE = -30^k$)

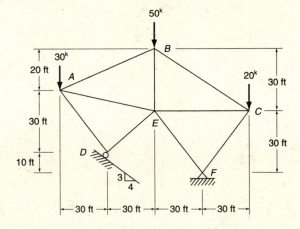

7.23

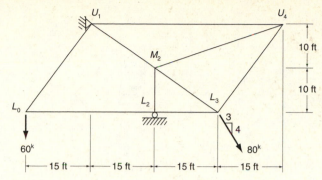

Compound Trusses

For Problems 7.24 through 7.31, determine the forces in all the members of these trusses.

7.24   (*Ans.* $M_1 L_2 = -42.4^k$, $U_2 M_3 = +17.7^k$, $L_5 L_6 = +87.5^k$)

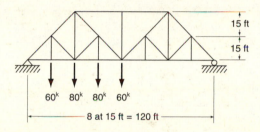

7.25

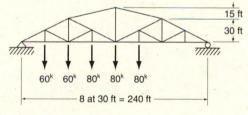

7.26   (*Ans.* $U_2 M_2 = +86.2^k$, $L_3 L_4 = +236.3^k$, $U_4 M_5 = -34.9^k$)

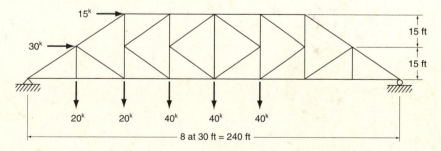

7.27

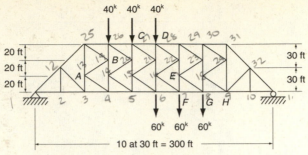

7.28   (*Ans.* $L_2U_3 = +10^k$, $U_3U_5 = -97.3^k$, $L_4L_{10} = +65^k$)

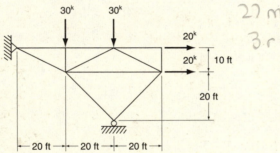

7.29

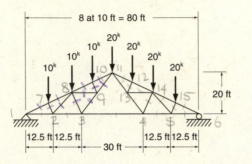

7.30   (*Ans. AB = +19.6^k, DE = −18.7^k, CF = −6.7^k*)

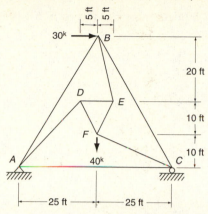

7.31

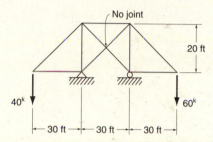

For Problems 7.32 through 7.36 determine directly the forces in each of the designated members using the method of moments.

7.32   Members U_2U_3, L_2L_3, and U_2L_3 of the truss of Problem 7.2.
(*Ans.* −158.1^k, +225^k, −90.1^k)

7.33   Members L_2L_3, U_3U_4, and U_2L_3 of the truss of Problem 7.6.

7.34   Members U_2U_3, L_2L_3, and U_2L_3 of the truss of Problem 7.7.
(*Ans.* −290^k, 348.2^k, −67.6^k)

7.35   Members L_0L_1, U_1U_2, and U_1U_2 of the truss of Problem 7.18.

7.36   Members L_0U_1, U_2L_2, and U_3L_4 (*Ans.* −27.8^k, − 45^k, + 19.1^k)

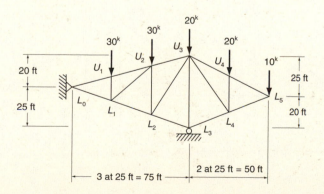

Complex Trusses

For Problems 7.37 through 7.39 determine the forces in all the members of these trusses.

7.37

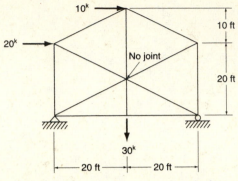

7.38   (*Ans.* $U_0 L_0 = +18.75^k$, $L_0 U_2 = -8.39^k$, $U_1 L_1 = 0$)

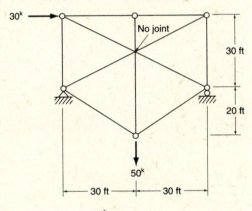

7.39

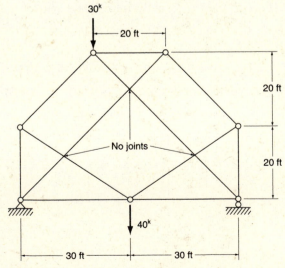

For Problems 7.40 through 7.47 use the zero-load test to determine if the members have critical form.

7.40   (*Ans.* unstable)

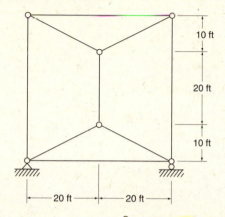

7.41

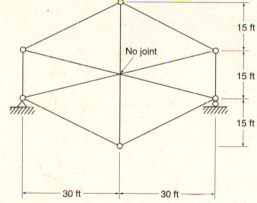

7.42   (*Ans.* stable)

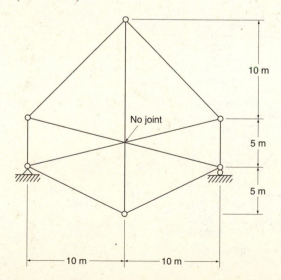

7.43

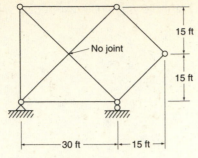

7.44   (*Ans.* unstable)

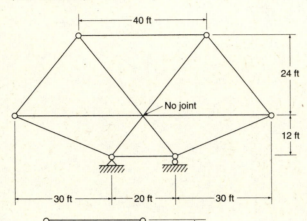

7.45

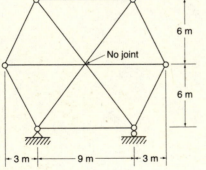

7.46   (*Ans.* stable)

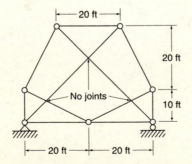

7.47

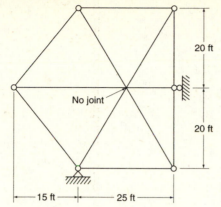

15 ft — 25 ft

Wichert Trusses

For Problems 7.48 through 7.49 determine the forces in all members.

7.48   (*Ans.* $U_1 U_2 = +6.67^k$, $L_4 L_5 = -113.1^k$, $L_8 L_9 = +73.3^k$)

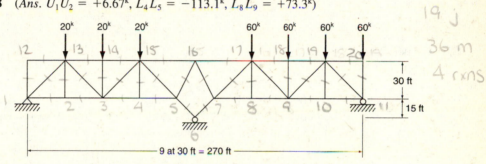

19 j
36 m
4 rxns

7.49

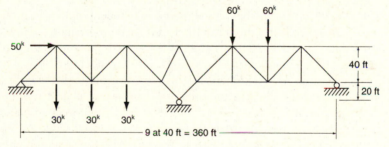

Computer Problems

For Problems 7.50 through 7.57 analyze the trusses using the SABLE program.

7.50   Problem 7.1 (*Ans.* $L_0 L_1 = +130^k$, $U_2 L_2 = +50^k$, $L_3 L_4 = +150^k$)

7.51   Problem 7.8

7.52   Problem 7.13 (*Ans.* $U_0 M_0 = +39^k$, $M_0 L_1 = +20.1^k$, $L_1 U_2 = -26.8^k$)

7.53   Problem 7.15

7.54   Problem 7.23 (*Ans.* $U_1 U_4 = +3^k$, $M_2 U_4 = -4.2^k$, $L_2 L_3 = -45^k$)

7.55   Problem 7.30

7.56   Problem 7.37 (*Ans.* $U_0 U_1 = -27{,}95^k$, $U_1 L_1 = +30^k$, $L_0 U_2 = +39.13^k$)

7.57   Problem 7.48

Chapter 8

Three-Dimensional or Space Trusses

8.1 GENERAL

A very brief and elementary discussion of small space trusses is presented in this chapter. It is hoped that this material will help the reader to develop some understanding of the behavior of space structures and to recognize that the equations of statics are as applicable in three dimensions as they are in two.

For all but the very smallest space structures, analysis by the methods of joints and moments described herein is completely unwieldy. As a result, a large percentage of colleges do not introduce the topic until students take a course in matrix analysis.

Nearly all structures are three-dimensional in nature. However, they may usually be broken down into separate systems, each lying in a single plane at right angles to the other. Figure 8.1(a) shows that an independent analysis of each of the systems is permissible. Two members, AB and BC, at right angles to each other, are shown in the figure. It is evident that a force in AB has no effect on the force in BC, because the component of a force at 90° is zero. Similarly, the forces in one truss have no effect on the forces in another truss framed into it at a right angle.

The members joining two systems together serve as members of both systems, and their total force is obtained by combining the forces developed as a part of each of the systems. The end posts of bridge trusses having end portals (see Figures 17.11 and 17.12) are one illustration. They serve as the end posts of the bridge trusses and as the columns of the portal.

Many towers, domes, and derricks are three-dimensional structures made up of members arranged so that it is impossible to divide them into different systems, each lying in a single plane, which may be handled individually. The difference is that the truss systems lie in planes that are not at right angles to one another. The forces in one truss framed into another at an angle other than 90° affect the forces in that truss, just as a force in member AB of Figure 8.1(b) causes a force in member BC. For trusses of this type we must analyze the whole structure as a unit, rather than consider the

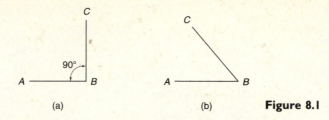

Figure 8.1

systems in various planes individually. This chapter is devoted to these types of trusses.

Structural engineers are so accustomed to visualizing structures in one plane that when they encounter space trusses they may make frequent mistakes because their minds still are operating on a single-plane basis. If the layout of a space truss is not completely clear, the construction of a small model will probably clarify the situation. Even the most simple models of paper, cardboard, or wire are helpful.

8.2 BASIC PRINCIPLES

Prior to introducing a method of analyzing space trusses, a few of the basic principles pertaining to such structures need to be considered. Three-dimensional trusses, as were two-dimensional trusses, are assumed to be made up of members subject to axial force only. In other words, the trusses are assumed to have members that are straight between joints, to have loads applied at joints only, to have members whose ends are free to rotate (note that for this situation to be true the members would have to be connected with universal joints, or at least with several frictionless pins), and so forth. Analyses based on these assumptions usually are quite satisfactory despite the welded and bolted connections used in actual practice.

A system of forces coming together at a single point, though not all in the same plane, may be combined into one resultant force. Similarly, an inclined force may have three coordinate components, and three reference planes will be used in handling them. The three planes used here are one horizontal and two vertical planes, each being perpendicular to the other. The intersections of these planes form the three coordinate axes used, X, Y, and Z. The force in any member inclined to these axes may be broken down into components along them, their magnitudes being proportional to their length projections on the axes. The inclined force F in Figure 8.2 is graphically broken down into components F_x, F_y, and F_z.

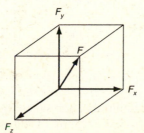

Figure 8.2

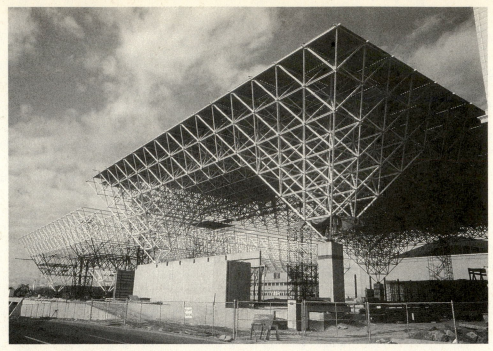

Denver, Colorado, Convention Center. (Courtesy Bethlehem Steel Corporation.)

The same values may be computed algebraically from the following relationship: The force in a member is to the length of the member as the X, Y, or Z component of force is to the corresponding X, Y, or Z component of length.

$$\frac{F}{\ell} = \frac{F_x}{\ell_x} = \frac{F_y}{\ell_y} = \frac{F_z}{\ell_z}$$

$$\ell^2 = \ell_x^2 + \ell_y^2 + \ell_z^2$$

$$F^2 = F_x^2 + F_y^2 + F_z^2$$

Space trusses may be either statically determinate or statically indeterminate; consideration is given here only to those that are statically determinate. The methods developed in later chapters for statically indeterminate structures apply equally to three-dimensional and two-dimensional trusses.

8.3 STATICS EQUATIONS

There are more statics equations available for determining the reactions of three-dimensional structures because there are two more axes to take moments about and one new axis along which to sum up forces. For equilibrium the sum of the forces along each of the three reference axes equals zero, as does the sum of the moments of all the forces about each of the axes. A total of six equations is available ($\Sigma X = 0$, $\Sigma Y = 0$, $\Sigma Z = 0$, $\Sigma M_x = 0$, $\Sigma M_y = 0$, and $\Sigma M_z = 0$), and six reaction components may be determined directly from them.

Should a structure have more than six reaction components, it is statically indeterminate externally; if less than six, it is unstable; and if equal to six, it is statically determinate externally. Many space trusses, however, have more than six reaction components and yet are statically determinate overall. Example 8.2 shows that reactions for this type of structure may be determined by solving them concurrently with the member forces using only statics equations.

The basic figure of the space truss is the triangle. A triangle can be extended into a space truss by adding three members and one joint. Each of the new members frames into one of the joints of the basic triangle, the other ends coming together to form a new joint. The elementary space truss formed has six members and four joints and is called a tetrahedron (a figure with four triangular surfaces). It may be enlarged by the addition of three members and one joint. For each of the joints of a space frame three equations ($\Sigma X = 0$, $\Sigma Y = 0$, and $\Sigma Z = 0$) are available to calculate the unknowns. Letting j be the number of joints, m the number of members, and r the number of reaction components, it can be seen that for a space truss to be statically determinate the following relation must hold:

$$3j = m - r$$

Should there be joints in the truss where the members are all in one plane, only two equations are available at each, and it is necessary to subtract one from the left-hand side of the equation for each such joint. The omission of one member for each reaction component in excess of six will cause this equation to be satisfied, and the structure will be statically determinate internally. When this situation occurs, it is possible to compute by three-dimensional statics the forces and reactions for the truss, although it is statically indeterminate externally.

8.4 STABILITY OF SPACE TRUSSES

The general rule for stability as regards outer forces is that the projection of the structure on any one of the three planes must itself be stable; therefore, as with two-dimensional structures, there must be at least three nonconcurrent reaction components in any one plane. The results of reaction computations will be inconsistent for any other case.

In the preceding paragraphs external stability and determinateness and internal stability and determinacy have been treated as though they were completely independent subjects. The two have been separated for clarity for the reader who has not previously encountered space trusses. The reader will learn, however, from the example problems in the pages that follow that it is impossible in a majority of cases to consider the two separately. For instance, many trusses are statically indeterminate externally and statically determinate internally and can be completely analyzed by statics. Few two-dimensional structures fall into this class.

In Section 8.3 it was stated that, for a space truss to be statically determinate, the following equation had to be satisfied:

$$3j = m - r$$

This equation is not sufficient to show whether a particular space truss is stable, however. Externally the reactions must be arranged so as to prevent movement of the structure; internally the members must be placed so as to keep the joints from moving with respect to each other.

For external stability the reactions must be placed so that they can resist translation along and rotation about each of the three axes x, y, and z. To achieve this goal there must be at least six nonparallel reactions, and they must not intersect a common axis.

Internal stability can be achieved if the truss is built up tetrahedron by tetrahedron, that is, by successively adding one joint and three members. In large space frames it may be quite difficult to see if this condition has been met. An analysis of the truss, however, will provide a clear statement as to stability. If we can obtain a *unique* solution the truss is stable. If not, the truss is unstable. The zero-load test (Section 7.9) also can be used to check stability.

8.5 SPECIAL THEOREMS APPLYING TO SPACE TRUSSES

From the principles of elementary statics there may be developed two theorems that are useful in analyzing space trusses. These are discussed in the following paragraphs.

1. The component of a force at 90° is zero, because no matter how large the force may be it equals zero when multiplied by the cosine of 90°. A force in one plane cannot have components in a plane normal to the original plane. Furthermore, a force in one plane cannot cause moment about any axis in its plane, because it will either intersect the axis or be parallel to it.

From the foregoing it is evident that if several members of a truss come together at a joint, all but one lying in the same plane, the component of force in the member normal to the plane of the other members must equal the sum of the components of the external forces at the joint normal to the same plane. If no external forces are present, the member has a force of zero.

2. The equations of statics clearly show that if there is a joint in a truss where no external loads are applied and where it has been proved that all but two of the members coming into the joint have no force, these two members must have zero force unless they happen to lie in a straight line.

8.6 TYPES OF SUPPORT

Trusses in one plane have been assumed to be supported with rollers or hinges that could supply one or two reaction components. For three-dimensional structures the same types of supports are used, but the number of reaction components may vary from one to three. These supports are described as follows and are illustrated in Figure 8.3.

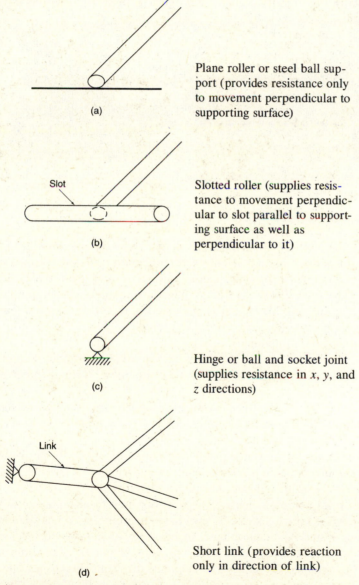

Plane roller or steel ball support (provides resistance only to movement perpendicular to supporting surface)

(a)

Slotted roller (supplies resistance to movement perpendicular to slot parallel to supporting surface as well as perpendicular to it)

(b)

Hinge or ball and socket joint (supplies resistance in x, y, and z directions)

(c)

Short link (provides reaction only in direction of link)

(d)

Figure 8.3 Types of supports for space trusses.

Broome County Veterans Memorial Arena, Binghamton, NY. (Courtesy Bethlehem Steel Corporation.)

1. The *plane roller* or *steel ball* or *flat plate* support provides resistance to movement perpendicular to the supporting surface. Thus, it has one reaction component which can be toward or away from the surface.
2. The *slotted roller* is free to move in one direction parallel to the supporting surface. Movement is prevented in the other direction parallel to the supporting surface as well as perpendicular to it, giving a total of two reaction components.
3. The *hinge* or *ball and socket* joint provides resistance to movement in the x, y, and z directions. Thus, it provides a total of three reaction components.
4. The *short link* provides resistance only in the direction of the link, having only one reaction component.

This discussion indicates that it is possible to select a type of support having three reaction components or one that may have one or two of the components eliminated. A little thought on the subject shows that the possibility of limiting the number of reaction components of a space truss is very advantageous. A truss that is statically indeterminate externally may have its total reaction components limited to six, making it statically determinate. (Advantages of statically determinate and statically indeterminate structures are discussed in Chapter 12.) For some structures it is desirable to eliminate the reaction components in certain directions. The most obvious example occurs when a space truss is supported on walls where a reaction or thrust perpendicular to the wall is undesirable.

The directions in which reaction components are possible are indicated herein by dark heavy lines at the support points, as shown in the diagrams of the frames analyzed in Examples 8.1 to 8.3.

8.7 ILLUSTRATIVE EXAMPLES

Examples 8.1 and 8.2 illustrate the application of the foregoing principles to elementary space trusses. Example 8.1 considers a structure supported at three points with six reaction components, which can be computed directly. The second example presents a space truss supported at four points with seven reaction components, which cannot be solved directly.

EXAMPLE 8.1

Determine the reactions and member forces in the space truss shown in Figure 8.4.

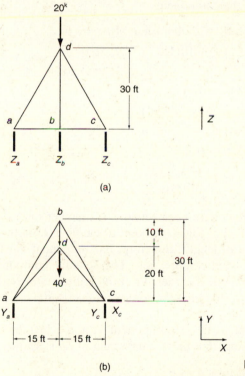

(a)

(b)

Figure 8.4

Solution.   The truss is statically determinate and stable externally because there is a total of six reaction components—three nonconcurrent ones in each plane. Internally it is statically determinate, as proved with the joint equation.

$$3j = m + r$$
$$12 = 6 + 6$$
$$12 = 12$$

For a truss with three vertical reaction components, moments may be taken about an axis through any two of them to find the third.

$$\Sigma M_x = 0 \text{ about } ac$$
$$(40)(30) - (20)(20) + 30Z_b = 0$$
$$Z_b = -26.7^k \downarrow$$
$$\Sigma M_y = 0 \text{ about line of action of } Y_a$$
$$(20)(15) + (26.7)(15) - 30Z_c = 0$$
$$Z_c = +23.3^k \uparrow$$
$$\Sigma Z = 0$$
$$-20 - 26.7 + 23.3 + Z_a = 0$$
$$Z_a = +23.4^k \uparrow$$

Similarly, where there are three unknown horizontal reaction components, moments may be taken about a vertical axis passing through the point of intersection of two of the components.

$$\Sigma M_z = 0 \text{ about line of action of } Z_c$$
$$-(40)(15) + 30Y_a = 0$$
$$Y_a = +20^k \uparrow$$
$$\Sigma Y = 0$$
$$20 - 40 + Y_c = 0$$
$$Y_c = +20^k \uparrow$$
$$\Sigma X = 0$$
$$0 + X_c = 0$$
$$X_c = 0$$

When the reactions have been found, the member forces can readily be computed by the method of joints. At joint a, member ad is the only member having a Z component of length; therefore, its component must be equal and opposite to Z_a, or 23.4-kip compression. The X and Y components of ad are proportional to its components of length in those directions. Setting up a table similar to the one shown simplifies the computation of components and resultant forces.

Considering joint a, the Y component of force in member ab can be determined by joints now that the Y component of ad is known.

$$\Sigma Y \text{ at joint } a = 0$$
$$20 - 15.6 - Y_{ab} = 0$$
$$Y_{ab} = -4.4 \text{ compression}$$

The other member forces are computed by joints and shown in Table 8.1.

∎

TABLE 8.1

| Member | Projection | | | | Component of force | | | |
	X	Y	Z	Length	X	Y	Z	Force
ab	15	30	0	33.5	− 2.2	− 4.4	0	− 4.92
ad	15	20	30	39.1	−11.7	−15.6	−23.4	−30.5
ac	30	0	0	30.0	+13.9	0	0	+13.9
bc	15	30	0	33.5	− 2.2	− 4.5	0	− 5.02
bd	0	10	30	31.6	0	+ 8.9	+26.7	+28.1
cd	15	20	30	39.1	−11.7	−15.5	−23.3	−30.3

EXAMPLE 8.2

Find all reactions and member forces of the space truss shown in Figure 8.5.

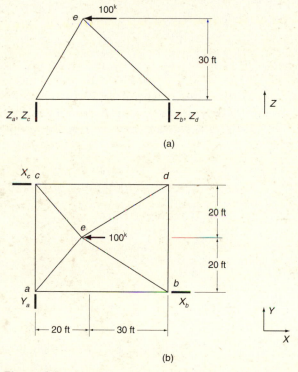

(a)

(b)

Figure 8.5

Solution. Examination of the truss shows it to be statically indeterminate externally because there are seven definite reaction components and only six equations of statics. Internally, however, it is statically determinate as shown, and the analysis may be handled by statics.

$$3j = m + r$$
$$15 = 8 + 7$$
$$15 = 15$$

Although the truss is statically indeterminate externally, there are only three unknown reaction components in the XY plane, and these may be determined immediately. The other four components will be solved in conjunction with the member forces.

$$\Sigma M_z = 0 \text{ about line of action of } z_a$$
$$-(100)(20) + 40X_c = 0$$
$$X_c = +50^k \rightarrow$$

$$\Sigma X = 0$$
$$+50 - 100 + X_b = 0$$
$$X_b = +50^k \rightarrow$$

$$\Sigma Y = 0$$
$$0 + Y_a = 0$$
$$Y_a = 0$$

If the value of one of the Z reaction components should be known, the values of the other three could be determined by statics. It is assumed that Z_d has a value of S downward, and the reaction components are computed in terms of S.

$$\Sigma M_y = 0 \text{ about } ac$$
$$-(100)(30) + 50S + 50Z_b = 0$$
$$Z_b = 60 - S$$

$$\Sigma M_x = 0 \text{ about } ab$$
$$+40S - 40Z_c = 0$$
$$Z_c = +S$$

$$\Sigma Z = 0$$
$$+S - S - (60 - S) + Z_a = 0$$
$$Z_a = 60 - S$$

Checking by $\Sigma M_x = 0$ about cd
$$(60 - S)(40) - 40Z_a = 0$$
$$Z_a = 60 - S$$

The calculation of member forces may now be started from the reaction components in terms of S. These computations are continued until the forces at both ends of one bar are determined in terms of S. The two values must be equal, and they are equated to give the correct value of S.

The Z component of force in de equals S and is in tension, whereas the Z component of force in be equals $60 - S$ and is also in tension. The Y component

TABLE 8.2

Member	Projection			Length	Component of force			Force
	X	Y	Z		X	Y	Z	
ab	50	0	0	50	+20	0	0	+20
ae	20	20	30	41.2	−20	−20	−30	−41.2
ac	0	40	0	40	0	+20	0	+20
be	30	20	30	46.9	+30	+20	+30	+46.9
bd	0	40	0	40	0	−20	0	−20
de	30	20	30	46.9	+30	+20	+30	+46.9
cd	50	0	0	50	−30	0	0	−30
ce	20	20	30	41.2	−20	−20	−30	−41.2

of force in de equals $(20/30)(S) = \frac{2}{3}S$, and the Y component of force in be is $(20/30)(60 - S) = 40 - \frac{2}{3}S$. By $\Sigma Y = 0$ at joint d, member bd is in compression with a force of $\frac{2}{3}S$. Similarly, by $\Sigma Y = 0$ at joint b, member bd is seen to have a compressive force of $40 - \frac{2}{3}S$. Equating the two expressions yields the value of S.

$$\frac{2}{3}S = 40 - \frac{2}{3}S$$
$$\frac{4}{3}S = 40$$
$$S = 30^{\text{k}}$$

The numerical values of the Z reaction components can now be found from S, and the forces in the truss can be determined by joints. The use of a table to work with length and force components is again convenient. The results are shown in Table 8.2. ■

8.8 MORE COMPLICATED SPACE TRUSSES

The method of joints was used to determine the forces in the elementary space trusses analyzed in Section 8.7. The analyst may have difficulty calculating any of the forces by the method of joints for more complicated space trusses, however, even if the reactions are available. The truss of Figure 8.6 on page 214 falls into this class.

It is possible to compute all reaction components immediately, but no forces may be determined by the method of joints as presented in Examples 8.1 and 8.2. The application of the zero-member principle of paragraph (1) of Section 8.5, however, will quickly prove that several members have no force. For example, at joint f, members fc, fb, and fe lie in the same plane; member fd does not, and it has a component of force perpendicular to the plane of the other three members. This component must be equal and opposite to all components perpendicular to the plane from external loads applied at the joint. Since there are no loads present, member fd has no force. A similar analysis can be made at joint e to show the force in member fe is zero.

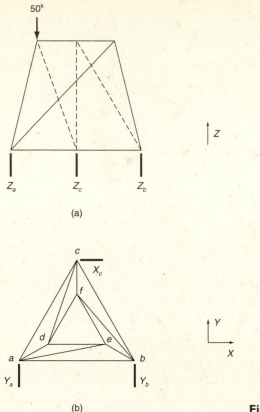

Figure 8.6

Two of the four members meeting at joint f have been shown to have no force. The remaining two members are not in a straight line and must have zero forces, since there are no external loads applied at the joint [paragraph (2) of Section 8.5]. It is now possible to compute the forces in the remaining members by joints for this particular truss.

For other structures the use of the zero-member principle will not be sufficient (if it is of any value at all) in determining forces. In this latter type of truss the use of moments will probably enable the reader to make the analysis. The truss of Figure 8.6 could be analyzed by taking moments. If an imaginary section were passed completely around joint e isolating it as a free body, moments could be taken about line ab to find the force in member ef. Forces in ae and eb intersect line ab, and the line of action of de is parallel to ab. Several moment equations of this type are used in Example 8.3.

As space trusses become more complicated, it may not be possible to find a single force by the methods introduced to this point. If a member is arbitrarily given a force, say F, it may be possible to compute the forces in several members in terms of F. Once the force at both ends of a member is known in terms of F, the two values may be equated to find F and the force analysis for the truss may be carried out.

Transmission towers for the country's first 345,000-volt transmission line, Chief Joseph-Snohomish Dam, Washington State. (Courtesy Bethlehem Steel Corporation.)

EXAMPLE 8.3

Analyze the space truss of Figure 8.7, shown on the following page.

Solution.   Although there are eight reaction components, the structure is statically determinate.

$$3j = m + r$$
$$(3)(8) = 16 + 8$$
$$24 = 24$$

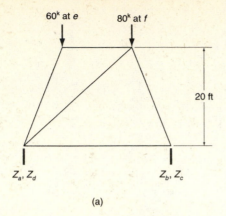

(a)

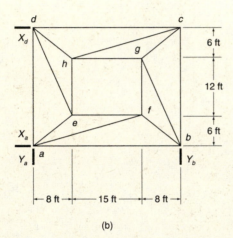

(b)   **Figure 8.7**

A brief examination of the structure will reveal four zero members. At joint h member eh has no force, and at joint g the same can be said for member gh; therefore, members hd and hc meeting at joint h have zero forces.

By removing joint e as a free body and taking moments about line ad, the force in ef is determined, and by taking moments about line ab, force de is found. The method of joints is then used to find the forces in ae and da.

$$\Sigma M_y = 0 \text{ about line } ad$$
$$(60)(8) - 20F_{ef} = 0$$
$$F_{ef} = 24^k \text{ compression}$$

$$\Sigma M_x = 0 \text{ about line } ab. \text{ (Note } eh \text{ has a zero force.)}$$
$$(60)(6) - 20Y_{de} - 6Z_{de} = 0$$
$$360 - 20Y_{de} - (6)(20/18)(Y_{de}) = 0$$

$$Y_{de} = 13.5^k \text{ compression}$$

$$\Sigma Y = 0 \text{ at joint } e$$
$$Y_{ae} = 13.5^{k} \text{ compression}$$
$$\Sigma Y = 0 \text{ at joint } d$$
$$Y_{da} = 13.5^{k} \text{ tension}$$

An imaginary section is passed around joint f and moments are taken about line ab to find the force in fg. By using the same free body, moments are taken about line bc to find the force in af.

$$\Sigma M_{x} = 0 \text{ about line } ab$$
$$(80)(6) - 20F_{jg} = 0$$
$$F_{jg} = 24^{k} \text{ compression}$$
$$\Sigma M_{y} = 0 \text{ about line } bc$$
$$-(80)(8) + (24)(20) + 20X_{af} + 8Z_{af} = 0$$
$$X_{af} = 5.9^{k} \text{ compression}$$

A similar moment procedure may be used, with joint g as the free body, to determine the force in member bg, and the values of all remaining forces and reactions can be determined by joints. A summary of the results is given in Table 8.3.

TABLE 8.3

Member	Projection			Length	Component of force			Force
	X	Y	Z		X	Y	Z	
ab	31	0	0	31	+22	0	0	+22
bc	0	24	0	24	0	+ 6	0	+ 6
cd	31	0	0	31	+ 8	0	0	+ 8
da	0	24	0	24	0	+13.5	0	+13.5
ae	8	6	20	22.4	−18	−13.5	−45	−50.4
af	23	6	20	31.1	− 5.9	− 1.5	− 5.1	− 7.9
bf	8	6	20	22.4	−30	−22.5	−75	−84
bg	8	18	20	28.1	+ 8	+18	+20	+28.1
cg	8	6	20	22.4	− 8	−6	−20	−22.4
ch	23	6	20	31.1	0	0	0	0
dh	8	6	20	22.4	0	0	0	0
de	8	18	20	28.1	− 6	−13.5	−15	−21.1
ef	15	0	0	15	−24	0	0	−24
fg	0	12	0	12	0	−24	0	−24
gh	15	0	0	15	0	0	0	0
he	0	12	0	12	0	0	0	0

Reactions:

$$Z_a = 50 \uparrow \qquad X_a = 2 \rightarrow$$
$$Z_b = 55 \uparrow \qquad X_d = 2 \leftarrow$$
$$Z_c = 20 \uparrow \qquad Y_a = 1.5 \uparrow$$
$$Z_d = 15 \uparrow \qquad Y_b = 1.5 \downarrow \quad \blacksquare$$

8.9 SIMULTANEOUS-EQUATION ANALYSIS

Three simultaneous equations ($\Sigma X = 0$, $\Sigma Y = 0$, and $\Sigma Z = 0$) may be written for the forces meeting at each joint of a space truss. If the truss is statically determinate the equations may be solved for both the member forces and the reaction components. Although the subject of matrix formulation for solving simultaneous equations is discussed in Chapters 20 through 21, the solution of the equations by the process of elimination is used in this chapter. Actually, this type of solution may be rather quick for small space trusses, despite the large number of equations, because of the small number of unknowns that appear in each equation.

Before considering the preparation of simultaneous equations for an entire structure, a single joint is considered. Should there be only three unknown forces at a joint, it may be possible to obtain their values by writing the three statics equations for that joint. (Frequently this procedure will not work because the equations may be linearly dependent.)

The preparation and solution of simultaneous equations for space trusses can be simplified by making use of *tension coefficients*.[1] The tension coefficient for a member equals its force divided by its length. In each of the following expressions for force components the value F/ℓ is replaced by T, the tension coefficient.

$$F_x = \frac{\ell_x}{\ell}F = \frac{F}{\ell}\ell_x = T\ell_x$$

$$F_y = \frac{\ell_y}{\ell}F = \frac{F}{\ell}\ell_y = T\ell_y$$

$$F_z = \frac{\ell_z}{\ell}F = \frac{F}{\ell}\ell_z = T\ell_z$$

The space truss of Figure 8.4 is redrawn in Figure 8.8 and, for convenience in preparing the equations, the members are numbered as shown. The following equations are written for joint d of this frame using tension coefficients and assuming all members are in tension. The reactions are assumed to be in the directions shown.

$$\Sigma X = -15T_{10} + 15T_{12} = 0$$
$$\Sigma Y = -20T_{10} + 10T_{11} - 20T_{12} = 40$$
$$\Sigma Z = -30T_{10} - 30T_{11} - 30T_{12} = 20$$

---

[1] R. V. Southwell, "Primary Stress Determination in Space Frames," *Engineering 109* (1920): 165.

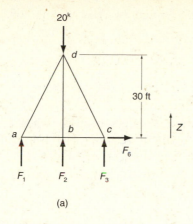

(a)

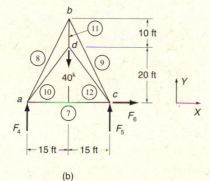

(b)   **Figure 8.8**

There should be no difficulty in solving these equations by the usual form of elimination. The Gauss-Jordan elimination procedure described here for solving the equations is advantageous when there are numerous equations.[2] First, the coefficients of both sides of the equations are written as follows:

$$
\begin{array}{rrrr}
-15 & 0 & 15 & 0 \\
-20 & 10 & -20 & 40 \\
-30 & -30 & -30 & 20
\end{array}
$$

Each equation is divided through by its leading nonzero constant with these results:

$$
\begin{array}{rrrr}
1 & 0 & -1 & 0 \\
1 & -0.5 & 1 & -2 \\
1 & 1 & 1 & -0.667
\end{array}
$$

The first equation is left as is and each of the other equations not starting with a zero value is subtracted from the first.

[2] G. L. Rogers and M. L. Causey, *Mechanics of Engineering Structures* (New York: Wiley, 1962), 27–31.

These equations are each divided through by their leading nonzero constants with the following results:

$$
\begin{array}{cccc}
1 & 0 & -1 & 0 \\
0 & 1 & -4 & 4 \\
0 & 1 & 2 & -0.667
\end{array}
$$

The process is repeated with all rows after the second one whose second element is not zero being subtracted from row 2, with the resulting rows being divided through by their leading nonzero element. The final result is as follows:

$$
\begin{array}{cccc}
1 & 0 & -1 & 0 \\
 & 1 & -4 & 4 \\
 & & 1 & -0.777
\end{array}
$$

These rows now have the form of a triangular matrix (see Appendix B) and correspond to the following equations:

$$
\begin{aligned}
T_{10} && -T_{12} &= 0 \\
& T_{11} & -4T_{12} &= 4 \\
&& T_{12} &= -0.777
\end{aligned}
$$

The value of T_{12} is calculated from the last equation; T_{11}, from the next to last equation; and so on. The results are $T_{12} = -0.777$; $T_{11} = 0.892$; and $T_{10} = -0.777$. The member forces are formed by multiplying the tension coefficients by the member lengths.

$$
\begin{aligned}
F_{10} &= (39.1)(-0.777) = -30.4^k \\
F_{11} &= (31.6)(\ 0.892) = +28.2^k \\
F_{12} &= (39.1)(-0.777) = -30.4^k
\end{aligned}
$$

In a similar manner the equations for the entire space truss of Figure 8.8 are written at the end of this paragraph. When a computer is not available and the simultaneous-equation procedure is used, it may be more practical to start at joints having only three unknowns and determine their values (if feasible) and to use the values so computed in later equations, rather than considering all the equations for an entire structure at one time. Nevertheless, all the equations for the truss of Figure 8.8 are considered as a group here.

$$
\text{joint } a
\begin{cases}
30T_7 + 15T_8 + 15T_{10} & = 0 \quad (7) \\
F_4 + 30T_8 + 20T_{10} & = 0 \quad (4) \\
F_1 + 30T_{10} & = 0 \quad (1)
\end{cases}
$$

$$
\text{joint } b
\begin{cases}
-15T_8 + 15T_9 & = 0 \quad (8) \\
-30T_8 - 30T_9 - 10T_{11} & = 0 \quad (9) \\
F_2 + 30T_{11} & = 0 \quad (2)
\end{cases}
$$

$$\text{joint } c \begin{cases} F_6 - 30T_7 - 15T_9 - 15T_{12} = 0 & (6) \\ F_5 + 30T_9 + 20T_{12} = 0 & (5) \\ F_3 + 30T_{12} = 0 & (3) \end{cases}$$

$$\text{joint } d \begin{cases} -15T_{10} + 15T_{12} = 0 & (12) \\ -20T_{10} + 10T_{11} - 20T_{12} = 40 & (11) \\ -30T_{10} - 30T_{11} - 30T_{12} = 20 & (10) \end{cases}$$

To facilitate the solution of the equations, it is very important to arrange them properly. If not arranged properly, the work involved in solving them can be enormous. They should be arranged as nearly as possible to the desired final triangular matrix. In Table 8.4 the equations are listed in the order given by the numbers to the

TABLE 8.4

F_1	F_2	F_3	F_4	F_5	F_6	T_7	T_8	T_9	T_{10}	T_{11}	T_{12}	Load
1									30			
	1									30		
		1									30	
			1				30		20			
				1			30				20	
					1	−30		−15			−15	
						1	0.5		0.5			
							1	−1				
							1	−1		0.33		
									1	1	1	−0.67
									1	−0.5	1	− 2
									1	−1		

TABLE 8.5

F_1	F_2	F_3	F_4	F_5	F_6	T_7	T_8	T_9	T_{10}	T_{11}	T_{12}	Load
1									30			
	1									30		
		1									30	
			1				30		20			
				1			30				20	
					1	−30		−15			−15	
						1	0.5		0.5			
							1	−1				
								−1		0.17		
									1	1	1	−0.67
										1		0.89
											1	−0.78

right of each of the preceding equations. Each row also is divided by its leading nonzero constant.

These equations already are almost in the desired final form. Following the procedure previously described, the results shown in Table 8.5 (page 221) can be obtained. These equations can be solved with the same results shown in Example 8.1.

8.10 COMPUTER EXAMPLE

EXAMPLE 8.4 _____

Repeat Example 8.3 using SABLE. Number the members and joints as shown in Figure 8.9.

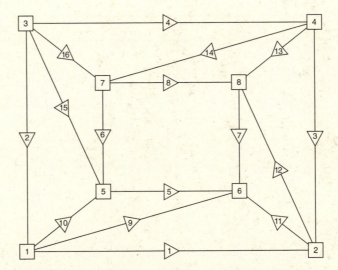

Figure 8.9 Numbering of joints and members for space truss of Example 8.4.

Solution.

Structural Data

Only part of input data are presented here in order to save space.

1. Nodal location and restraint data

| | Coordinates | | | Restraints | | | | | |
Node	X	Y	Z	T-X	T-Y	T-Z	R-X	R-Y	R-Z
1	0.000E+00	0.000E+00	0.000E+00	Y	Y	Y	Y	Y	Y
2	3.100E+01	0.000E+00	0.000E+00	N	Y	Y	Y	Y	Y
3	0.000E+00	2.400E+01	0.000E+00	Y	N	Y	Y	Y	Y
4	3.100E+01	2.400E+01	0.000E+00	N	N	Y	Y	Y	Y
5	8.000E+00	6.000E+00	2.000E+01	N	N	N	Y	Y	Y

	Coordinates			Restraints					
Node	X	Y	Z	T-X	T-Y	T-Z	R-X	R-Y	R-Z
----	----------	----------	----------	----	----	----	----	----	---
6	2.300E+01	6.000E+00	2.000E+01	N	N	N	Y	Y	Y
7	8.000E+00	1.800E+01	2.000E+01	N	N	N	Y	Y	Y
8	2.300E+01	1.800E+01	2.000E+01	N	N	N	Y	Y	Y

2. Applied joint loads

Node	Case	Force-X Moment-X	Force-Y Moment-Y	Force-Z Moment-Z
-----	-----	----------	----------	----------
5	1	0.000E+00	0.000E+00	−6.000E+01
		0.000E+00	0.000E+00	0.000E+00
6	1	0.000E+00	0.000E+00	−8.000E+01
		0.000E+00	0.000E+00	0.000E+00

3. Member properties and joint locations omitted

Results

1. Sample beam end forces

Beam	Case	End	Axial Torsion	Shear-Y Shear-Z	Moment-Z Moment-Y
-----	-----	----	----------	----------	----------
1	1	i	−2.194E+01	0.000E+00	0.000E+00
			0.000E+00	0.000E+00	0.000E+00
		j	2.194E+01	0.000E+00	0.000E+00
			0.000E+00	0.000E+00	0.000E+00
2	1	i	−1.350E+01	0.000E+00	0.000E+00
			0.000E+00	0.000E+00	0.000E+00
		j	1.350E+01	0.000E+00	0.000E+00
			0.000E+00	0.000E+00	0.000E+00
3	1	i	−6.000E+00	0.000E+00	0.000E+00
			0.000E+00	0.000E+00	0.000E+00
		j	6.000E+00	0.000E+00	0.000E+00
			0.000E+00	0.000E+00	0.000E+00
4	1	i	−8.000E+00	0.000E+00	0.000E+00
			0.000E+00	0.000E+00	0.000E+00
		j	8.000E+00	0.000E+00	0.000E+00
			0.000E+00	0.000E+00	0.000E+00

2. Sample calculated nodal forces

Node	Case	Force-X Moment-X	Force-Y Moment-Y	Force-Z Moment-Z
-----	-----	----------	----------	----------
1	1	2.000E+00	1.548E+00	5.016E+01
		0.000E+00	0.000E+00	0.000E+00
2	1	0.000E+00	−1.548E+00	5.484E+01
		0.000E+00	0.000E+00	0.000E+00
3	1	−2.000E+00	0.000E+00	1.500E+01
		0.000E+00	0.000E+00	0.000E+00	■

PROBLEMS

For Problems 8.1 through 8.8, compute the reaction components and the member forces for the space trusses.

8.1

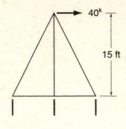
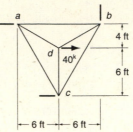

8.2   (*Ans. ac* = -43.16^k, *bd* = $+50.31^k$)

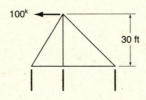
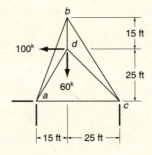

8.3

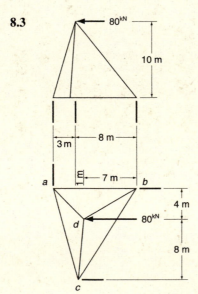

8.4   (*Ans. ac* = $+6.67^k$, *ce* = -15.64^k, *ab* = $+20.00^k$)

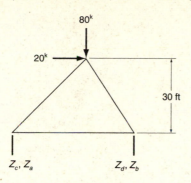

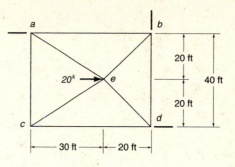

8.5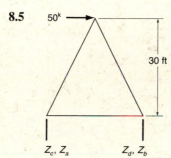

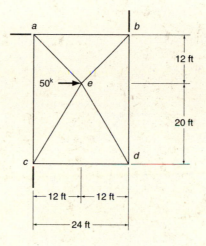

8.6   (*Ans. cd* = $+12.5^k$, *gh* = -50^k, *bf* = 0, *ch* = $+63.7^k$, Z_c = 50^k ↓, Y_d = 12.5^k ↑)

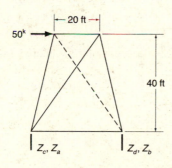

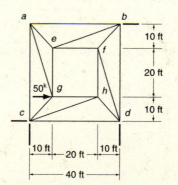

8.7

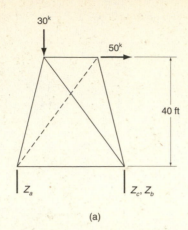

(a)

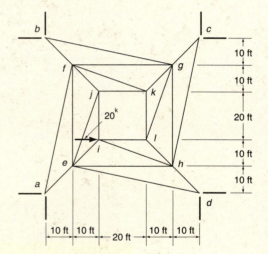

(b)

8.8   (*Ans.* $de = -14.8^k$, $eh = +15.0^k$, $fj = -8.3^k$, $hi = -32.7^k$, $X_b = 3.33^k \rightarrow$, $Y_a = 6.67^k \downarrow$)

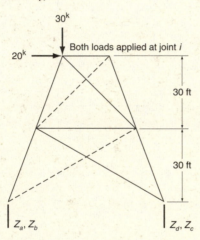

8.9   Solve Problem 8.1 using simultaneous equations for the entire truss.

Repeat the problems listed using the enclosed computer diskette.

8.10   Problem 8.2 (*Ans.* $ac = -43.2^k$, $bd = +50.3^k$, $cd = +89.8^k$)

8.11   Problem 8.7

8.12   Problem 8.8 (*Ans.* $de = -14.8^k$, $ej = +10.9^k$, $hi = -32.7^k$)

Chapter 9

Influence Lines

9.1 INTRODUCTION

Structures supporting groups of loads fixed in one position have been discussed in the foregoing chapters. Whether beams, frames, or trusses were being considered and whether the functions sought were shears, reactions, or member forces, the loads were stationary. In practice, however, the engineer rarely deals with structures supporting only fixed loads. Nearly all structures are subject to loads moving back and forth across their spans. Perhaps bridges with their vehicular traffic are the most noticeable examples, but industrial buildings with traveling cranes, office buildings with furniture and human loads, and frames supporting conveyor belts are in the same category.

Each member of a structure must be designed for the most severe conditions that can possibly develop in that member. The designer places the live loads at the positions where they will produce these conditions. The critical positions for placing live loads will not be the same for every member. For example, the maximum force in one member of a bridge truss may occur when there is a line of trucks from end to end of the bridge, whereas the maximum force in some other member may occur when the trucks extend only from that member to one end of the bridge. The maximum forces in certain beams and columns of a building will occur when the live loads are concentrated in certain portions of the building, whereas the maximum forces in other beams and columns will occur when the loads are placed elsewhere.

On some occasions it is possible by inspection to determine where to place the loads to give the most critical forces, but on many other occasions it is necessary to resort to certain criteria or diagrams to find the locations. The most useful of these devices is the influence line.

Tennessee-Tombigbee Waterway Bridge in Mississippi. (Courtesy of the Mississippi State Highway Department.)

9.2 THE INFLUENCE LINE DEFINED

The influence line, which was first used by Professor E. Winkler of Berlin in 1867, shows graphically how the movement of a unit load across a structure influences some function of the structure.[1] The functions that may be represented include reactions, shears, moments, forces, and deflections.

An influence line may be defined as a diagram whose ordinates show the magnitude and character of some function of a structure as a unit load moves across the structure. Each ordinate of the diagram gives the value of the function when the load is at that ordinate.

Influence lines are primarily used to determine where to place live loads to cause maximum forces. They may also be used to compute those forces. The procedure for drawing the diagrams is simply the plotting of values of the function under study as ordinates for various positions of the unit load along the span and the connecting of those ordinates. You should mentally picture the load moving across the span and try to imagine what is happening to the function in question during the movement. The study of influence lines can immeasurably increase your knowledge of what happens to a structure under different loading conditions.

Study of the following sections should fix clearly in your mind what an influence line is. The actual mechanics of developing the diagrams are elementary, once the

[1] J. S. Kinney, *Indeterminate Structural Analysis* (Reading, Mass.: Addison-Wesley, 1957), Chap. 1.

definition is completely understood. No new fundamentals are introduced here; rather, a method of recording information in a convenient and useful form is given.

9.3 INFLUENCE LINES FOR SIMPLE BEAM REACTIONS

Influence lines for the reactions of a simple beam are given in Figure 9.1. The variation of the left-hand reaction V_L as a unit load moves from left to right across the beam is considered initially. When the load is directly over the left support, $V_L = 1$; when it is 2 ft to the right of the left support, $V_L = 18/20$, or 0.9; when it is 4 ft to the right, $V_L = 16/20$, or 0.8; and so on.

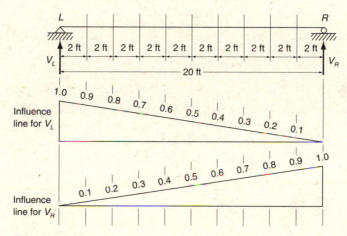

Figure 9.1

Values of V_L are shown at 2-ft intervals as the unit load moves across the span. These values lie in a straight line because they change uniformly for equal intervals of the load. For every 2-ft interval the ordinate changes 0.1. The values of V_R, the right-hand reaction, are plotted similarly for successive 2-ft intervals of the unit load. For each position of the unit load the sum of the ordinates of the two diagrams at any point equals (and for equilibrium certainly must equal) the unit load.

9.4 INFLUENCE LINES FOR SIMPLE BEAM SHEARS

Influence lines are plotted in Figure 9.2 on the following page for the shear at two sections in a simple beam. The following sign convention for shear is used: Positive shear occurs when the sum of the transverse forces to the left of a section is up or when the sum of the forces to the right of the section is down.

Placing the unit load over the left support causes no shear at either of the two sections. Moving the unit load 2 ft to the right of the left support results in a left-hand reaction of 0.9, and the sum of the forces to the left of section 1-1 is 0.1 down, or a shear of −0.1. When the load is 4 ft to the right of the left support and an infinitesimal distance to the left of section 1-1, the shear to the left is −0.2. If the load is moved

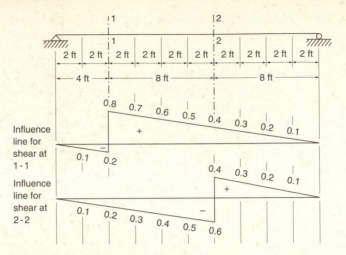

Figure 9.2

a very slight distance to the right of section 1-1, the sum of the forces to the left of the section becomes 0.8 up, or a +0.8 shear. Continuing to move the load across the span toward the right support results in changing values of the shear at section 1-1. These values are plotted for 2-ft intervals of the unit load. The influence line for shear at section 2-2 is developed in the same manner.

Note that the slope of the shear influence line to the left of the section in question must equal the slope of the influence line to the right of the section. In Figure 9.2, for instance, for the influence line at section 1-1 the slope to the left is 0.2/4 = 0.05,

Precast-concrete Kalihiwai Bridge near Kilauea, Kauai, Hawaii. (Courtesy of the Hawaii Department of Transportation.)

while the slope to the right is $0.8/16 = 0.05$. This information is very useful in drawing other shear influence lines.

9.5 INFLUENCE LINES FOR SIMPLE BEAM MOMENTS

Influence lines are plotted in Figure 9.3 for the moment at the same sections of the beam used in Figure 9.2 for the shear illustrations. To review, a positive moment causes tension in the bottom fibers of a beam and occurs at a particular section when the sum of the moments of all the forces to the left is clockwise, or when the sum to the right is counterclockwise. Moments are taken at each of the sections for 2-ft intervals of the unit load.

The major difference between shear and moment diagrams as compared with influence lines should now be clear. *A shear or moment diagram shows the variation of shear or moment across an entire structure for loads fixed in one position. An influence line for shear or moment shows the variation of that function at one section in the structure caused by the movement of a unit load from one end of the structure to the other.*

Influence lines for functions of statically determinate structures consist of a set of straight lines. An experienced analyst will be able to compute values of the function under study at a few critical positions and connect the plotted values with straight lines. A person beginning his or her study, however, must be very careful to compute the value of the function for enough positions of the unit load. The shapes of influence lines for forces in truss members often are deceptive in their seeming simplicity. It is obviously better to plot ordinates for several extra positions of the load than to fail to plot one essential value.

Several influence lines for moment, shear, and reactions for an overhanging beam are plotted in Figure 9.4 on the following page.

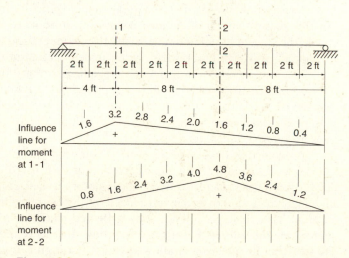

Figure 9.3

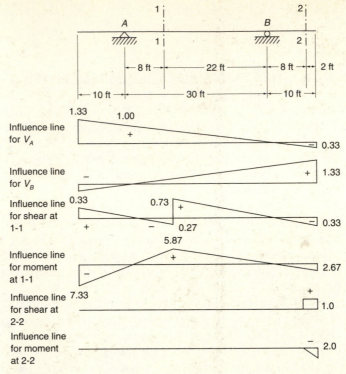

Figure 9.4

9.6 QUALITATIVE INFLUENCE LINES

The average student initially has a great deal of difficulty in drawing influence lines. For this reason qualitative influence lines are introduced at this time, as they will enable the student to obtain immediately the correct shape of the desired figures and, it is hoped, will give him or her a better understanding of these useful devices.

The influence lines drawn in the preceding sections for which numerical values were computed are referred to as *quantitative influence lines*. It is possible, however, to make rough sketches of these diagrams with sufficient accuracy for many practical purposes without computing any numerical values. These latter diagrams are referred to as *qualitative influence lines*.

A detailed discussion of the principle on which these sketches are made is given in Chapter 14 together with a consideration of their usefulness. Such a discussion is delayed until the student has had some exposure to deflections. Qualitative influence lines are based on a principle introduced by the German Professor Heinrich Müller-Breslau. This principle follows: **The deflected shape of a structure represents to some scale the influence line for a function such as reaction, shear, or moment if the function in question is allowed to act through a unit displacement.** In other words, the structure draws its own influence line when the proper displacement is applied. This principle is derived in section 14.2.

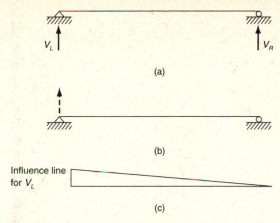

(a)

(b)

Influence line
for V_L

(c)

Figure 9.5

As a first example, the qualitative influence line for the left reaction of the beam of Figure 9.5(a) is considered. The constraint at the left support is removed and a displacement is introduced there in the direction of the reaction as shown in part (b) of the figure. When the left end of the beam is pushed up, the area between the original and final position of the beam is the influence line for V_L to some scale.

In a similar manner the influence lines for the left and right reactions of the beam of Figure 9.6 are sketched.

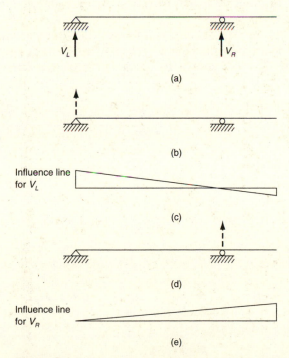

(a)

(b)

Influence line
for V_L

(c)

(d)

Influence line
for V_R

(e)

Figure 9.6

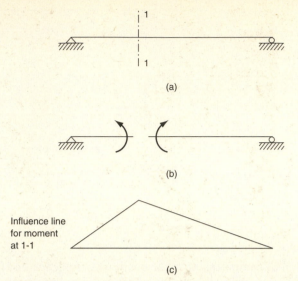

Figure 9.7

As a third example, the influence line for moment at section 1-1 in the beam of Figure 9.7 is considered. This diagram can be obtained by cutting the beam at the point in question and applying moments just to the left and just to the right of the cut section as shown. It can be seen in the figure that the moment on each side of the section is positive with respect to the segment of the beam on that side of the section. The resulting deflected shape of the beam is the qualitative influence line for moment at section 1-1.

To draw a qualitative influence line for shear, the beam is assumed to be cut at the point in question and a vertical force of the nature required to give positive shear applied to the beam on each side of the section [see Fig. 9.8(b)]. To understand the

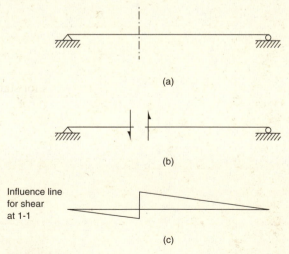

Figure 9.8

Overpass, Boise, Idaho. (Courtesy of the American Concrete Institute.)

direction used for these forces, the reader should note that they are applied to the left and to the right of the cut section so as to produce a positive shear for each segment. In other words, the force on the left segment is in the direction of a positive shear force applied from the right side (↓) and vice versa. Additional examples are presented for qualitative influence lines in Figure 9.9 on the following page.

Müller-Breslau's principle is useful for sketching influence lines for statically determinate structures, but its greatest value is for statically indeterminate structures. Though the diagrams are drawn exactly as before, the reader should note that they consist of curved lines instead of straight lines, as was the case for statically determinate structures. On page 236, Figure 9.10 shows several such examples.

9.7 USES OF INFLUENCE LINES; CONCENTRATED LOADS

Influence lines are the plotted values of functions of structures for various positions of a unit load. Having an influence line for a particular function of a structure makes the value of the function for a concentrated load at any position on the structure immediately available. The beam of Figure 9.1 and the influence line for the left reaction are used to illustrate this statement. A concentrated 1-kip load 4 ft to the right

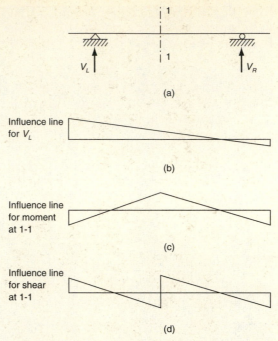

Figure 9.9

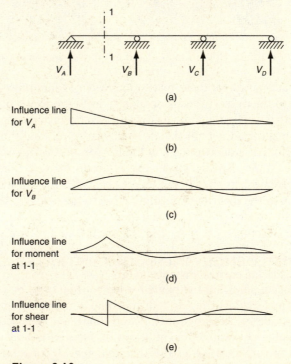

Figure 9.10

of the left support would cause V_L to equal 0.8 kip. Should a concentrated load of 175 kips be placed in the same position, V_L would be 175 times as great, or 140 kips.

The value of a function due to a series of concentrated loads is quickly obtained by multiplying each concentrated load by the corresponding ordinate of the influence line for that function. Loads of 150 kips 6 ft to the right of L in Figure 9.1 and 200 kips 16 ft to the right of L would cause V_L to equal $(150)(0.7) + (200)(0.2)$, or 145 kips.

Influence lines for the left reaction and the centerline moment are shown for a simple beam in Figure 9.11, and the values of these functions are calculated for the several loads supported by the beam.

$$V_L = (20)(0.8) + (30)(0.4) + (30)(0.1) = 31^k$$
$$M_{\mathbb{C}} = (20)(5.0) + (30)(10.0) + (30)(2.5) = 475'^k$$

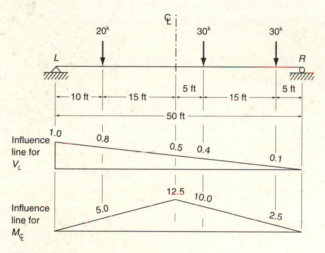

Figure 9.11

9.8 USES OF INFLUENCE LINES; UNIFORM LOADS

The value of a certain function of a structure may be obtained from the influence line, when the structure is loaded with a uniform load, by multiplying the area of the influence line by the intensity of the uniform load. The following discussion proves this statement to be correct.

A uniform load of intensity w lb/ft (pound per foot) is equivalent to a continuous series of smaller loads of (w) $(\frac{1}{10})$ lb on each $\frac{1}{10}$ ft, or $w\,dx$ lb on each dx distance. Considering each dx distance to be loaded with a concentrated load of $w\,dx$, the value of the function under study for one of these small loads is $(w\,dx)(y)$, or $wy\,dx$, where y is the ordinate of the influence line at that point. The effect of all of these concentrated loads is equal to $\int wy\,dx$. This expression shows that the effect of a uniform load on some function of a structure equals the intensity of the uniform load (w) times the area of the influence line $(\int y\,dx)$ along the section of the structure covered by the uniform load.

Assuming the beam of Figure 9.11 to be loaded with a uniform load of 3 klf for the entire span, the values of V_L and $M_{\mathrm{C\!\!\!L}}$ would be as follows:

$$V_L = (3)(\tfrac{1}{2} \times 1.0 \times 50) = 75^{\mathrm{k}}$$

$$M_{\mathrm{C\!\!\!L}} = (3)(\tfrac{1}{2} \times 12.5 \times 50) = 937.5'^{\mathrm{k}}$$

If the uniform load extended only from the left end to the centerline of the beam, the values of V_L and $M_{\mathrm{C\!\!\!L}}$ would be

$$V_L = (3)\left(\frac{1.0 + 0.5}{2} \times 25\right) = 56.25^{\mathrm{k}}$$

$$M_{\mathrm{C\!\!\!L}} = (3)(\tfrac{1}{2} \times 12.5 \times 25) = 468.75'^{\mathrm{k}}$$

Should a structure support uniform and concentrated loads, the value of the function under study can be found by multiplying each concentrated load by its respective ordinate on the influence line and the uniform load by the area of the influence line opposite the section covered by the uniform load.

9.9 COMMON SIMPLE BEAM FORMULAS FROM INFLUENCE LINES

Several useful expressions for moments in simple beams can be determined with influence lines. Formulas may be developed for moment at the centerline of a simple beam in Figure 9.12(a), the beam being loaded first with a uniform load and second with a concentrated load at the centerline. From Figure 9.12(b), formulas may be developed for moment at any point in a simple beam loaded with a uniform load and for moment at any point where a concentrated load is located. These formulas follow:

(a) Loaded with a uniform load	(b) Loaded with a uniform load
$M_{\mathrm{C\!\!\!L}} = (w)\left(\dfrac{1}{2} \times \ell \times \dfrac{\ell}{4}\right) = \dfrac{w\ell^2}{8}$	$M_{1\text{-}1} = (w)\left(\dfrac{1}{2} \times \dfrac{ab}{\ell} \times \ell\right) = \dfrac{wab}{2}$
Loaded with a concentrated load P at centerline:	Loaded with a concentrated load P at section 1-1:
$M_{\mathrm{C\!\!\!L}} = \dfrac{P\ell}{4}$	$M_{1\text{-}1} = \dfrac{Pab}{\ell}$

9.10 PLACING LIVE LOADS TO CAUSE MAXIMUM VALUES USING INFLUENCE LINES

Beams must be designed to support satisfactorily the largest values of shears and moments caused by the loads to which they are subjected. As an illustration the beam of Figure 9.9 and its influence line for moment at section 1-1 are redrawn in Figure 9.13(a) and (b) on page 240. It is desired to obtain the maximum possible positive moment at section 1-1 for a uniform dead load and a uniform live load plus impact.

Massachusetts Eye and Ear Infirmary in Boston. (Courtesy Bethlehem Steel Corporation.)

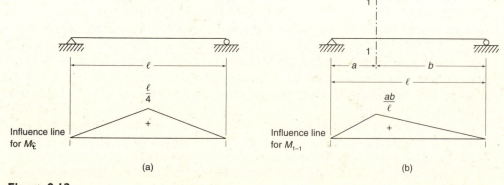

Figure 9.12

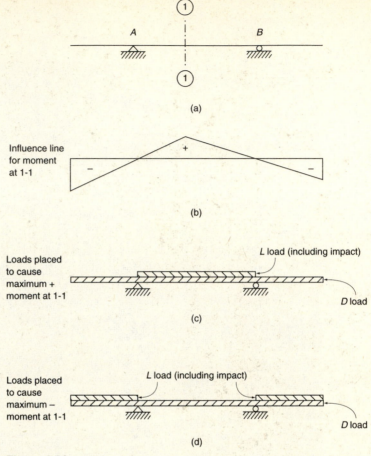

Figure 9.13

The uniform dead load, which is the weight of the structure, will be located from end to end of the beam as shown in part (c) of the figure. The unit load caused positive moment at section 1-1 only when it was located between the supports A and B. Thus, in part (c) of the figure the uniform live load increased by the percent of impact is placed from A to B only. If there had been a concentrated live load, it, together with its impact increase, would have been placed at section 1-1 (as the unit load caused the greatest positive moment when it was located there). Now that the loads have been placed to cause the maximum positive moment at section 1-1, that moment can be calculated using the ordinates on the influence line or by statics equations.

In a similar fashion it is assumed that the maximum negative moment at section 1-1 is desired. For this case the loads would be placed as shown in Figure 9.13(d). If there had been a live concentrated load, it would have been placed at the left or right end of the beam—whichever had the largest negative ordinate on the influence line.

9.11 PLACING LIVE LOADS TO CAUSE MAXIMUM VALUES BASED ON MAXIMUM CURVATURE

In the preceding section an influence line was used to determine the critical positions for placing live loads to cause maximum moments. The same results can be obtained (and perhaps more easily in many situations) by considering the deflected shape or curvature of a member under load. If the live loads are placed so they cause the greatest curvature at a particular point, they will have bent the member the greatest amount at that point, which means the greatest moment will have been caused.

For the first illustration it is desired to cause the greatest positive moment at section 1-1 in the beam of Figure 9.14 due to the same loads considered in the last section. In part (b) of the figure the deflected shape of the beam is sketched at section 1-1 as it would be when a positive moment occurs there. Then the rest of the beam's deflected shape is sketched as shown by the dotted lines. The dead load is placed all across the beam, while the live load increased by impact is placed from A to B so that it will magnify the deflected shape at section 1-1.

A similar situation is shown in part (c) of the figure, where it is desired to obtain maximum negative moment at 1-1. The deflected shape of the beam is sketched as it would be when a negative moment occurs at 1-1 and the rest of the beam's deflected shape is drawn. To exaggerate this negative or upward bending at 1-1, the live loads need to be placed on the parts of the beams outside the supports.

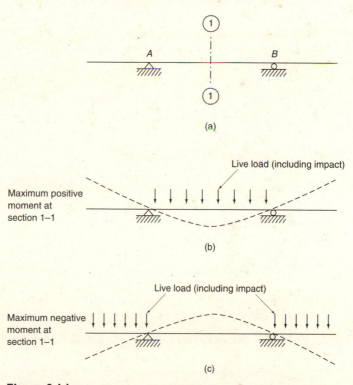

Figure 9.14

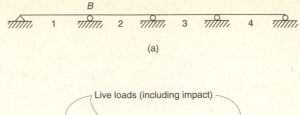

(a)

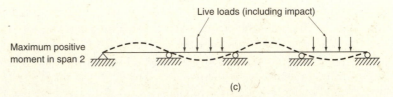

(b)

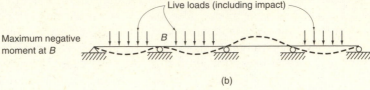

(c)

Figure 9.15

For the four-span beam of Figure 9.15 it is desired to place live loads so as to cause maximum negative moment at support B and maximum positive moment in span 2. In part (b) of the figure the deflected shape of the beam at B is sketched as it would be when bent with a negative moment. Then the rest of the beam's deflected shape is sketched with a dotted line. It can be seen that live loads should be placed in spans 1, 2, and 4 to exaggerate negative bending at B. In part (c) of the figure it is desired to obtain maximum positive moment in span 2. A similar procedure is followed and it is found that live loads should be placed in spans 2 and 4.

9.12 INFLUENCE LINES FOR TRUSSES

The variation of forces in truss members due to moving loads is of great importance. Influence lines may be drawn and used for making force calculations, or they may be sketched roughly without computing the values of the ordinates and used only for placing the moving loads to cause maximum or minimum forces.

The procedure used for preparing influence lines for trusses is closely related to the one used for beams. The exact manner of application of loads to a bridge truss from the bridge floor is described in the following section. A similar discussion could be made for the application of loads to roof trusses.

9.13 ARRANGEMENT OF BRIDGE FLOOR SYSTEMS

The arrangement of the members of a bridge floor system should be carefully studied so that the manner of application of loads to the truss will be fully understood. Probably the most common type of floor system consists of a concrete slab supported

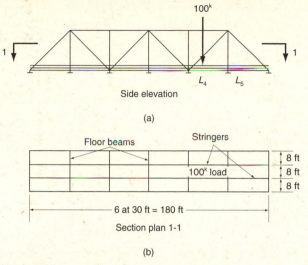

Side elevation

(a)

Section plan 1-1

(b)

Figure 9.16

by steel stringers running parallel to the trusses. The stringers run the length of each panel and are supported at their ends by floor beams that run transverse to the roadway and frame into the panel points or joints of the truss (Figure 9.16).

The foregoing discussion apparently indicates that the stringers rest on the floor beams and the floor beams on the trusses. This method of explanation has been used to emphasize the manner in which loads are transferred from the pavement to the trusses, but the members usually are connected directly to each other on the same level. Stringers are conservatively assumed to be simply supported, but actually there is some continuity in their construction.

A 100-kip load is applied to the floor slab in the fifth panel of the truss of Figure 9.16. The load is transferred from the floor slab to the stringers, thence to the floor beams, and finally to joints L_4 and L_5 of the supporting trusses. The amount of load going to each stringer depends on the position of the load between the stringers: if halfway, each stringer would carry half. Similarly, the amount of load transferred from the stringers to the floor beams depends on the longitudinal position of the load.

Figure 9.17 on page 244 illustrates the calculations involved in figuring the transfer of the 100-kip load to the trusses. The final reactions shown for the floor beams represent the downward loads applied at the truss panel points. The computation of truss loads usually is a much simpler process than the one described here.

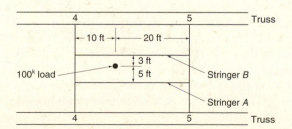

Load transferred to each stringer,

$$A = \tfrac{3}{8} \times 100 = 37.5^k$$
$$B = \tfrac{5}{8} \times 100 = 62.5^k$$

Load transferred from stringer A to floor beams,

$$4\text{-}4 = \tfrac{20}{30} \times 37.5 = 25^k$$
$$5\text{-}5 = \tfrac{10}{30} \times 37.5 = 12.5^k$$

Load transferred from stringer B to floor beams,

$$4\text{-}4 = \tfrac{20}{30} \times 62.5 = 41.67^k$$
$$5\text{-}5 = \tfrac{10}{30} \times 62.5 = 20.83^k$$

Floor beams loaded as follows:

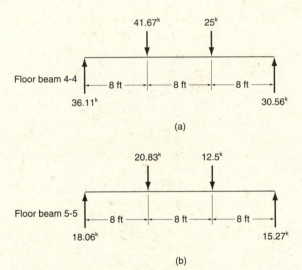

Figure 9.17

9.14 INFLUENCE LINES FOR TRUSS REACTIONS

Influence lines for reactions of simply supported trusses are used to determine the maximum loads that may be applied to the supports. Although their preparation is elementary, they offer a good introductory problem in learning the construction of influence lines for truss members.

Influence lines for the reactions at both supports of an overhanging truss are given in Figure 9.18. Loads can be applied to the truss only by the floor beams at the panel points, and floor beams are assumed to be present at each of the panel points, including the end ones. A load applied at the very end of the truss opposite the end panel point will be transferred to that panel point by the end floor beam.

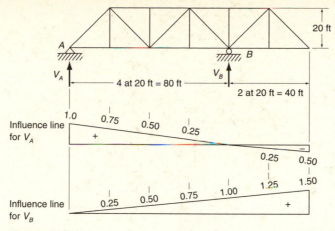

Figure 9.18

9.15 INFLUENCE LINES FOR MEMBER FORCES OF PARALLEL-CHORD TRUSSES

Influence lines for forces in truss members may be constructed in the same manner as those for various beam functions. The unit load moves across the truss, and the ordinates for the force in the member under consideration may be computed for the load at each panel point. In most cases it is unnecessary to place the load at every panel point and calculate the resulting value of the force, because certain portions of influence lines can readily be seen to consist of straight lines for several panels.

One method used for calculating the forces in a chord member of a truss consists of passing an imaginary section through the truss, cutting the member in question and taking moments at the intersection of the other members cut by the section. The resulting force in the member is equal to the moment divided by the lever arm; therefore, the influence line for a chord member is the same shape as the influence line for moment at its moment center.

The truss of Figure 9.19 is used to illustrate this point on the following page. The force in member $L_1 L_2$ is determined by passing section 1-1 and taking moments at U_1. An influence line is shown for the moment at U_1 and for the force in $L_1 L_2$, the ordinates of the latter figure being those of the former divided by the lever arm. Similarly, section 2-2 is passed to compute the force in $U_4 U_5$, and influence lines are shown for the moment at L_4 and for the force in $U_4 U_5$.

The forces in the diagonals of parallel-chord trusses may be calculated from the shear in each panel. The influence line for the shear in a panel is of the same shape as the influence line for the force in the diagonal, because the vertical component of force in the diagonal is equal numerically to the shear in the panel. On page 246, Figure 9.20 illustrates this fact for two of the diagonals of the same truss considered in Figure 9.19. For some positions of the unit loads the diagonals are in compression and for others they are in tension.

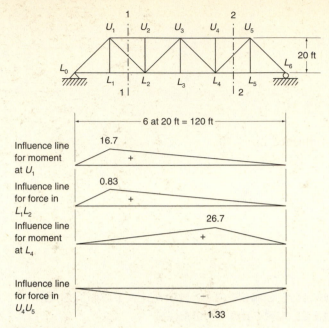

Figure 9.19

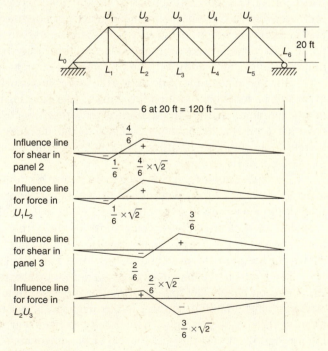

Figure 9.20

The vertical components of force in the diagonals can be converted into the actual forces from their slopes. The sign convention for positive and negative shears is the same as the one used previously.

9.16 INFLUENCE LINES FOR MEMBERS OF NONPARALLEL-CHORD TRUSSES

Influence-line ordinates for the force in a chord member of a "curved-chord" truss may be determined by passing a vertical section through the panel and taking moments at the intersection of the diagonal and the other chord. Several such influence lines are drawn for chords of a Parker truss in Figure 9.21.

The ordinates for the influence line for force in a diagonal may be obtained by passing a vertical section through the panel and taking moments at the intersection of the two chord members, as illustrated in Figure 9.21, where the force in U_1L_2 is obtained by passing section 1-1 and taking moments at the intersection of chords U_1U_2 and L_1L_2 at point x. The influence line is drawn for the vertical component of force in the inclined member. In the following pages several more influence lines are

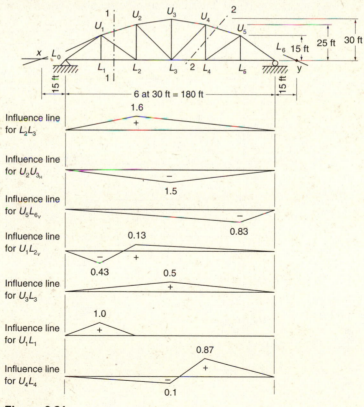

Figure 9.21

drawn for either the vertical or horizontal components of force for inclined members. Force components obtained from the diagrams may be quickly adjusted to resultant forces from slopes of the members.

The influence line for the midvertical $U_3 L_3$ is obtained indirectly by computing the vertical components of force in $U_2 U_3$ and $U_3 U_4$. The ordinates for $U_3 L_3$ are found by summing up these components. The influence lines for the other verticals are more easily drawn. Member $U_1 L_1$ can have a force only when the unit load lies between L_0 and L_2. It has no force if the load is at either of these joints, but a tension of unity occurs when the load is at L_1. Influence lines for verticals such as $U_4 L_4$ can be drawn by two methods. A section such as 2-2 may be passed through the truss and moments taken at the intersection of the chords at point y, or if the influence diagram for $L_4 U_5$ is available, its vertical components may be used to calculate the ordinates for $U_4 L_4$.

9.17 INFLUENCE LINES FOR K TRUSS

Figure 9.22 shows influence lines for several members of a K truss. The calculations necessary for preparing the diagrams for the chord members are equivalent to those used for the chords of trusses previously considered. The values needed to plot the diagrams for vertical and diagonal members are slightly more difficult to obtain.

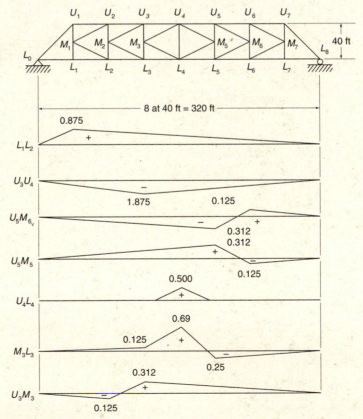

Figure 9.22

The forces in the two diagonals of each panel may be obtained from the shear in the panel. By knowing that the horizontal components are equal and opposite, the relationship between their vertical components can be found from their slopes. If the slopes are equal, the shear to be carried divides equally between the two. The influence lines for the verticals, such as $U_5 M_5$, may be determined from the influence lines for the adjoining diagonals, if available. On the other hand, the ordinates may be computed independently for various positions of the unit load. The reader should make a careful comparison of the influence lines for the upper and lower verticals, such as those given for $M_3 L_3$ and $U_3 M_3$ in the figure.

The Influence line for the midvertical $U_4 L_4$ can be developed by computing the vertical components of force in $M_3 L_4$ and $L_4 M_5$, or in $M_3 U_4$ and $U_4 M_5$, for each position of the unit load. The vertical components of force in each of these pairs of members will cancel each other unless the shear in panel 4 is unequal to the shear in panel 5, which is possible only when the unit load is at L_4.

9.18 DETERMINATION OF MAXIMUM FORCES

Truss members are designed to resist the maximum forces that may be caused by any combination of the dead, live, and impact loads to which the truss may be subjected. The live load probably consists of a series of moving concentrated loads representing the wheel loads of the vehicles using the structure, but for convenience in force analysis an approximately equivalent uniform live load with only one or two concentrated loads often is used in their place. Live loads for which highway and railroad bridges are designed and common impact expressions are discussed in detail in Sections 9.20 through 9.22.

The dead load, representing the weight of the structure and permanent attachments, extends for the entire length of the truss, but the uniform and concentrated live loads are placed at the points on the influence line that cause maximum force of the character being studied. If tension is being studied, the live uniform load is placed along the section of the truss corresponding to the positive or tensile section of the influence line, and the live concentrated loads are placed at the maximum positive tensile ordinates on the diagram.

Members whose influence lines have both positive and negative ordinates may possibly be in tension for one combination of loads and in compression for another. A member subject to *force reversal* must be designed to resist both the maximum compressive and maximum tensile forces.

In the next few paragraphs the maximum possible forces in several members of the truss of Figure 9.23 (page 250) are determined due to the following loads:

1. Dead uniform load of 1.5 klf
2. Live uniform load of 2 klf
3. Moving concentrated load of 20 kips
4. Impact of 24.4%

The influence lines are drawn, and the forces are computed by the exact method as described in the following paragraphs.

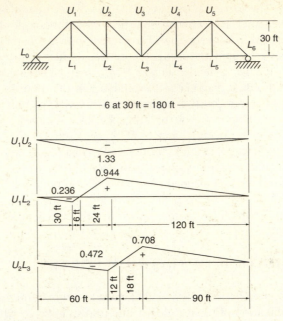

Figure 9.23

$U_1 U_2$ The member is in compression for every position of the unit load; therefore, the dead uniform load and the live uniform load are placed over the entire span. The moving concentrated load of 20 kips is placed at the maximum compression ordinate on the influence line. The impact factor is multiplied by the live load forces and added to the total.

$$
\begin{aligned}
\text{DL} &= (1.5)(180)(-1.33)(\tfrac{1}{2}) &&= -180.0 \\
\text{LL} &= (2)(180)(-1.33)(\tfrac{1}{2}) &&= -240.0 \\
&+ (20)(-1.33) &&= -\ \ 26.7 \\
I &= (0.244)(-240.0 - 26.7) &&= -\ \ 65.1 \\
\text{total force} & &&= -511.8^{\text{k}} \text{ compression}
\end{aligned}
$$

$U_1 L_2$ Examination of the influence line for $U_1 L_2$ shows that for some positions of the unit load the member is in compression, whereas for others it is in tension. The live loads should be placed over the positive portion of the diagram and the dead loads across the entire structure to obtain the largest possible tensile force. Similarly, the live loads should be placed over the negative portion of the diagram and the dead loads over the entire structure to obtain the largest possible compressive force.

Maximum tension:
$$
\begin{aligned}
\text{DL} &= (1.5)(144)(+0.944)(\tfrac{1}{2}) &&= +102.0 \\
&+ (1.5)(36)(-0.236)(\tfrac{1}{2}) &&= -\ \ 6.4
\end{aligned}
$$

$$LL = (2)(144)(+0.944)(\tfrac{1}{2}) \qquad = +136.0$$
$$+ (20)(+0.944) \qquad\qquad = + \ 18.9$$
$$I = (0.244)(+136.0 + 18.9) = + \ 37.8$$
$$\text{total force} \qquad\qquad\qquad = +288.3^k \text{ tension}$$

Maximum compression:
$$DL = (1.5)(144)(+0.944)(\tfrac{1}{2}) \qquad = +102.0$$
$$+ (1.5)(36)(-0.236)(\tfrac{1}{2}) \qquad = - \ \ 6.4$$
$$LL = (2)(36)(-0.236)(\tfrac{1}{2}) \qquad\quad = - \ \ 8.5$$
$$+ (20)(-0.236) \qquad\qquad\quad = - \ \ 4.7$$
$$I = (0.244)(-8.5 - 4.7) \qquad = - \ \ 3.2$$
$$\text{total force} \qquad\qquad\qquad\quad = + \ 79.2^k \text{ tension}$$

$U_2 L_3$

The calculations for $U_1 L_2$ proved it could have only tensile forces, regardless of the positioning of the live loads given. The following calculations show force reversal may occur in member $U_2 L_3$.

Maximum tension:
$$DL = (1.5)(108)(+0.708)(\tfrac{1}{2}) \ = + \ 57.3$$
$$+ (1.5)(72)(-0.472)(\tfrac{1}{2}) = - \ 25.5$$
$$LL = (2)(108)(+0.708)(\tfrac{1}{2}) \ \ = + \ 76.4$$
$$+ (20)(+0.708) \qquad\quad = + \ 14.2$$
$$I = (0.244)(+76.4 + 14.2) = + \ 22.1$$
$$\text{total force} \qquad\qquad = +144.5^k \text{ tension}$$

Maximum compression:
$$DL = (1.5)(108)(+0.708)(\tfrac{1}{2}) \ = +57.3$$
$$+ (1.5)(72)(-0.472)(\tfrac{1}{2}) = -25.5$$
$$LL = (2.0)(72)(-0.472)(\tfrac{1}{2}) \ \ = -34.0$$
$$+ (20)(-0.472) \qquad\quad = - \ 9.4$$
$$I = (0.244)(-34.0 - 9.4) \ = -10.6$$
$$\text{total force} \qquad\qquad = -22.2^k \text{ compression}$$

9.19 COUNTERS IN BRIDGE TRUSSES

The fact that a member in compression is in danger of bending or buckling reduces its strength and makes its design something of a problem. The design of a 20-ft member for a tensile force of 100 kips usually will result in a smaller section than is required for a member of the same length subject to a compressive force of the same magnitude. The ability of a member to resist compressive loads depends on its stiffness, which is measured by the *slenderness ratio*. The slenderness ratio is the ratio of the length of a member to its least radius of gyration. As a section becomes longer, or as its slenderness ratio increases, the danger of buckling increases, and a larger section is required to withstand the same load.

This discussion shows there is a considerable advantage in keeping the diagonals of a truss in tension if possible. If a truss supported only dead load, it would be a simple matter to arrange the diagonals so that they were all in tension. All of the diagonals of the Pratt truss of Figure 9.24(a) would be in tension for a uniform dead load extending over the entire span. The calculations in Section 9.18, however, have shown that live loads may cause the forces in some of the diagonals of a bridge truss to alternate between tension and compression. The constant passage of trains or trucks back and forth across a bridge probably will cause the forces in some of the diagonals to change continually from tension to compression and back to tension.

The possibilities of force reversal are much greater in the diagonals near the center of a truss. The reason for this situation can be seen by referring to the truss of Figure 9.23, where a positive shear obviously causes tension in members $U_1 L_2$ and $U_2 L_3$. The positive dead-load shear is much smaller in panel 3 than in panel 2, and it is more likely for the live load to be in a position to cause a negative shear large enough to overcome the positive shear and produce compression in the diagonal.

A few decades ago, when it was common for truss members to be pin-connected, the diagonals were actually eyebars that were capable of resisting little compression. The same force condition exists in trusses erected today, with diagonals consisting of a pair of small steel angles or other shapes of little stiffness. It was formerly common to add another tension-resisting diagonal to the panels where force reversal could occur, the new diagonal running across the first one and into the previously uncon-nected corners of the panel. These members, called *counters* or *counter diagonals,* can be seen in many older bridges across the country, but rarely in new ones.

Figure 9.24(b) shows a Pratt truss to which counters have been added in the middle four panels, the counters being represented by dotted lines. When counters have been added in a panel, both diagonals may consist of relatively slender and light members, neither being able to resist appreciable compression. With light and slender diagonals the entire shear in the panel is assumed to be resisted by the diagonal that would be in tension for that type of shear, whereas the other diagonal is relaxed or without stress. The two diagonals in a panel may be thought of as cables that can resist

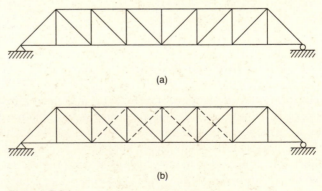

(a)

(b)

Figure 9.24

no compression whatsoever. If compression were applied to one of the cables, it would become limp, whereas the other one would be stretched. A truss with counters is actually statically indeterminate unless the counter is adjusted to have zero force under dead load.

Today's bridges are designed with diagonals capable of resisting force reversals. In fact, all bridge truss members, whether subject to force reversal or not, must be capable of withstanding the large force changes that occur when vehicles move back and forth across those structures. A member that is subject to frequent force changes even though the character of the force does not change (as $+50$ to $+10$ to $+50$ kips, etc.) is in danger of a fatigue failure unless it is specifically designed for that situation. For structural steel members, tension must be involved for fatigue to be a problem.

Modern steel specifications provide a maximum permissible stress range (from high to low) for each truss member. The stress range is defined as the algebraic difference between the maximum and minimum stresses. For this calculation tensile stress is given an algebraic sign that is opposite to that of compression stress. The AASHTO and AISC specifications provide a permissible stress range that is dependent on the estimated number of cycles of stress, on the type and location of a particular member, and on its type of connection. Obviously, the more critical the situation, the smaller is the permissible stress range.

9.20 LIVE LOADS FOR HIGHWAY BRIDGES

Until about 1900, bridges in the United States were "proof loaded" before they were considered acceptable for use. Highway bridges were loaded with carts filled with stone or pig iron, while railway bridges were loaded with two locomotives in tandem. Such procedures were probably very useful in identifying poor designs and/or poor workmanship, but were no guarantee against overloads and fatigue stress situations.[2]

As discussed in *America's Highways 1776–1976,* during much of the 19th century, highway bridges were designed to support live loads of approximately 80 to 100 psf applied to the bridge decks. These loads supposedly represented large, closely spaced crowds of people moving across the bridges. In 1875 the ASCE recommended that highway bridges be designed to support live loads varying from 40 to 100 psf. The smaller values were to be used for very long spans. In 1913, the Office of Public Roads published a circular recommending that highway bridges be designed for a live loading consisting of a series of electric cars or a 15-ton road roller plus a uniform live load on the rest of the bridge deck.

Although highway bridges must support several different types of vehicles, the heaviest possible loads are caused by a series of trucks. In 1931 the AASHTO Bridge Committee issued its first printed edition of the AASHTO Standard Specification for Highway Bridges. A very important part of these specifications was the use of the truck system of live loads. The truck loads were designated as H-20, H-15, and H-10, representing two-axle design trucks of 20, 15, and 10 tons, respectively. Each lane of

---

[2]U.S. Department of Transportation, Federal Highway Administration, *America's Highways 1776–1976* (Superintendent of Documents, U.S. Government Printing Office, 1976), 429–432.

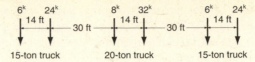

Figure 9.25   The old AASHTO truck loads.

a bridge was to have an H-truck placed in it and was to be preceded and followed by a series of trucks weighing 3/4ths as much as the basic truck.[3] This loading is shown in Figure 9.25.

Today the AASHTO specifies that highway bridges be designed for lines of motor trucks occupying 10-ft-wide lanes. Only one truck is placed in each span for each lane. The truck loads specified are designated with an H prefix (or M if SI units are being used) followed by a number indicating the total weight of the truck in tons (10^4 newtons). The weight may be followed by another number indicating the year of the specifications. For example, an H20-44 loading indicates a 20-ton truck and the 1944 specifications. A sketch of the truck and the distances between axle centers, wheel centers, and so on, is shown in Figure 9.26.

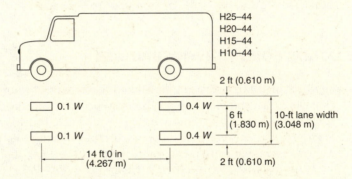

Figure 9.26   Today's AASHTO loadings.

The selection of the particular truck loading to be used in design depends on the bridge location, anticipated traffic, and so on. These loadings may be broken down into three groups as follows.

Two-Axle Trucks: H10, H15, H20, and H25

The weight of an H truck is assumed to be distributed two-tenths to the front axle (for example, 4 tons, or 8 kips, for an H20 loading) and eight-tenths to the rear axle. The axles are spaced 14 ft, 0 in. on center, and the center-to-center lateral spacing of the wheels is 6 ft, 0 in. Should a truck loading varying in weight from these be desired, one that has axle loads in direct proportion to the standard ones listed here may be used. A loading as small as the H10 may be used only for bridges supporting the lightest traffic.

[3] See note 2.

Two-Axle Trucks Plus One-Axle Semitrailer: HS15-44, HS20-44, and HS25-44

For today's highway bridges carrying a great amount of truck traffic, the two-axle truck loading with a one-axle semitrailer weighing 80% of the main truck load is commonly specified for design. The DOT (Department of Transportation) for many states today requires that their bridges be designed for the HS25-44 trucks. This truck has 5 tons on the front axle, 20 tons on the rear axle, and 20 tons on the trailer axle. The distance from the rear truck axle to the semitrailer axle is varied from 14 to 30 ft, depending on which spacing will cause the most critical conditions.

Uniform Lane Loadings

Computation of forces caused by a series of concentrated loads, whether they represent two-axle trucks or two-axle trucks with semitrailers, is a tedious job with a hand calculator; therefore, a lane loading that will produce approximately the same forces frequently is used. The lane loading consists of a uniform load plus a single moving concentrated load. This load system represents a line of medium-weight traffic with a heavy truck somewhere in the line. The uniform load per foot is equal to 0.016 times

John F. Fitzgerald Expressway, Mystic River Bridge, Boston, Massachusetts. (Courtesy of the American Institute of Steel Construction, Inc.)

the total weight of the truck to which the load is to be roughly equivalent. The concentrated load equals 0.45 times the truck weight for moment calculations and 0.65 times the truck weight for shear calculations. These values for an H20 loading would be as follows: 0.016 × 20 tons equals 640 lb/ft of lane; concentrated load for moment 0.45 × 20 tons equals 18 kips; and concentrated load for shear 0.65 × 20 tons equals 26 kips.

For continuous spans another concentrated load of equal weight is to be placed in one of the other spans in such a position as to cause maximum negative moment. For positive moment only one concentrated load is to be used per lane, with the uniform load placed in as many spans as necessary to produce the maximum positive value.

The lane loading is more convenient to handle, but it should not be used unless it produces moments or shears equal to or greater than those produced by the corresponding H loading. Based on the information presented later in this chapter, calculations can be made that will show that the equivalent lane loading for the HS20-44 will produce greater moments in simple spans of 145 ft and above and greater shears for simple spans of 128 ft and above. Appendix A of the AASHTO specifications contains tables that give the maximum shears and moments in simple spans for the various H loadings or for their equivalent lane loadings, whichever controls.

The possibility of having a continuous series of heavily loaded trucks in every lane of a bridge that has more than two lanes does not seem as great that for a bridge that has only two lanes. The AASHTO therefore permits the values caused by full loads in every lane to be reduced by a certain factor if the bridge has more than two lanes.

Interstate Highway System Loading

Another loading system can be used instead of the HS20-44 in the design of structures for the Interstate Highway System. This alternate system, which consists of a pair of 24-kip axle loads spaced 4 ft on center, is critical for short spans only. It is possible to show that this loading will produce maximum moments for simple spans from 11.5 to 37 ft and maximum shears for spans from 6 to 22 ft. For other spans the HS20-44 loading or its equivalent lane loading will be critical.

9.21 LIVE LOADS FOR RAILWAY BRIDGES

Railway bridges are commonly analyzed for a series of loads devised by Theodore Cooper. His loads, referred to as E loadings, represent two locomotives followed by a line of freight cars. A series of concentrated loads is used for the locomotives, and a uniform load represents the freight cars. Mr. Cooper introduced his loading system in 1894; it was the so-called E-40 load, which is pictured in Figure 9.27. The train is assumed to have a 40-kip load on the driving axle of the engine. Since his system was introduced, the weights of trains have been increased considerably, until at the present time bridges are designed on the basis of loads in the vicinity of an E-72 loading, and the use of E-80 and E-90 loadings is not uncommon.

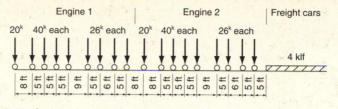

E-40 Cooper loading

Figure 9.27

   Various tables that are easily obtainable present detailed information pertaining to Cooper's loadings such as axle loads, moments, and shears. If information is available for one E loading, the information for any other E loading can be obtained by direct proportion. The axle loads of an E-75 are 75/40 those for an E-40; those for an E-60 are 60/72 of those for an E-72; and so on. Tables used in conjunction with the maximum criteria presented later in this chapter greatly reduce the computations.

   Cooper's loadings do not accurately picture today's trains, but they still are in general use despite the availability of several more modern and more realistic loadings, such as Dr. D. B. Steinman's M-60 loading.[4]

9.22 IMPACT LOADINGS

The truck and train loads applied to highway and railroad bridges are applied not gently and gradually, but rather violently, which causes forces to increase. Additional loads, called *impact loads,* must be considered. They are taken into account by increasing the live-load forces by some percentage, the percentage being obtained from purely empirical expressions. Numerous formulas have been presented for estimating impact. One example is the following AASHTO formula for highway bridges, in which I is the percent of impact and L is the length of the span, in feet, over which live load is placed to obtain a maximum stress. The AASHTO says that it is unnecessary to use an impact percentage greater than 30%, regardless of the value given by the formula. Note that the longer the span length becomes, the smaller becomes the impact.

$$I = \frac{50}{L + 125} \quad \text{or} \quad \frac{15.24}{L + 38} \quad \text{if } L \text{ is in meters}$$

   Impact factors or percentages for railroad bridges are higher than those for highway bridges because of the much greater vibrations caused by the wheels of a train as compared to the relatively soft rubber-tired vehicles on a highway bridge. A person need only stand near a railroad bridge for a few seconds while a fast-moving and heavily loaded freight train passes over to see the difference. Tests have shown the impact on railroad bridges will often run as high as 100% or more. Not only does a

[4]"Locomotive Loadings for Railway Bridges," *Transactions of the American Society of Civil Engineers* 86 (1923): 606–636.

train have a direct vertical impact, or bouncing up and down, but it also has a lurching or swaying back-and-forth type of motion. Some AREA impact formulas are as follows:

Direct vertical effect for beams, girders, floor beams, and so on:

$$I = 60 - \frac{L^2}{500} \qquad \text{for } L < 100 \text{ ft}$$

$$I = \frac{1800}{L - 40} + 10 \qquad \text{for } L = 100 \text{ ft or more}$$

Direct vertical effect for trusses:

$$I = \frac{4000}{L + 25} + 15$$

The AISC specification states that, unless otherwise specified, live loads shall be increased by certain percentages. Some of these values are 100% for elevators, 33% for hangers supporting floors and balconies, not less than 50% for supports of reciprocating machinery or power-driven units, and so on.

9.23 MAXIMUM VALUES FOR MOVING LOADS

In the preceding pages of this chapter it has been repeatedly indicated that to design beams, girders, trusses, or any other structures supporting moving loads, the designer must be able to determine which positions of these loads cause maximum shear, moment, and so on, at various points in the structure. If one can place the loads at the positions causing maximums, one need not worry about any other positions the loads might take on the structure. Should a structure be loaded with a uniform live load and not more than one or two moving concentrated loads, the critical positions for placing the loads will be obvious from the influence lines.

If, however, the structure is to support a series of concentrated loads of varying magnitudes, such as groups of truck or train wheels, the problem is not as simple. The influence line will, of course, indicate the approximate positions for placing the loads, because it is reasonable to assume that the heaviest loads should be grouped in the vicinity of the largest ordinates of the diagram.

Space is not taken herein to consider all of the possible situations that might be faced in structural analysis. It is felt, however, that the determination of the absolute maximum moment caused in a beam by a series of concentrated loads is so frequently encountered by the designer that it should be included.

The absolute maximum moment in a simple beam usually is thought of as occurring at the beam centerline. Maximum moment does occur at the centerline if the beam is loaded with a uniform load or a single concentrated load. A beam, however, may be required to support a moving series of varying concentrated loads such as the wheels of a train, and the absolute maximum moment will in all probability occur at some position other than the centerline.

The largest possible moment should be determined, because the beam must be capable of withstanding the worst possible conditions. To calculate the moment, it is necessary to find the point where it occurs and the position of the loads causing it. Assuming the largest moment to be developed at the centerline of long-span beams is reasonable, but for short-span beams this assumption may be considerably in error. It is therefore necessary to have a definite procedure for determining absolute maximum moment.

The moment diagram for a simple beam loaded with a group of concentrated loads will consist of a set of straight lines regardless of the position of the loads; therefore, the absolute maximum moment occurring during the movement of these loads across the span will occur at one of the loads, usually the one nearest the center of gravity of the group. The beam in Figure 9.28 and the series of loads P_1, P_2, P_3, and so on are studied in the following paragraphs. The load P_3 is assumed to be the one nearest the center of gravity of the loads on the span, and it is located a distance ℓ_1 from P_R (the resultant of all the loads on the span) and a distance ℓ_2 from P_{1-2} (the resultant of loads P_1 and P_2). The left reaction R_L is located a distance x from P_R. In the following paragraphs maximum moment is assumed to occur at P_3, and a definite method is developed for placing this load to cause the maximum.

The moment at P_3 may be written as follows:

$$M = R_R(\ell - x - \ell_1) - (P_{1-2})(\ell_2)$$

Substituting the value of R_R, $P_R x / \ell$, gives

$$M = \left(\frac{P_R x}{\ell}\right)(\ell - x - \ell_1) - (P_{1-2})(\ell_2)$$

It is desired to find the value of x for which the moment at P_3 will be a maximum. Maximum moment at P_3, which occurs when the shear is zero, may be found by differentiating the moment expression with respect to x, equating the result to zero, and solving for x.

$$\frac{dM}{dx} = \ell - 2x - \ell_1 = 0$$

$$x = \frac{\ell}{2} - \frac{\ell_1}{2}$$

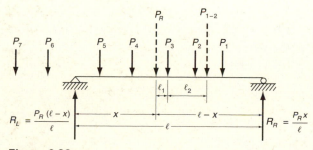

Figure 9.28

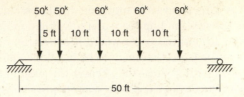

Figure 9.29

From the preceding derivation a general rule for absolute maximum moment may be stated as follows: **Maximum moment in a beam loaded with a moving series of concentrated loads usually will occur at the load nearest the center of gravity of the loads on the beam when the center of gravity is the same distance on one side of the centerline of the beam as the load nearest the center of gravity of the loads is on the other side.**

Should the load nearest the center of gravity of the loads be a relatively small one, the absolute maximum moment may occur at some other load nearby. Occasionally two or three loads have to be considered to find the greatest value; however, the problem is not a difficult one because another moment criteria not described herein—*average load to left equals average load to right*—must be satisfied, and there will be little trouble in determining which of the nearby loads will govern. (Actually, it can be shown that the absolute maximum moment occurs under the load that would be placed at the centerline of the beam to cause maximum moment there, when that wheel is placed as far on one side of the beam centerline as the center of gravity of all the loads is on the other.[5]

As an example it is desired to determine the absolute maximum moment that can occur in the 50-ft simple beam of Figure 9.29 as the series of concentrated loads shown moves across the span. The center of gravity of the loads is determined.

$$\frac{(50)(5) + (60)(15 + 25 + 35)}{280} = 16.96 \text{ ft from left load}$$

Then the loads are placed as follows and the shear and moment diagrams are drawn. The absolute largest moment that can occur is $1980'^k$.

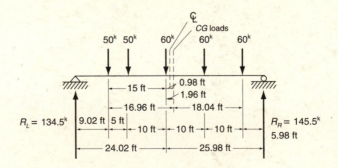

[5]A. A. Jakkula, and H. K. Stephenson, *Fundamentals of Structural Analysis* (New York: Van Nostrand, 1953), 241–242.

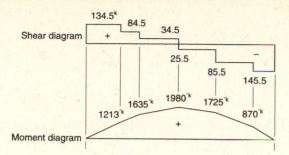

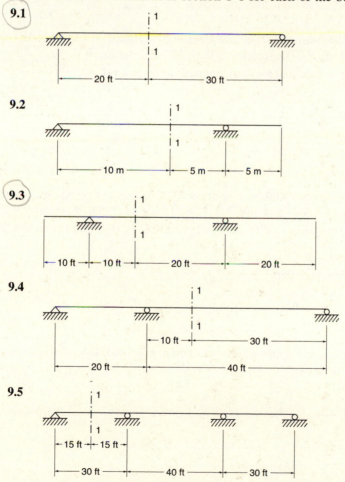

PROBLEMS

For Problems 9.1 through 9.6, draw qualitative influence lines for all of the reactions and for shear and moment at section 1-1 for each of the beams.

9.6

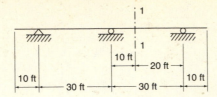

For Problems 9.7 through 9.18, draw quantitative influence lines for the situations listed.

9.7   Both reactions and shear and moment at section 1-1

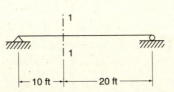

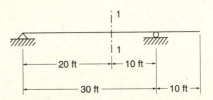

9.8   Both reactions and shear and moment at section 1-1. (*Ans.* Load at free end: $V_L = 0.33 \downarrow$, $V_R = 1.33 \uparrow$, $V_{1-1} = -0.33$, $M_{1-1} = -6.67$)

9.9   Both reactions and shear and moment at section 1-1

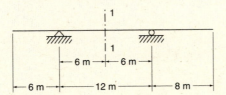

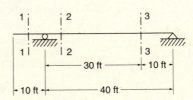

9.10   Both reactions and shear at sections 1-1, 2-2 (just to left and right of left support), and 3-3 (*Ans.* Load @ section 3-3: $V_L = +0.25$, $V_R = +0.75$, shear = 0 @ section 1-1)

9.11   Both reactions, shear at sections 1-1 and 2-2, and moment at section 2-2

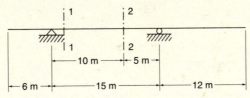

9.12   Vertical reaction and moment reaction at fixed end, shear and moment at section 1-1 (*Ans.* Load @ free end: $V_L = +1.0$, $M_L = -15$, M @ section 1-1 = -10)

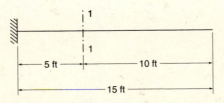

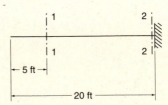

9.13   Shear and moment at sections 1-1 and 2-2

9.14 Both reactions as load moves from A to D (*Ans.* Load @ left end: $V_B = +1.50$, $V_E = -0.50$)

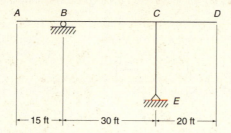

9.15 All reactions

9.16 Vertical reactions at supports A and B (*Ans.* Load @ left end $= +1.40$, $V_B = -0.40$)

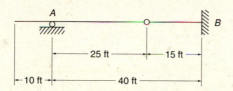

9.17 All vertical reactions, moment, and shear at section 1-1

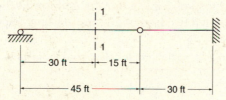

9.18 Reactions at supports A and B (*Ans.* Load @ left unsupported hinge: $V_A = -0.33$, $V_B = +1.33$)

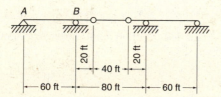

9.19 Draw influence lines for both reactions and for shear just to the left of the 16-kip load, and for moment at the 16-kip load. Determine the magnitude of each of these functions using the influence lines for the loads fixed in the positions shown in the accompanying illustration.

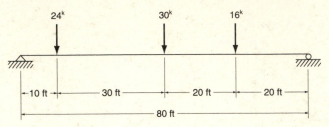

9.20 Draw influence lines for tension force in cable, for shear in beam just to left of B, and for moment at point C as a unit load moves from A to D. (*Ans.* Load @ B: Vertical component of cable tension $= +1.0$, M @ $C = 0$)

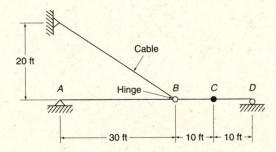

For Problems 9.21 through 9.27 using influence lines determine the quantities requested for a uniform dead load of 2 klf, a moving uniform live load of 3 klf, and a moving or floating concentrated live load of 20 kips. Assume impact $= 25\%$ in each live load case.

9.21 Maximum left reaction and maximum plus shear and moment at section 1-1

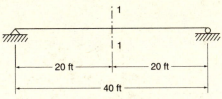

9.22 Maximum positive values of left reaction and shear and moment at section 1-1 (*Ans.* $+149.37^k$, -2.29^k, $+ 1393.7$ ft-k)

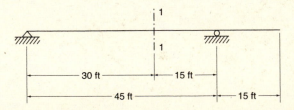

9.23 Maximum negative values of shear and moment at section 1-1, shear just to left of support and moment at the support for the beam of Problem 9.13.

9.24 Maximum negative shear and moment at section 1-1 (*Ans.* -24.66^k, -180ft-k)

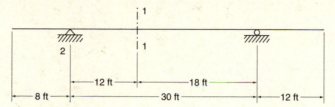

9.25 Maximum upward value of reactions at A and B and maximum negative shear and moment at section 1-1

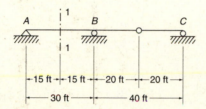

9.26 Maximum positive and negative values of shear and moment at section 3-3 in the beam of Problem 9.10 (*Ans.* Maximum positive values $= -6.87^k$, $+1025$ ft-k)

9.27 Maximum positive shear at unsupported hinge, maximum negative moment at support B, and maximum downward value of reaction at support A for the beam of Problem 9.25

For Problems 9.28 through 9.45 draw influence lines for the members indicated.

9.28 $L_0 U_1$, $L_0 L_1$, $U_1 L_1$, $U_1 U_2$ (*Ans.* Values @ L_1: $L_0 U_{1v} = -0.75$, $L_0 L_1 = +0.75$, $U_1 L_1 = +1.0$, $U_1 U_2 = -0.5$)

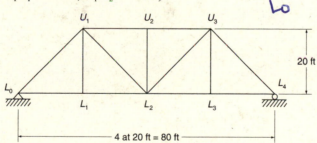

9.29 $L_0 L_1$, $U_1 U_2$, $L_1 U_2$

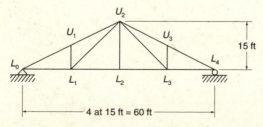

9.30 $U_1 U_2$, $L_1 L_2$, $U_1 L_2$, $L_2 U_3$ (*Ans.* Load @ L_2: $U_1 U_2 = -2.0$, $L_1 L_2 = +1.0$, $U_1 L_{2_v} = +0.667$)

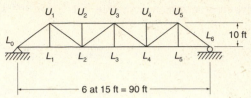

9.31 $L_1 L_2$, $U_2 L_2$, $U_2 L_3$, $L_3 U_4$ as unit load moves across top of truss

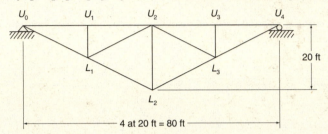

9.32 $U_1 U_2$, $U_2 L_3$, $U_3 L_3$, $L_4 L_5$ (*Ans.* Load @ L_3: $U_1 U_{2_H} = -0.416$, $U_2 L_{3_v} = +0.31$, $U_3 L_3 = +0.375$)

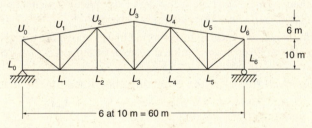

9.33 $L_0 U_1$, $L_1 L_2$, $U_1 U_2$

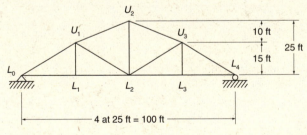

9.34   Members $L_3 L_4$, $U_1 L_2$, and $U_2 L_2$ of the truss of Prob. 9.33 (*Ans.* Load @ L_2: $L_3 L_4 = +0.833$, $U_1 L_{2_v} = +0.10$, $U_2 L_2 = +0.80$)

9.35 $U_1 U_3$, $L_2 L_4$, $L_2 U_3$, $L_4 U_5$

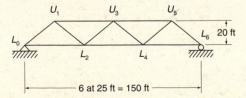

9.36 U_0L_1, U_1U_2, U_1L_2, L_3U_4 as unit load moves across top of truss (*Ans.* Load @ L_1: $U_0L_{1v} = -0.83$, $U_1U_2 = -0.89$, $U_1L_{2v} = -0.167$)

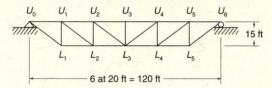

9.37 L_1L_2, U_1L_1, U_1L_2, U_3L_4 as unit load moves across top of truss

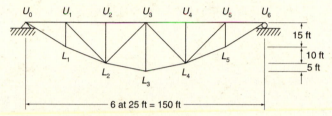

9.38 L_0U_1, U_1L_2, U_2U_3 (*Ans.* Load @ L_4: $L_0L_{1v} = +1.00$, $U_1L_{2v} = -1.00$, $U_2U_3 = +3.33$)

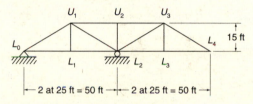

9.39 L_0U_1, U_1L_2, U_3L_4, U_5U_6

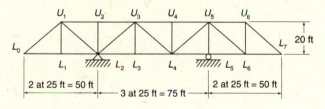

9.40 L_0U_1, U_1L_2, U_3L_4, U_5U_6 (*Ans.* Load @ L_0: $L_0U_{1v} = +1.0$, Load @ L_3: $U_3L_{4v} = -0.375$)

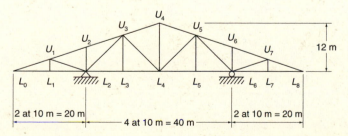

9.41 U_1L_2, L_2L_3, U_3L_4

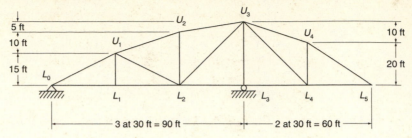

9.42 U_1U_2, U_2L_3, U_3L_3 (*Ans.* Load @ L_2: $U_1U_{2_H} = -1.60$, $U_2L_{3_V} = 0.50$, $U_3L_3 = +0.333$)

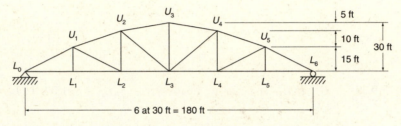

9.43 U_2U_4, L_3L_5, U_4L_5, L_5U_6 as unit load moves across top of truss

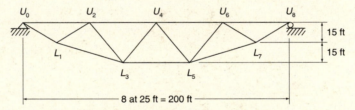

9.44 U_1U_2, U_1L_1, L_2U_3. Assume unit load moves across top of truss. (*Ans.* Load @ U_2: $U_1U_2 = -2.00$, $U_1L_1 = -0.583$, $L_2U_{3_V} = +0.583$)

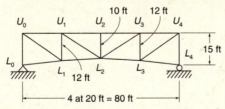

9.45 U_2U_3, M_1L_2, M_2L_2, U_3L_3

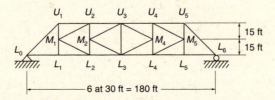

9.46   Compute the maximum and minimum forces in $L_2 U_3$ of the truss of Problem 9.30 for a uniform dead load of 1 klf, a moving uniform load of 2 klf, a moving concentrated load of 20 kips, and an impact factor of 27%. (*Ans.* $+29.20^k$ and -98.21^k)

9.47   Determine if force reversal is possible in member $U_1 L_2$ of the truss of Problem 9.41 for the loads and impact factor used in Problem 9.46.

9.48   Draw influence lines for the vertical and horizontal reactions at the left support of this three-hinged arch. (*Ans.* Load @ crown hinge: $V_L = 0.5 \uparrow$, $H_L = 1.0 \rightarrow$)

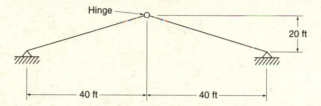

For Problems 9.49 through 9.52 draw influence lines for the members indicated.

9.49 $U_1 U_2$, $M_3 L_4$, $U_4 L_4$, $U_6 L_6$

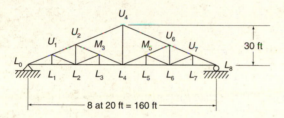

9.50 $U_1 U_2$, $L_2 L_3$, $L_4 U_5$ as unit load moves across top of span (*Ans.* $U_1 U_2 - 1.5$ at U_1; $L_2 L_{3_H}$ $+1.0$ at U_1: $L_4 U_{5_H} = -1.0$ at U_3)

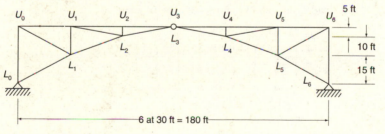

9.51   Members a, b, and c.

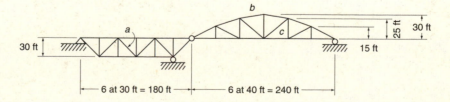

9.52　$U_2 U_3$, $U_2 M_2$, $M_2 L_2$, $U_3 L_3$ (*Ans.* $U_2 U_{3H} = -1.6$ at L_2, $U_2 M_2 = +0.06$ at L_1, $M_2 L_2 = +0.889$ at L_2)

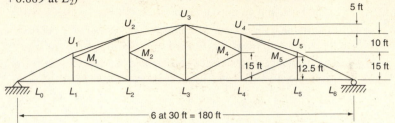

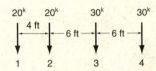

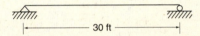

9.53　Determine the absolute maximum shear and moment possible in a 30-ft simple beam due to the load system shown.

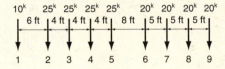

9.54　A simple beam of 20-m span supports a pair of 60-kN moving concentrated loads 4 m apart. Compute the maximum possible moment at the centerline of the beam and the absolute maximum moment in the beam. (*Ans.* 480 kN · m at ₵, 486 kN · m absolute maximum)

9.55　What is the maximum possible moment that can occur in an 80-ft simple beam as the load system shown moves across the span?

Chapter 10

Deflection and Angle Changes—Geometric Methods

10.1 INTRODUCTION

This chapter and the next are concerned with the elastic deformations of structures. Both the linear displacements of points (deflections) and the rotational displacements of lines (slopes) are considered. The word *elastic* is used to mean that stresses are proportional to strains, that there is a straight line variation of stress from the neutral axis of a beam to its extreme fibers, and that the members will return to their original shapes and lengths after loads are removed.

The deformations of structures are caused by bending moments and by axial and shear forces. For beams and frames the largest values are caused by moments, while for trusses the largest values are caused by axial forces. Shear deflections are neglected in this text, as they are quite small in almost all beamlike structures. (Shear deflections as a percentage of beam deflections increase as the ratio of beam depth to span increases. For the usual depth/span ratio of 1/12 to 1/6, the percentages of shear deflections to bending deflections vary from about 1% to 8%. For a ratio of 1/4 the percentages can be as high as 15% to 18%.)[1]

In this chapter displacements are computed using the moment-area and conjugate beam procedures. Often, these are referred to as *geometric methods* because the deformations are obtained directly from the strains in the structure. In other words, the displacements (deflections and angle changes or slopes) at a particular point are obtained by summing up the effects of the strains in the structure. In Chapter 11 displacements are determined with the *energy methods,* which are based on the conservation of energy principle. Both the geometric and energy procedures will provide identical results.

[1] C. K. Wang, *Intermediate Structural Analysis* (New York: McGraw-Hill Book Company, 1983), 750.

10.2 SKETCHING DEFORMED SHAPES OF STRUCTURES

The analyst should visualize or even make rough sketches of the anticipated deformed shapes of structures under load before making actual calculations. Such a practice will give the analyst a feel for the behavior of the structure and will provide a rough check of the magnitudes and directions of displacements.

In Figure 10.1 the deflected shapes of several loaded structures are approximated. In each case the member weight effects are neglected. You should note that the corners of the frame of part (e) of the figure are free to rotate, but the angles between the members meeting there are assumed to be constant. If the moment diagrams have previously been prepared, they can be helpful in making the sketches where we have both positive and negative moments.

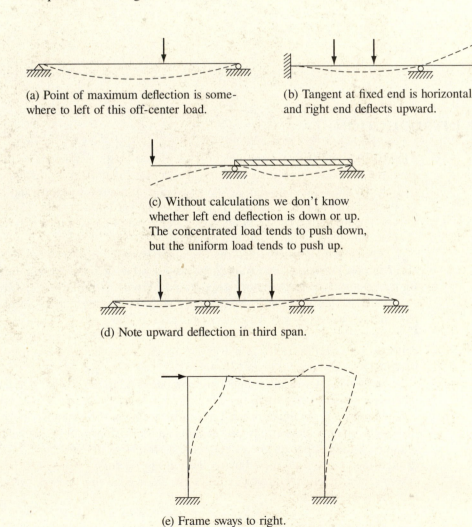

(a) Point of maximum deflection is somewhere to left of this off-center load.

(b) Tangent at fixed end is horizontal and right end deflects upward.

(c) Without calculations we don't know whether left end deflection is down or up. The concentrated load tends to push down, but the uniform load tends to push up.

(d) Note upward deflection in third span.

(e) Frame sways to right.

Figure 10.1   Sketches of estimated deformed shapes of some structures under load.

10.3 REASONS FOR COMPUTING DEFLECTIONS

The members of all structures are made up of materials that deform when loaded. If deflections exceed allowable values, they may detract from the appearance of the structures and the materials attached to the members may be damaged. For example, a floor joist that deflects too much may cause cracks in the ceiling below, or if it supports concrete or tile floors, it may cause cracks in the floors. In addition, the use of a floor supported by beams that "give" appreciably does not inspire confidence, although the beams may be perfectly safe. Excessive vibration may occur in a floor of this type, particularly if it supports machinery.

Houston Ship Channel Bridge, Houston, Texas. (Courtesy of the Texas State Department of Highways and Public Transportation.)

Standard American practice is to limit deflections caused by live load to 1/360 of the span lengths. This figure probably was originated for beams supporting plastered ceilings and was thought to be sufficient to prevent plaster cracks (although a large part of the deflections in a building are due to dead load, which will have taken place substantially before plaster is applied).

The 1/360 deflection is only one of many maximum deflection values in use because of different loading situations, different designers, and different specifications. For situations in which precise and delicate machinery is supported, maximum deflections may be limited to 1/1500 or 1/2000 of the span lengths. The 1992 AASHTO specifications limit deflections in steel beams and girders due to live load and impact to 1/800 of the span length. The value, which is applicable to both simple and continuous spans, is preferably reduced to 1/1000 for bridges in urban areas that are used in part by pedestrians. Corresponding AASHTO values for cantilevered arms are 1/300 and 1/375.

Members subject to large downward deflections often are unsightly and may even cause users of the structure to be frightened. Such members may be *cambered* so their displacements do not appear to be so large. The members are constructed of such a shape that they will become straight under some loading condition (usually dead load). A simple beam would be constructed with a slight convex bend so that under gravity loads it would become straight as assumed in the calculations. Some designers take into account both dead and live loads in figuring the amount of camber.

Deflection computations may be used for computing the reactions for statically indeterminate beams and trusses as well as the forces in the members of redundant trusses. Despite the importance of deflections, it is rarely necessary—even for statically indeterminate structures—to compute structure deformations for the purpose of modifying the original dimensions on which computations are based. The deformations of the materials used in ordinary work are quite small as compared to the overall dimensions of the structure. For example, the strain that occurs in a steel section that has a modulus of elasticity of 29×10^6 psi (pounds per square inch) when the stress is 20,000 psi is only

$$\epsilon = \frac{f}{E} = \frac{20 \times 10^3}{29 \times 10^6} = 0.000690$$

or 0.0690% of the member length.

Quite a few methods are available for determining deflections. It is desirable for the structural engineer to be familiar with several of these. For some structures one method may be easier to apply; for others another method is more satisfactory. In addition, the ability to handle any structural problem by more than one method is of great importance for checking results.

In this chapter and the next, the following methods of computing slopes and deflections are presented:

1. Moment-area theorems
2. Conjugate-beam procedure
3. Virtual work
4. Castigliano's theorem, Part II

10.4 THE MOMENT-AREA THEOREMS

The first method presented for the calculation of deflections is the very interesting and valuable moment-area method presented by Charles E. Greene of the University of Michigan about 1873. Under changing loads the neutral axis of a member changes in shape according to the positions and magnitudes of the loads. The elastic curve of a member is the shape the neutral axis takes under temporary loads. Professor Greene's theorems are based on the shape of the elastic curve of a member and the relationship between bending moment and the rate of change of slope at a point on the curve.

To develop the theorems, the simple beam of Figure 10.2 is considered. Under the loads P_1 to P_4 it deflects downward as indicated in the figure.

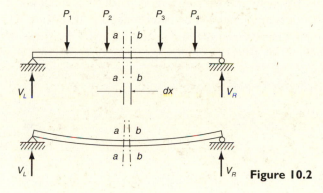

Figure 10.2

The dx section bounded on its ends by sections a–a and b–b is shown in Figure 10.3. The size, degree of curvature, and distortion of the segment are tremendously exaggerated so that the slopes and deflections to be discussed can be seen easily. Line ac lies along the neutral axis of the beam and is unchanged in length. Line

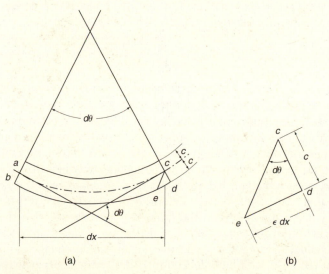

(a)   (b)

Figure 10.3

ce is drawn parallel to *ab*; therefore, *be* equals *ac* and *de* represents the lengthening of the bottom fiber of the *dx* section. Figure 10.3(b) shows an enlarged view of triangle *cde* and the angle *dθ*, which is the change in slope of the tangent to the elastic curve at the left end of the section from the tangent at the right end. Sufficient information is now available to determine *dθ*. In the derivation to follow it is to be remembered that the angle *dθ* being considered is very small, and for a very small angle the sine, tangent, and the angle in radians are identical, which permits their values to be used interchangeably. It is worthwhile to check a set of natural trigonometry tables to see the range of angles for which the functions almost coincide.

The bending moments developed by the external loads are positive and cause shortening of the upper beam fibers and lengthening of the lower fibers. The changes in fiber dimensions have caused the change in slope *dθ*. The modulus of elasticity is known, and the stress at any point can be determined by the flexure formula; therefore, the strain in any fiber can be found because it equals the stress divided by the modulus of elasticity. The value of *dθ* may be expressed as follows:

$$\tan d\theta = d\theta \text{ in radians (rad)} = \frac{\text{strain}}{c} = \frac{ed}{cd}$$

$$d\theta = \frac{\epsilon \, dx}{c}$$

By substituting the value of ϵ,

$$d\theta = \frac{(f/E) \, dx}{c}$$

But f is equal to Mc/I, and

$$d\theta = \frac{(Mc/EI) \, dx}{c} = \frac{M \, dx}{EI}$$

The change in slope in a *dx* distance is equal to *M dx/EI*, and the total change in slope from one point *A* in the beam to another point *B* can be expressed as the summation of all the *dθ* changes in the *dx* distances between the two points.

$$\theta_{AB} = \int_A^B \frac{M \, dx}{EI}$$

If the *M/EI* diagram is drawn for the beam, the above expression will be seen to equal each *dx* distance, from *A* to *B*, multiplied by its respective *M/EI* ordinate. The summation of these multiplications is the area of the diagram between the two points. From this discussion the first moment-area theorem may be expressed as follows: **The change in slope between the tangents to the elastic curve at two points on a member is equal to the area of the *M/EI* diagram between the two points.**

Once we have a method by which changes in slopes between tangents to the elastic curve at various points may be determined, it is only a brief step to a method for computing deflections between the tangents. In a *dx* distance the neutral axis changes in direction by an amount *dθ*. The deflection of one point on the beam with respect to the tangent at another point due to this angle change is equal to *x* (the distance from the point at which deflection is desired to the particular differential distance) times *dθ*.

Stick-welding decking on a shopping center mall in Charlotte, North Carolina. (Courtesy Lincoln Electric Company.)

$$d\delta = x\, d\theta$$

The value of $d\theta$ from the first theorem is substituted in this expression:

$$d\delta = x\frac{M\, dx}{EI} = \frac{Mx\, dx}{EI}$$

To determine the total deflection from the tangent at one point A to the tangent at another point B on the beam, it is necessary to obtain a summation of the products of each $d\theta$ angle (from A to B) times the distance to the point where deflection is desired. The preceding sentence is a statement of the second moment-area theorem.

$$\delta_{AB} = \int_A^B \frac{Mx\, dx}{EI}$$

The deflection of a tangent to the elastic curve of a beam with respect to a tangent at another point is equal to the moment of the M/EI diagram between the two points, taken about the point at which deflection is desired.

10.5 APPLICATION OF THE MOMENT-AREA THEOREMS

In the paragraphs to follow it will be shown that moment area is most conveniently used for determining slopes and deflections for beams in which the direction of the tangent to the elastic curve at one or more points is known, such as cantilevered beams, where the tangent at the fixed end does not change in slope. The method is applied quite easily to beams loaded with concentrated loads, because the moment diagrams consist of straight lines. These diagrams can be broken down into single triangles and rectangles, which facilitates the mathematics. Beams supporting uniform loads or uniformly varying loads may be handled, but the mathematics is slightly more difficult.

The properties of several figures given in Figure 10.4 are useful in handling the M/EI diagrams.

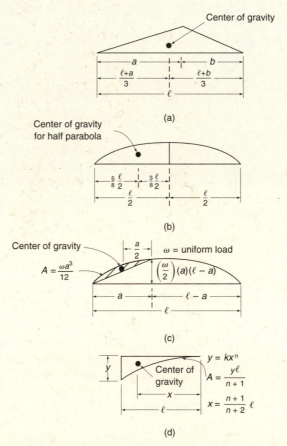

Figure 10.4   Frequently used area properties.

Examples 10.1 to 10.6 illustrate the application of the moment-area theorems. It may occasionally be possible to simplify the mathematics by drawing the moment diagram and making the calculations in terms of symbols, such as P for a concentrated load, w for a uniform load, or ℓ for span length, as illustrated by Examples 10.1 and 10.3. The numerical values of each of the symbols are substituted in the final step to obtain the slope or deflection desired.

Care must be taken to use consistent units in the calculations. The procedure here is to use all of the distances in feet and all of the loads and reactions in kilopounds. At the end of each problem the kips are changed to pounds and the feet to inches. The resulting deflections will be in inches and the slopes in radians. (There are 2π rad in 360°.)

To prevent mistakes in the application of moment-area theory, it is emphasized that the slopes and deflections that are obtained are with respect to tangents to the elastic curve at the points being considered. The theorems do not directly give the slope or deflection at a point in the beam as compared to the horizontal (except in one or two special cases); they give the change in slope of the elastic curve from one point to another or the deflection of the tangent at one point with respect to the tangent at another point.

If a beam or frame has several loads applied, the M/EI diagram may be inconvenient to handle. The calculations may be simplified by drawing a separate diagram for each of the loads and determining the slopes and deflections for each diagram separately. The final results for a particular point can be found by adding the values for all of the loads. The principle of superposition applies to the moment-area method and to any of the other methods of determining slopes and deflections discussed in subsequent pages.

Should a beam or frame be loaded with both concentrated loads and uniform loads, it is advisable to separate the moment diagrams for the uniform loads from the diagrams for the concentrated loads. Such a separation is desirable because of difficulty in determining the properties (areas and centers of gravities) of the resulting combined diagrams, which are necessary in applying the moment-area theorems as well as the conjugate-beam procedure presented later in this chapter.

Example 10.6 shows that moment area is one method that may be used to determine the moments at the ends of a fixed-ended beam, which is statically indeterminate to the third degree.

EXAMPLE 10.1 ———

Determine the slope and deflection of the right end of the cantilevered beam shown in Figure 10.5.

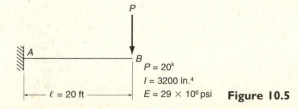

$P = 20^k$
$I = 3200$ In.^{4}
$E = 29 \times 10^6$ psi   **Figure 10.5**

Solution.   A tangent to the elastic curve at the fixed end is horizontal; therefore, the changes in slope and deflection of a tangent at the free end with respect to a tangent at the fixed end are the slope and deflection of that point.

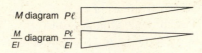

Slope at *B* equals the area of the *M/EI* diagram from *A* to *B*

$$\theta_B = \left(\frac{1}{2}\right)(\ell)\left(\frac{P\ell}{EI}\right) = \frac{P\ell^2}{2EI}$$

$$= \frac{(20 \times 1000)(20 \times 12)^2}{(2)(29 \times 10^6)(3200)} = 0.00621 \text{ rad} = 0.36°$$

Deflection at *B* equals the moment of the *M/EI* diagram from *A* to *B* about *B*

$$\delta_B = \left(\frac{1}{2}\right)(\ell)\left(\frac{P\ell}{EI}\right)\left(\frac{2}{3}\ell\right) = \frac{P\ell^3}{3EI}$$

$$= \frac{(20 \times 1000)(20 \times 12)^3}{(3)(29 \times 10^6)(3200)} = 0.99 \text{ in.} \quad \blacksquare$$

EXAMPLE 10.2 _____

Determine the slope and deflection of the beam at point *B*, 10 ft from the left end of the structure shown in Figure 10.6.

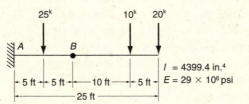

I = 4399.4 in.⁴
E = 29 × 10⁶ psi

Figure 10.6

Solution.   The left end is again fixed; the slope at *B* equals the area of the *M/EI* diagram from *A* to *B*; and the deflection at *B* equals the moment of the *M/EI* diagram from *A* to *B* taken about *B*. The diagram is broken down into convenient triangles as shown for making the calculations.

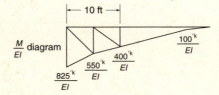

Slope:

$$\theta_B = \frac{(\frac{1}{2})(825)(5) + (\frac{1}{2})(550)(5) + (\frac{1}{2})(550)(5) + (\frac{1}{2})(400)(5)}{EI} = \frac{5812.5 \text{ ft}^2\text{-k}}{EI}$$

$$= \frac{(5812.5)(12 \times 12)(1000)}{(29 \times 10^6)(4399.4)} = 0.00656 \text{ rad} = 0.38°$$

Deflection:

$$\delta_B = \frac{(\frac{1}{2})(825)(5)(8.33) + (\frac{1}{2})(550)(5)(6.67) + (\frac{1}{2})(550)(5)(3.33) + (\frac{1}{2})(400)(5)(1.67)}{EI}$$

$$= \frac{32,600 \ (\text{ft}^3\text{-k})}{EI} = \frac{(32,600)(12 \times 12 \times 12)(1000)}{(29 \times 10^6)(4399.4)} = 0.442 \text{ in.} \quad \blacksquare$$

EXAMPLE 10.3

Determine the slope and deflection at the free end of the cantilevered beam shown in Figure 10.7.

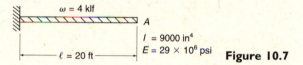

$\omega = 4$ klf

A

$I = 9000$ in^4

$E = 29 \times 10^6$ psi

$\ell = 20$ ft

Figure 10.7

Solution.   Carefully note the units used in solving the slope and deflection equations developed. To substitute in the equations the units inches and pounds are used herein. The value of w, the uniform load, is then 4000/12 lb/in. and not just 4000 lb/ft. If the reader is not careful with the units for w, he or she can easily miss the slope or deflection for a particular problem by a multiple of 12.

$\frac{M}{EI}$ diagram

$\frac{w\ell^2}{EI}$

Slope:

$$\theta_A = \left(\frac{1}{3}\right)(\ell)\left(\frac{w\ell^2}{2EI}\right) = \frac{w\ell^3}{6EI}$$

$$\theta_A = \frac{(4000/12)(20 \times 12)^3}{(6)(29 \times 10^6)(9000)} = 0.00294 \text{ rad} = 0.17°$$

Deflection:

$$\delta_A = \left(\frac{1}{3}\right)(\ell)\left(\frac{w\ell^2}{2EI}\right)\left(\frac{3}{4}\ell\right) = \frac{w\ell^4}{8EI}$$

$$= \frac{(4000/12)(20 \times 12)^4}{(8)(29 \times 10^6)(9000)} = 0.530 \text{ in.} \quad \blacksquare$$

EXAMPLE 10.4

Compute the slope and deflection at the free end of the cantilevered beam shown in Figure 10.8.

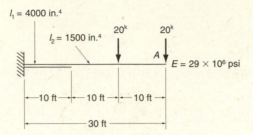

Figure 10.8

Solution.   The moment of inertia of the beam has been increased near the support where bending moment is greatest. The *M/EI* diagram is drawn by keeping the constant *E* as a symbol but dividing the ordinates by the proper moments of inertia. The resulting figure is conveniently divided into traingles and the computations made as before.

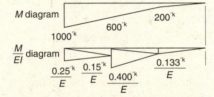

Slope:

$$\theta_A = \frac{(\frac{1}{2})(10)(0.25) + (\frac{1}{2})(10)(0.15) + (\frac{1}{2})(10)(0.400) + (\frac{1}{2})(10)(0.133)(2)}{E}$$

$$= \frac{5.333 \text{ ft}^2\text{-k}}{E} = \frac{(5.333)(144)(1000)}{29 \times 10^6} = 0.0265 \text{ rad} = 1.52°$$

Deflection:

$$\delta_A = [(\tfrac{1}{2})(10)(0.25)(26.67) + (\tfrac{1}{2})(10)(0.15)(23.33) + (\tfrac{1}{2})(10)(0.400)(16.67)$$
$$+ (\tfrac{1}{2})(10)(0.133)(13.33) + (\tfrac{1}{2})(10)(0.133)(6.67)]/E$$

$$= \frac{97.47 \text{ ft}^3\text{-k}}{E} = \frac{(97.47)(1728)(1000)}{29 \times 10^6} = 5.81 \text{ in.} \quad ■$$

EXAMPLE 10.5

Compute the deflection at the centerline of the uniformly loaded simple beam shown in Figure 10.9.

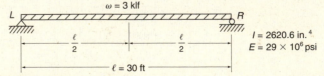

Figure 10.9

Solution.   The tangents to the elastic curve at each end of the beam are inclined. It is a simple matter to determine the deflection between a tangent at the center-line and one of the end tangents, but the result is not the actual deflection at the centerline of the beam. To obtain the correct deflection, a somewhat roundabout procedure is used.

1. The deflection δ_1 of the tangent at the right end R from the tangent at the left end L is found.
2. The deflection of a tangent at the centerline from a tangent at L, δ_2, is found.
3. By proportions the distance from the original chord between L and R and the tangent at L, δ_3, can be computed. The difference between δ_3 and δ_2 is the centerline deflection.

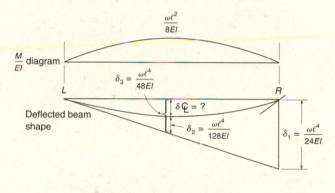

$$\delta_1 = \left(\frac{2}{3}\right)(\ell)\left(\frac{w\ell^2}{8EI}\right)\left(\frac{\ell}{2}\right) = \frac{w\ell^4}{24EI}$$

$$\delta_2 = \left(\frac{2}{3}\right)\left(\frac{\ell}{2}\right)\left(\frac{w\ell^2}{8EI}\right)\left(\frac{3}{8}\right)\left(\frac{\ell}{2}\right) = \frac{w\ell^4}{128EI}$$

$$\delta_3 = \frac{1}{2} \times \delta_1 = \frac{1}{2} \times \frac{w\ell^4}{24EI} = \frac{w\ell^4}{48EI}$$

$$\delta_{\mathrm{C\!L}} = \frac{w\ell^4}{48EI} - \frac{w\ell^4}{128EI} = \frac{5w\ell^4}{384EI}$$

$$\delta_{\mathrm{C\!L}} = \frac{(5)(3000/12)(30 \times 12)^4}{(384)(29 \times 10^6)(2620.6)} = 0.719 \text{ in.}$$

Note: The procedure followed in this example was a general one, applicable to many problems; however, the calculations of this particular problem could have been appreciably shortened by computing the deflection from the tangent at the centerline of the beam (which is horizontal due to symmetry) to the tangent at one of the supports by taking moments at that support as follows:

$$\delta_{\mathrm{C\!L}} = \left(\frac{w\ell^2}{8EI}\right)\left(\frac{\ell}{2}\right)\left(\frac{2}{3}\right)\left(\frac{5}{8}\right)\left(\frac{\ell}{2}\right) = \frac{5w\ell^4}{384EI} \quad \blacksquare$$

EXAMPLE 10.6

Determine the moments at the ends of the fixed-ended beam shown in Figure 10.10 for which E and I are constant.

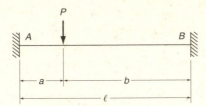

Figure 10.10

Solution. Examination of the beam reveals no change of slope and no deflection of the tangent at A from the tangent at B; therefore, the total area of the M/EI diagram from A to B is zero, and the moment of the M/EI diagram about either end is zero.

The M/EI diagram may be drawn in two parts: the simple beam moment diagram, the ordinates of which are known, and the moment diagram due to the unknown end moments M_A and M_B. For each of the latter moments, a triangular-shaped diagram may be drawn and the two combined into one trapezoid. The moment-area theorems are written to express the change in slope and deflection from B to A. Each of the two equations contains the two unknowns M_A and M_B, and the equations are solved simultaneously.

Westinghouse Rapid Transit, Pittsburgh, Pennsylvania. (Courtesy Bethlehem Steel Corporation.)

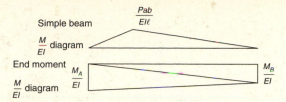

Theorem 1:

$$\left(\frac{1}{2}\right)\left(\frac{Pab}{EI\ell}\right)(\ell) + \left(\frac{1}{2}\right)\left(\frac{M_A}{EI}\right)(\ell) + \left(\frac{1}{2}\right)\left(\frac{M_b}{EI}\right)(\ell) = 0$$

$$\frac{Pab}{2EI} + \frac{M_A\ell}{2EI} + \frac{M_B\ell}{2EI} = 0 \qquad (1)$$

Theorem 2 (taking moments about left end A):

$$\left(\frac{Pab}{2EI}\right)\left(\frac{\ell + a}{3}\right) + \left(\frac{M_A\ell}{2EI}\right)\left(\frac{1}{3}\ell\right) + \left(\frac{M_B\ell}{2EI}\right)\left(\frac{2}{3}\ell\right) = 0$$

$$\frac{Pab\ell + Pa^2b}{6EI} + \frac{M_A\ell^2}{6EI} + \frac{M_B\ell^2}{3EI} = 0 \qquad (2)$$

By solving equations (1) and (2) simultaneously for M_A and M_B,

$$M_A = -\frac{Pab^2}{\ell^2} \qquad M_B = -\frac{Pa^2b}{\ell^2} \quad \blacksquare$$

10.6 THE METHOD OF ELASTIC WEIGHTS

A careful study of the procedure used in applying the area-moment theorems will reveal a simpler and more practical method of computing slopes and deflections for most beams. In reviewing this procedure the beam and M/EI diagram of Figure 10.11 are considered.

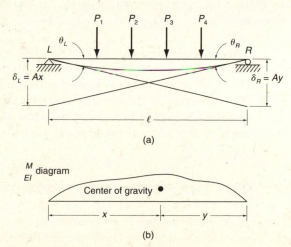

Figure 10.11

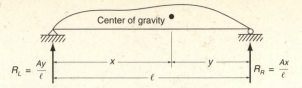

Figure 10.12

By letting A equal the area of the M/EI diagram, the deflection of the tangent at R from the tangent at L equals Ay, and the change in slope between the two tangents is A. An imaginary beam is loaded with the M/EI diagram, as shown in Figure 10.12, and the reactions R_L and R_R are determined. They equal Ay/ℓ and Ax/ℓ, respectively.

In Figure 10.11 the slopes of the tangents to the elastic curve at each end of the beam (θ_L and θ_R) are equal to the deflections between the tangents at each end divided by the span length, as follows:

$$\theta_L = \frac{\delta_R}{\ell} \qquad \theta_R = \frac{\delta_L}{\ell}$$

The values of δ_L and δ_R have previously been found to equal Ax and Ay, respectively, and may be substituted in these expressions.

$$\theta_L = \frac{Ay}{\ell} \qquad \theta_R = \frac{Ax}{\ell}$$

The end slopes are exactly the same as the reactions for the beam in Figure 10.12. At either end of the fictitious beam the shear equals the reaction and thus the slope in the actual beam. Further experiments will show that the shear at any point in the beam loaded with the M/EI diagram equals the slope at that point in the actual beam.

A similar argument can be made concerning the computation of deflections, and it will be found that the deflection at any point in the actual beam equals the moment at that point in the fictitious beam. In detail, the two theorems of elastic weights may be stated as follows:

1. **The slope of the elastic curve of a simple beam at a point, measured with respect to a chord between the supports, equals the shear at that point if the beam is loaded with the M/EI diagram.**
2. **The deflection of the elastic curve of a simple beam at a point, measured with respect to a chord between the supports, equals the moment at that point if the beam is loaded with the M/EI diagram.**

10.7 APPLICATION OF THE METHOD OF ELASTIC WEIGHTS

The method of elastic weights in its present form is applicable only to beams simply supported at each end. It will be found in using the method that maximum deflections in the actual beam occur at points of zero shear in the imaginary beam. The reasoning is the same as that presented for shear and moment diagrams in Section 5.5, where maximum moments were found to occur at points of zero shear.

Consideration has not been given to the subject of sign conventions for either the area-moment or the elastic-weight methods. With little difficulty the reader can see the directions of slopes and deflections by study of the shears and moments on the fictitious beam. A positive shear in the fictitious beam shows the left side is being pushed up with respect to the right side, or the beam is sloping downward from left to right. Similarly, a positive moment (see Figure 5.7) indicates downward deflection.

Examples 10.7 to 10.10 illustrate the application of elastic weights.

EXAMPLE 10.7

Determine the deflection at the centerline of the beam shown in Figure 10.13.

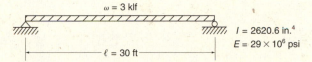

Figure 10.13

Solution.

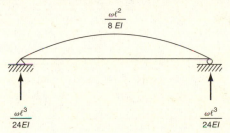

Deflection at centerline:

$$\delta_{\mathbb{C}} = \text{moment}_{\mathbb{C}} \left(\frac{w\ell^3}{24EI} \right)\left(\frac{\ell}{2} \right) - \left(\frac{2}{3} \right)\left(\frac{\ell}{2} \right)\left(\frac{w\ell^2}{8EI} \right)\left(\frac{3}{8}\frac{\ell}{2} \right)$$

$$= \frac{5w\ell^4}{384EI} = \frac{(5)(3000/12)(30 \times 12)^4}{(384)(29 \times 10^6)(2620.6)} = 0.719 \text{ in.} \quad \blacksquare$$

EXAMPLE 10.8

Determine the slope and deflection at the centerline of the beam shown in Figure 10.14.

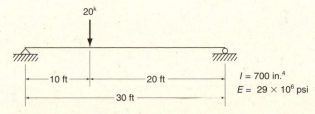

Figure 10.14

Solution. Loading the fictitious beam with the $\dfrac{M}{EI}$ diagram and computing the reactions.

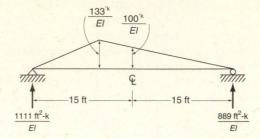

Deflection

$$\delta_{C\!L} = \frac{(889)(15) - (\frac{1}{2})(100)(15)(5)}{EI} = \frac{9585 \text{ ft}^3\text{-k}}{EI}$$

$$= \frac{(9585)(1728)(1000)}{(29 \times 10^6)(700)} = 0.816 \text{ in.}$$

Slope:

$$\theta_{C\!L} = \frac{-889 + (\frac{1}{2})(100)(15)}{EI} = -\frac{139 \text{ ft}^2\text{-k}}{EI}$$

$$= -\frac{(139)(144)(1000)}{(29 \times 10^6)(700)} = -0.000986 \text{ rad}$$

$$= -0.056^6 \text{ (negative slope } / \text{)} \quad \blacksquare$$

EXAMPLE 10.9 _____

Compute the maximum deflection for the beam shown in Figure 10.15.

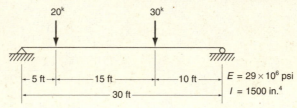

Figure 10.15

Solution.

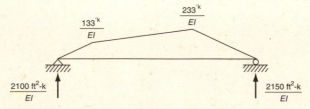

(a)

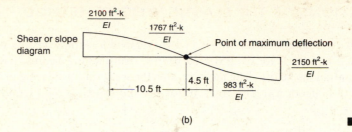

(b)

■

$$\delta_{max} = \frac{(2100)(15.5) - (\frac{1}{2})(5)(133)(12.17) - (10.5)(133)(5.25) - (\frac{1}{2})(10.5)(70)(3.5)}{EI}$$

$$= \frac{19{,}890 \text{ ft}^3\text{-k}}{EI} = \frac{(19{,}890)(1728)(1000)}{(29 \times 10^6)(1500)} = 0.790 \text{ in.}$$

EXAMPLE 10.10

Compute the centerline deflection for the simple beam shown in Figure 10.16.

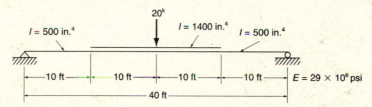

Figure 10.16

Solution.

$$\delta_{\mathcal{C}} = \left(\frac{2.07}{E}\right)(20) - \left(\frac{1}{2}\right)(10)\left(\frac{0.2}{E}\right)(13.33) - (10)\left(\frac{0.0715}{E}\right)(5)$$

$$- \left(\frac{1}{2}\right)(10)\left(\frac{0.0715}{E}\right)(3.33)$$

$$= \frac{23.3}{E} = \frac{(23.3)(1728)(1000)}{29 \times 10^6} = 1.39 \text{ in.} \quad ■$$

10.8 LIMITATIONS OF THE ELASTIC-WEIGHT METHOD

The method of elastic weights was developed for simple beams, and in its present form will not work for cantilevered beams, overhanging beams, fixed-ended beams, and continuous beams. The moment-area theorems are used to determine the correct slope and deflection at the free end of the uniformly loaded cantilevered beam of Figure 10.17.

$$\theta_B = \left(\frac{1}{3}\right)(\ell)\left(\frac{w\ell^2}{2EI}\right) = \frac{w\ell^3}{6EI}$$

$$\delta_B = \left(\frac{1}{3}\right)(\ell)\left(\frac{w\ell^2}{2EI}\right)\left(\frac{3}{4}\ell\right) = \frac{w\ell^4}{8EI}$$

If the elastic-weight method was used in an attempt to find the slope and deflection at the ends of the same beam, the result would be slopes and deflections of zero at the free end and $w\ell^3/6EI$ and $w\ell^4/24EI$ at the fixed end, as shown in Figure 10.18.

The slope and deflection at the fixed end A must be zero; however, application of elastic weights to the beam results in both shear and moment, falsely indicating slope and deflection.

If the fixed end of the beam was moved to the free end and the resulting beam loaded with the M/EI diagram, the shears and moments would correspond exactly to the slopes and deflections on the actual beam as found by the moment-area method.

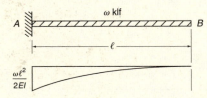

Figure 10.17

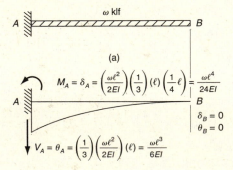

(a)

$$M_A = \delta_A = \left(\frac{\omega\ell^2}{2EI}\right)\left(\frac{1}{3}\right)(\ell)\left(\frac{1}{4}\ell\right) = \frac{\omega\ell^4}{24EI}$$

$$\delta_B = 0$$
$$\theta_B = 0$$

$$V_A = \theta_A = \left(\frac{1}{3}\right)\left(\frac{\omega\ell^2}{2EI}\right)(\ell) = \frac{\omega\ell^3}{6EI}$$

(b)

Figure 10.18

10.9 CONJUGATE-BEAM METHOD

The conjugate-beam method makes use of an "analogous" or "conjugate" beam to be handled by elastic weights in place of the actual beam, to which it cannot be correctly applied. The shear and moment in the imaginary beam, loaded with the M/EI diagram, must correspond exactly with the slope and deflection of the actual beam.

The correct mathematical relationship is obtained for a beam simply supported if it is loaded "as is" with the M/EI diagram. If the elastic-weight method is applied to other types of beams, the largest moments due to the M/EI loading occur at the supports, incorrectly indicating that the largest deflections occur at those points. For elastic weights to be applied correctly, use must be made of substitute beams or conjugate beams that have the supports changed so the correct relationships are obtained.

The loads and properties of the true beam have no effect on the manner in which the conjugate beam is supported. The only factors affecting the supports of the imaginary beam are the supports of the actual beam. The lengths of the two beams are equal. In the following paragraphs is a discussion of what the various types of beam support must become in the conjugate beam so that the elastic-weight method will apply. The mathematical proof of these relationships is explained in detail in books on strength of materials.

Free End

The free end of a beam slopes and deflects when the beam is loaded. The conjugate beam must have both shear and moment at that end when it is loaded with the M/EI diagram. The only type of end support having both shear and moment is the fixed end. *A free end in the actual beam becomes a fixed end in the conjugate beam.*

Fixed End

A similar discussion in reverse order can be made for a fixed end. No slope or deflection can occur at a fixed end, and there must not be any shear or moment in the conjugate beam at that point. *A fixed end in the actual beam becomes a free end in the conjugate beam.*

Simple End Support

A simple end slopes but does not deflect when the beam is loaded. The imaginary beam will have shear but no moment at that point, a situation that can occur only at a simple support. *A simple end support in the actual beam remains a simple end support in the conjugate beam.*

Simple Interior Support

There is no deflection at either a simple interior or a simple end support. Both types may slope when the beam is loaded, but the situations are somewhat different. The slope at a simple interior support is continuous across the support; that is, no sudden

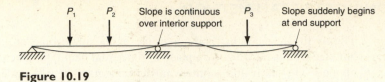

Figure 10.19

change of slope occurs. This condition is not present at a simple end support where the slope suddenly begins. (See the deflection curve for the beam of Figure 10.19.) If there is no change of slope at a simple interior support, there can be no change of shear at the corresponding support in the conjugate beam. Any type of external support at this point would cause a change in the shear; therefore, an internal pin (or unsupported hinge) is required. *A simple interior support in the actual beam becomes an unsupported internal hinge in the conjugate beam.*

Internal Hinge

At an unsupported internal hinge there is both slope and deflection, which means that the corresponding support in the conjugate beam must have shear and moment. *An internal hinge in the actual beam becomes a simple support in the conjugate beam.*

Summary

Figure 10.20 shows several types of common beams and their corresponding conjugates.

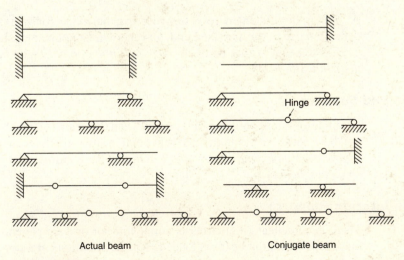

Actual beam   Conjugate beam

Figure 10.20

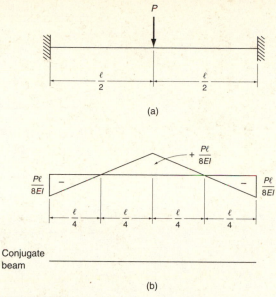

Figure 10.21

Equilibrium

The reactions, moments, and shears of the conjugate beam are easily computed by statics because the conjugate beam is always statically determinate, even though the real beam may be statically indeterminate. Sometimes the conjugate beam may appear to be completely unstable. The most conspicuous example is the conjugate beam for the fixed-end beam (Figure 10.21), which has no supports whatsoever. On second glance, the areas of the M/EI diagram are seen to be so precisely balanced between downward and upward loads (positive and negative areas of the diagram, respectively) as to require no supports. Any supports seemingly required would have zero reactions, and the proper shears and moments are supplied to coincide with the true slopes and deflections. Even a real beam continuous over several simple supports has a conjugate that is simply end-supported.

10.10 SUMMARY OF BEAM RELATIONS

A brief summary of the relations that exist between loads, shears, moments, slope changes, slopes, and deflections is presented in Figure 10.22 on the following page. The relations are shown for a uniformly loaded beam but are applicable to any type of loading. For the two sets of curves shown, the ordinate on one curve equals the slope at that point on the following curve. It is obvious from these figures that the same mathematical relations that exist between load, shear, and moment hold for M/EI loading, slope, and deflection.

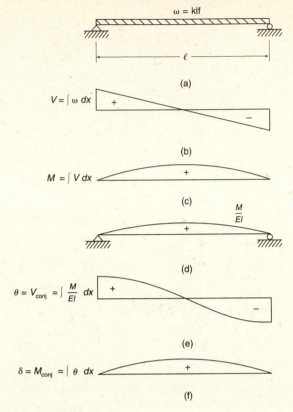

Figure 10.22

10.11 APPLICATION OF THE CONJUGATE METHOD TO BEAMS

Examples 10.11 and 10.12 illustrate the conjugate method of calculating slopes and deflections for beams. The procedure as to symbols and units used in applying the method is in general the same as that used for the moment-area and elastic-weight methods. Maximum deflections occur at points of zero shear on the conjugate structure. For example, the point of zero shear in the beam of Figure 10.21 is the centerline. The deflection is as follows:

$$\delta_{\mathcal{C}} = \text{moment}_{\mathcal{C}}$$
$$= \left(\frac{1}{2}\right)\left(\frac{\ell}{4}\right)\left(\frac{P\ell}{8EI}\right)\left(\frac{5}{12}\ell\right) - \left(\frac{1}{2}\right)\left(\frac{\ell}{4}\right)\left(\frac{P\ell}{8EI}\right)\left(\frac{\ell}{12}\right)$$
$$= \frac{P\ell^3}{192EI}$$

EXAMPLE 10.11

Determine the slope and deflection of point A in Figure 10.23.

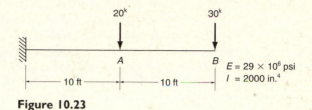

Figure 10.23

Solution.

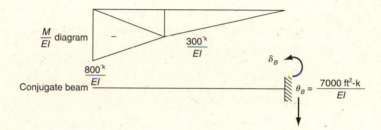

Slope:

$$\theta_A = \frac{(\frac{1}{2})(300)(10) + (\frac{1}{2})(800)(10)}{EI} = \frac{5500 \text{ ft}^2\text{-k}}{EI}$$

$$= \frac{(5500)(144)(1000)}{(29 \times 10^6)(2000)} = 0.0136 \text{ rad} = 0.78°\backslash$$

Deflection:

$$\delta_A = \frac{(\frac{1}{2})(300)(10)(3.33) + (\frac{1}{2})(800)(10)(6.67)}{EI} = \frac{31,667 \text{ ft}^2\text{-k}}{EI}$$

$$= \frac{(31,667)(1728)(1000)}{(29 \times 10^6)(2000)} = 0.943 \text{ in.} \downarrow \quad \blacksquare$$

EXAMPLE 10.12

Determine deflections at points A and B in the overhanging beam of Figure 10.24.

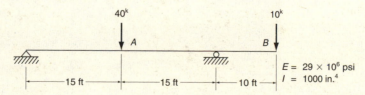

Figure 10.24

Solution. The *M/EI* diagram is drawn and placed on the conjugate beam, which has an interior hinge. The reactions are determined as they were for the cantilever-type structures of Chapter 4. The portion of the beam to the left of the hinge is considered as a simple beam, and its reactions are determined. The reaction at the hinge is applied as a concentrated load acting at the end of the cantilever to the right of the hinge in the opposite direction, and the reactions at the fixed end are determined. To simplify the mathematics, a separate moment diagram is drawn for each of the concentrated loads.

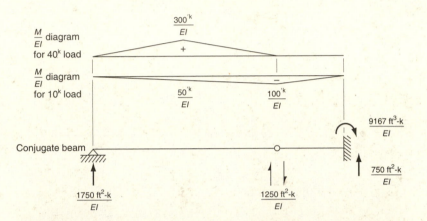

Harrison Avenue Bridge, Beaumont, Texas. (Courtesy Bethlehem Steel Corporation.)

$$\delta_A = \frac{(1750)(15) - (\frac{1}{2})(15)(300)(5) + (\frac{1}{2})(15)(50)(5)}{EI} = \frac{16{,}875 \text{ ft}^3\text{-k}}{EI}$$

$$= 1.01 \text{ in.} \downarrow$$

$$\delta_B = \frac{-(1250)(10) + (\frac{1}{2})(10)(100)(6.67)}{EI} = -\frac{9167 \text{ ft}^3\text{-k}}{EI} = 0.546 \text{ in.} \uparrow \quad \blacksquare$$

10.12 LONG-TERM DEFLECTIONS

Under sustained loads concrete will continue to deform for long periods of time. This additional deformation is called *creep,* or *plastic flow*. If a compressive load is applied to a concrete member, an immediate or elastic shortening occurs. If the load is left in place for a long time, the member will continue to shorten over a period of several years and the final deformation may be as much as two or three (or more) times the initial deformation. Creep is dependent on such items as humidity, temperature, curing conditions, age of concrete at time of loading, ratio of stress to strength, and other items.

When sustained loads are applied to reinforced-concrete beams, their compression sides will become shorter and shorter over time and the result will be larger deflections. The American Concrete Institute[2] states that the total long-term deflection in a particular member should be estimated by: (1) calculating the instantaneous deflection caused by all the loads; (2) computing the part of the instantaneous deflection that is caused by the sustained loads; (3) multiplying this value by an empirical factor from the ACI Code, which is dependent on the time elapsed; and (4) adding this value to the instantaneous deflection.

The sustained loads for a building include the dead load plus some percentage of the live load. For an apartment house or for an office building perhaps only 20 to 25% of the live load should be considered as being sustained, while as much as 70% to 80% of the live load of a warehouse might fall into this category.

A similar discussion is possible for timber structures. Seasoned timber members subjected to long-term loads develop a permanent deformation or sag approximately equal to twice the deflection computed for short-term loads of the same magnitude.[3]

10.13 APPLICATION OF THE CONJUGATE METHOD TO FRAMES

The rigid frames of Figure 10.25 on the following page consist of members rigidly connected at their joints. The joints are moment resisting and prevent the frame members from having freedom of rotation.

[2] *Building Code Requirements for Reinforced Concrete*, ACI 318–95 (Detroit: American Concrete Institute), Sect. 9.5.

[3] W. F. Scofield and W. H. O'Brien (revised by W. A. Oliver), *Modern Timber Engineering* (New Orleans, La.: Southern Pine Association, 1963), 97.

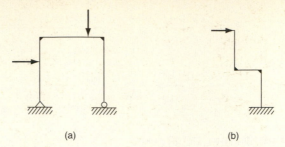

Figure 10.25

The slopes and deflections of frames such as these can be computed by the conjugate method as they were for beams. The procedure, however, is confusing and *the authors do not recommend its use for the usual frame*. The virtual-work method described in Chapter 11 provides a simpler and more logical method for frames.

Should the conjugate method be applied to frames, there will be additional factors (from those encountered in beams) that must be included in the calculations.

Consider the frame and loads of Figure 10.26(a) and let the deflection of point E be desired. Figure 10.26(b) shows a sketch of the estimated deformed shape of the frame. It is seen that there are two factors affecting the total deflection at E, these being the deflection at E if joint B were prevented from moving laterally plus the effect of the actual movement to the right of joint B. The frame sways to the right, which causes E to deflect an additional amount. Similarly, to find the total change in slope or the rotation of a particular joint in a complicated frame, it may be necessary to consider the rotations of several joints.

The foregoing discussion has shown that the calculation of deflections for a rigid frame by the usual conjugate-beam procedure may involve the taking of moments, the structure being loaded with the M/EI diagram, plus a complicated consideration of joint rotations. Slope and deflection are determined for an elementary frame in Example 10.13 in this manner.

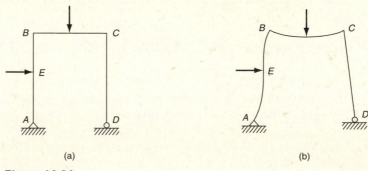

Figure 10.26

EXAMPLE 10.13

Determine the deflections at points B and D of the frame shown in Figure 10.27.

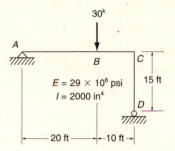

Figure 10.27

Solution.   A sketch of the estimated deflected shape of the structure and the M/EI diagram are drawn. The deflection of point B is easily determined by the usual procedure. The deflected shape of the structure shows that member CD remains straight, because it has no moment, and it inclines outward at an angle equal to θ_C. The deflection at D equals the length of CD times θ_C.

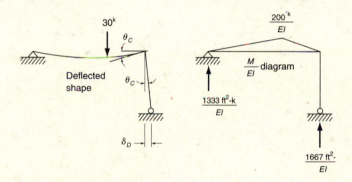

$$\delta_B = \frac{(1333)(20) - (\frac{1}{2})(20)(200)(6.67)}{EI} = \frac{13.333 \text{ ft}^3\text{-k}}{EI} = 0.397 \text{ in. } \downarrow$$

$$\theta_C = \frac{1667 \text{ ft}^2\text{-k}}{EI} = 0.00414 \text{ rad}$$

$$\delta_D = (15 \times 12)(0.00414) = 0.745 \text{ in. } \rightarrow$$

The conjugate method for frames is quite popular with some engineers. Readers who are interested in the topic should refer to Kinney,[4] where it is fully described. ∎

[4] J. S. Kinney, *Indeterminate Structural Analysis* (Reading, Mass.: Addison-Wesley), 143–162.

PROBLEMS

Moment-Area Problems

Using the moment-area method, determine the quantities asked for each of Problems 10.1 through 10.16.

10.1 θ_A, δ_A: $E = 29 \times 10^6$ psi. $I = 1000$ in.^{4}

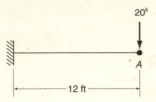

Problem 10.1

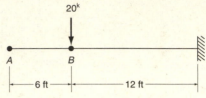

Problem 10.2

10.2 $\theta_A, \theta_B, \delta_A, \delta_B$: $E = 29 \times 10^6$ psi. $I = 1500$ in.^{4} (*Ans.* $\theta_A = \theta_B = 0.00477$ rad/, $\delta_A = 0.801$ in. ↓, $\delta_B = 0.458$ in. ↓)

10.3 $\theta_A, \theta_B, \delta_A, \delta_B$: $E = 29 \times 10^6$ psi. $I = 4100$ in.^{4}

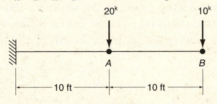

Problem 10.3

Problem 10.4

10.4 θ_A, δ_A: $E = 29 \times 10^6$ psi. $I = 843$ in.^{4} (*Ans.* $\theta_A = 0.00295$ rad/, $\delta_A = 0.265$ in. ↓)

10.5 $\theta_A, \theta_B, \delta_A, \delta_B$: $E = 29 \times 10^6$ psi. $I = 1140$ in.^{4}

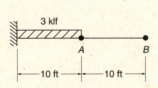

Problem 10.5

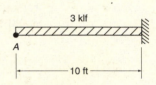

Problem 10.6

10.6 $\theta_A, \theta_B, \delta_A, \delta_B$: $E = 29 \times 10^6$ psi. $I = 4000$ in.^{4} (*Ans.* $\theta_A = 0.00430$ rad\, $\theta_B = 0.00502$ rad\, $\delta_A = 0.300$ in. ↓, $\delta_B = 0.760$ in. ↓)

10.7 $\theta_A, \theta_B, \delta_A, \delta_B$: $E = 29 \times 10^6$ psi. $I = 5200$ in.^{4}

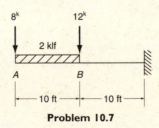

Problem 10.7

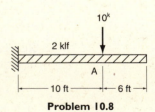

Problem 10.8

10.8 θ_A, δ_A, $E = 29 \times 10^6$ psi. $I = 3000$ in.^{4} (*Ans.* $\theta_A = 0.00297$ rad\, $\delta_A = 0.231$ in. ↓)

10.9 $\theta_A,\ \theta_B,\ \theta_C,\ \delta_A,\ \delta_B,\ \delta_C$: $E = 29 \times 10^6$ psi. $I = 1330$ in.^{4}

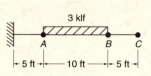

Problem 10.9

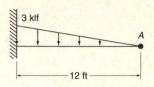

Problem 10.10

10.10 $\theta_A,\ \delta_A$: $E = 5 \times 10^6$ psi. $I = 6000$ in.^{4} (*Ans.* 0.00104 rad\, 0.119 in. ↓)

10.11 $\theta_A,\ \theta_B,\ \delta_A,\ \delta_B$: $E = 29 \times 10^6$ psi

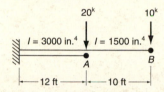

Problem 10.11

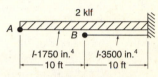

Problem 10.12

10.12 $\theta_A,\ \theta_B,\ \delta_A,\ \delta_B$: $E = 29 \times 10^6$ psi. (*Ans.* 0.00426 rad/, 0.00331 rad/, 0.724 in. ↓, 0.241 in. ↓)

10.13 $\theta_A,\ \delta_A$: $E = 200\ 000$ MPa, $I = 3.0 \times 10^8$ mm^4

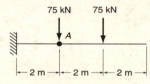

Problem 10.13

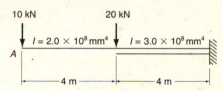

Problem 10.14

10.14 $\theta_A,\ \delta_A$: $E = 200\ 000$ MPa (*Ans.* $\theta_A = 0.008\ 67$ rad/, $\delta_A = 48$ mm)

10.15 $\theta_A,\ \delta_A$: $E = 29 \times 10^6$ psi. $I = 1500$ in.^{4}

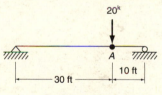

Problem 10.15

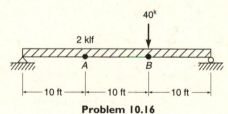

Problem 10.16

10.16 δ_A: $E = 29 \times 10^6$ psi. $I = 3200$ in.^{4} (*Ans.* 0.631 in. ↓)

For Problems 10.17 through 10.21 compute the fixed-end moments for the beams: E and I constant except as shown.

10.17

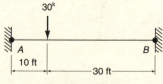

Problem 10.17

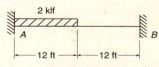

Problem 10.18

10.18   (*Ans.* $M_A = -66$ ft-k, $M_B = -30$ ft-k)

10.19

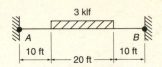

Problem 10.19

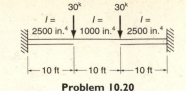

Problem 10.20

10.20   (*Ans. $M_A = M_B = -233$ ft-k*)

10.21

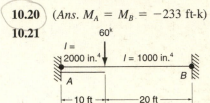

Conjugate-Beam Problems

Use the conjugate-beam method for solving Problems 10.22 through 10.54.

10.22 θ_A, δ_A: $E = 29 \times 10^6$ psi. $I = 1800$ in.^{4} (*Ans.* 0.00123 rad\, 0.294 in. ↓)

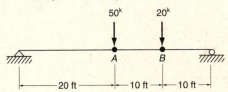

10.23   Determine the maximum deflection in the beam of Problem 10.22.

10.24 θ_A, θ_B, δ_A, δ_B: $E = 29 \times 10^6$ psi. $I = 4000$ in.^{4} (*Ans.* 0.000310 rad\, 0.00590 rad/, 1.27 in. ↓, 0.906 in. ↓)

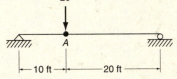

10.25   Determine the maximum deflection in the beam of Problem 10.24.

10.26   Determine the slope and deflection at points *A* and *B* for the beam of Problem 10.16. (*Ans.* $\theta_A = 0.0034$ rad\, $\theta_B = 0.00306$ rad/, $\delta_A = 0.631$ in. ↓, $\delta_B = 0.673$ in. ↓)

10.27 θ_A, δ_A: $E = 29 \times 10^6$ psi. $I = 2370$ in.^{4}

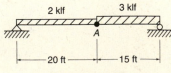

Problem 10.27

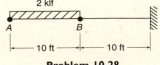

Problem 10.28

10.28 θ_A, θ_B, δ_A, δ_B: $E = 29 \times 10^6$ psi. $I = 1820$ in.^{4} (*Ans.* 0.00637 rad/, 0.00546 rad/, 1.12 in. ↓, 0.382 in. ↓)

10.29 θ_A, δ_A: $E = 200\,000$ MPa. $I = 6.0 \times 10^6$ mm^4

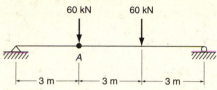

10.30 θ_A, δ_A: $E = 29 \times 10^6$ psi. $I = 2500$ in.^{4} (*Ans.* 0.000419 rad/, 0.779 in. ↓)

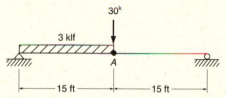

10.31 θ_A, θ_B, δ_A, δ_B: $E = 29 \times 10^6$ psi. $I = 5000$ in.^{4}

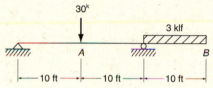

10.32 θ_A, δ_A: $E = 4 \times 10^6$ psi. $I = 6200$ in.^{4} (*Ans.* 0.000697 rad/, 0.902 in. ↓)

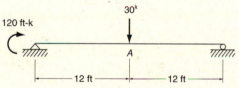

10.33 θ_A, θ_B, δ_A, δ_B: $E = 29 \times 10^6$ psi. $I = 1800$ in.^{4}

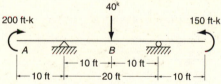

10.34 θ_A, δ_A, δ_B: $E = 29 \times 10^6$ psi. $I = 3400$ in.^{4} (*Ans.* 0.000730 rad/, 0.197 in. ↓, 1.15 in.↓)

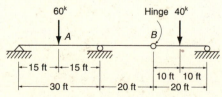

10.35 θ_A, δ_A: $E = 200\,000$ MPa

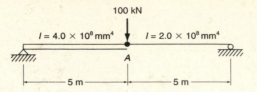

10.36 θ_A, δ_A: $E = 29 \times 10^6$ psi. $I = 1600$ in.⁴ (*Ans.* 0.00414 rad/, 0.795 in. ↓)

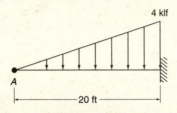

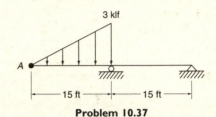

Problem 10.36   **Problem 10.37**

10.37 θ_A, δ_A: $E = 29 \times 10^4$ psi. $I = 513$ in.⁴

10.38   Repeat Problem 10.16 if the beam consists of reinforced concrete with $I = 12{,}400$ in.⁴ $E = 3.16 \times 10^6$ psi. Assume that the uniform load is a long-term or sustained load and the 40-kip concentrated load is a short-term live load. Assume that the sustained load multiplier is 2.0. (*Ans.* 0.0120 rad\, 2.30 in. ↓)

10.39 θ_A, θ_B, δ_A, δ_B: $E = 29 \times 10^6$ psi

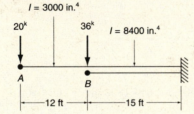

10.40 θ_A, δ_A: $E = 29 \times 10^6$ psi (*Ans.* 0.00808 rad\, 1.63 in. ↓)

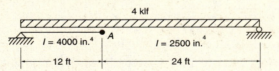

10.41 θ_A, δ_A: $E = 29 \times 10^6$ psi

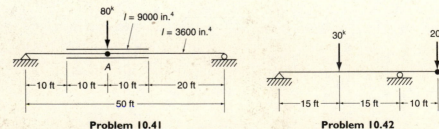

Problem 10.41   **Problem 10.42**

10.42 θ_A, δ_A: $E = 29 \times 10^6$ psi. $I = 1750$ in.⁴ (*Ans.* 0.00372 rad/, 0.334 in. ↓)

10.43 θ_A, θ_B, δ_A, δ_B: $E = 29 \times 10^6$ psi. $I = 1820$ in.^{4}

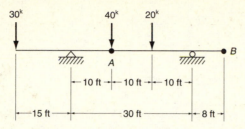

10.44 θ_A, δ_A: $E = 29 \times 10^6$ psi. $I = 1500$ in.^{4} (*Ans.* 0.00234 rad $\seardrow$, 0.157 in. $\downarrow$)

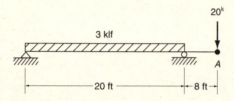

10.45 θ_A, θ_B, δ_A, δ_B: $E = 29 \times 10^6$ psi

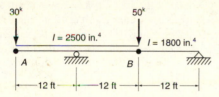

10.46 δ_A, θ_B: $E = 29 \times 10^6$ psi. $I = 1600$ in.^{4} (*Ans.* 0.122 in. $\downarrow$, 0.00194 rad\)

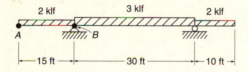

10.47 θ_A, θ_B, δ_A, δ_B: $E = 29 \times 10^6$ psi. $I = 1500$ in.^{4}

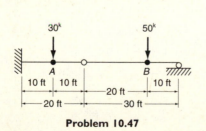

Problem 10.47

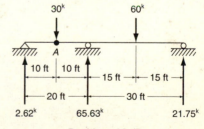

Problem 10.48

10.48 θ_A, δ_A: Reactions given, $E = 29 \times 10^6$ psi. $I = 2690$ in.^{4} (*Ans.* $\theta_A = 0.000380$ rad/, $\delta_A = 0.0263$ in. $\uparrow$)

10.49 θ_A, δ_B: $E = 29 \times 10^6$ psi. $I = 2830$ in.^{4}

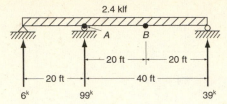

10.50   Determine slope and deflection at the 20 k load. Beam width is 12 in. (Suggestion in preparing M/EI diagram: Divide beam width into 2-ft sections and use the I at the center of each of the sections.) $E = 3.12 \times 10^6$ psi. (*Ans.* 0.00665 rad\, 0.283 in. ↓)

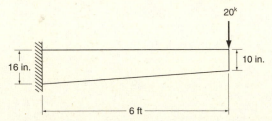

10.51   Determine slope and deflection at the 100-kip load in the beam shown in the accompanying illustration. The beam has a constant width of 12 in. and its depth varies as shown. (Suggestion in drawing M/EI diagram: Take the part of the beam for which the depth varies and divide it into 2-ft sections and use the I at the center of each of the sections.) $E = 3.12 \times 10^6$ psi.

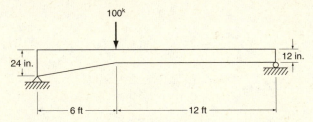

10.52   Determine deflection at the 20^k load if the beam is 12 in. wide and $E = 3.6 \times 10^6$ psi. Break the beam into the 2-ft sections shown and use the I at the center of each for the depths given. (*Ans.* 0.166 in. ↓)

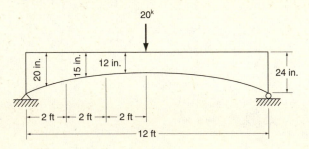

10.53 θ_A, δ_A: $E = 29 \times 10^6$ psi. $I = 1200$ in.^{4}

10.54 δ_{vert} at A, δ_{horiz} at B: $E = 29 \times 10^6$ psi. $I = 1400$ in.^{4} (*Ans.* $\delta_A = 1.21$ in. $\downarrow$, $\delta_B = 0.382$ in. $\leftarrow$)

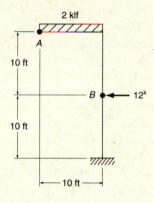

Chapter 11

Deflection and Angle Changes—Energy Methods

11.1 INTRODUCTION TO ENERGY METHODS

In Chapter 10 deflections and angle changes were computed using geometric methods (moment area and the conjugate method). These procedures are satisfactory for many structures, including some rather complicated ones.

In this chapter it will be shown that the same deflections and angle changes may be determined using the principle of conservation of energy. Two energy methods, virtual work and Castigliano's theorem, will be introduced. For some complicated structures, these methods may be more desirable than the geometric methods because of the simplicity with which the expressions may be set up for making the solutions. In addition, the energy methods are applicable to more types of structures.

11.2 CONSERVATION OF ENERGY PRINCIPLE

Before the conservation of energy principle is introduced, a few comments are presented concerning the concept of work. For this discussion *work* is defined as the product of a force and its displacement in the direction in which the force is acting. Should the force be constant during the displacement, the work will equal the force times the total displacement as follows:

$$W = F\Delta$$

On many practical occasions the force changes in magnitude as does the deformation. For such a situation it is necessary to sum up the small increments of work for which the force can be assumed to be constant, that is

$$W = \int F \, d\Delta$$

The *conservation of energy principle* is the basis of all energy methods. When a set of external loads is applied to a deformable structure the points of load application move, with the result that the members or elements making up the structure become deformed. According to the conservation of energy principle, the work done by the external loads, W_e, equals the work done by the internal forces acting on the elements of the structure, W_i. Thus

$$W_e = W_i$$

As the deformation of a structure takes place, the internal work—commonly referred to as strain energy—is stored within the structure as potential energy. If the elastic limit of the material is not exceeded, the elastic strain energy will be sufficient to return the structure to its original undeformed state when the loads are removed. Should a structure be subjected to more than one load the total energy stored in the body will equal the sum of the energies stored in the structure by each of the loads.

It is to be clearly noted that the conservation of energy principle will be applicable only when static loads are applied to elastic systems. If the loads are not applied gradually, acceleration may occur and some of the external work will be transferred into kinetic energy. If inelastic strains are present, some of the energy will be lost in the form of heat.

11.3 VIRTUAL WORK OR COMPLEMENTARY VIRTUAL WORK METHOD

With the virtual work principle, a system of forces in equilibrium is related to a compatible system of displacements in a structure. The word *virtual* means not in fact, but equivalent. Thus, the virtual quantities discussed in this chapter do not exist in a real sense. When we speak of a virtual displacement we are speaking of a fictitious displacement imposed on a structure. *The work performed by a set of real forces during a virtual displacement is called virtual work.*

We will see in the discussion to follow that for each virtual work theorem there is a corresponding theorem that is based on complementary virtual work. As a result the method is sometimes referred to as complementary virtual work.

Virtual work is based on the principle of virtual velocities, which was introduced by Johann Bernoulli of Switzerland in 1717. The virtual work theorem can be stated as follows:

If a displacement is applied to a deformable body that is in equilibrium under a known load or loads, the external work performed by the existing load or loads due to this new displacement will equal the internal work performed by the stresses existing in the body that were caused by the original load or loads.

The virtual work or complementary work method also is often referred to as the method of work or the dummy unit-load method. The name *dummy unit-load method* is used because a fictitious or dummy unit load is used in the solution.

Virtual work is based on the law of conservation of energy. To make use of this law in the derivation to follow, it is necessary to make the following assumptions:

Figure 11.1   Axially loaded straight bar.

1. The external and internal forces are in equilibrium.
2. The elastic limit of the material is not exceeded.
3. There is no movement of the supports.

Should a load F be gradually applied to the bar shown in Figure 11.1 so that the load and the bar's deformation (or increase in length) increase together from zero to their total values (F and Δ), the external work done will equal the average load $F/2$ times the total displacement Δ.

$$W_e = \frac{F}{2}\Delta$$

If a load F_1 is gradually applied to the bar, it will elongate an amount Δ_1. Therefore, the forces that undergo displacement during this elongation will perform external work equal to $W_e = 1/2F_1\Delta_1$, as shown by the triangular area with horizontal cross-hatched lines in Figure 11.2.

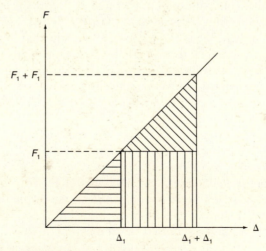

Figure 11.2   Load-deformation curve for a linear member.

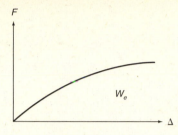

Figure 11.3 Load-deformation curve for a nonlinear member.

Should another gradually applied force F_2 be added to the bar causing an additional displacement Δ_2, additional external work will be performed. The force F_1 will be present for all of the Δ_2 displacement and the work will be $F_1\Delta_2$. This work is represented in Figure 11.2 by the rectangle with vertical lines. The gradually applied force F_2 will perform an additional amount of work equal to $1/2F_2\Delta_2$. Thus, the total external work is

$$W_e = \frac{1}{2}F_1\Delta_1 + F_1\Delta_2 + \frac{1}{2}F_2\Delta_2$$

Another type of work is complementary work. It is represented by the area above the curve and W^* here. Complementary work does not really have a physical meaning as does the external work W_e, but we can see that for a gradually applied load

$$W_e + W^* = F\Delta$$

Thus, W^* is the complement of the work W_e because it completes the rectangle, and as long as Hooke's law applies, $W_e = W^*$. Should the bar material's behavior not be linear, however, the load deformation diagram will be a curve and W_e and W^* will be unequal, as seen in Figure 11.3.

11.4 TRUSS DEFLECTIONS BY VIRTUAL WORK

In this section the virtual work method is used to determine truss deflections, making use of the dummy unit-load procedure. To determine displacements by this procedure two loading systems are considered. The first system consists of the structure subjected to the force or forces for which the deflections are to be calculated. The second system consists of the structure subjected only to a fictitious unit or dummy load acting at the point and in the direction in which displacement is desired.

The dummy structure with the unit load in place is now subjected to the deformations of the real truss when it is subjected to the real forces, after which the resulting internal and external work are equated. For this calculation the only external load that performs work is the unit load, and as it remains constant, the external work equals $1 \times \Delta$.

Next we need to write an expression for the internal work. The truss of Figure 11.4 on the following page will be considered for this discussion. Loads P_1 to P_3 are applied to the truss as shown and cause forces in the truss members. Each member of the truss shortens or lengthens depending on the character of its force. These internal deformations cause external deflections, and each of the external loads

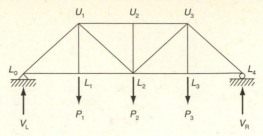

Figure 11.4

moves through a short distance. The law of conservation of energy as it applies to the truss may now be stated in detail. The external work performed by the loads P_1 to P_3 as they move through their respective truss deflections equals the internal work performed by the member forces as they move through their respective changes in length.

To write an expression for the internal work performed by a truss member, it is necessary to develop an expression for the deformation of the member. For this purpose the bar of Figure 11.5 is considered.

The force applied to the bar causes it to elongate by an amount $\Delta\ell$. The elongation may be computed from the properties of the bar. The strain or elongation per unit length, ϵ, is equal to the total elongation divided by the length of the bar and also is equal to the stress intensity divided by the modulus of elasticity. An expression for $\Delta\ell$ may be developed as follows:

$$E = \frac{f}{\epsilon} = \frac{F/A}{\Delta\ell/\ell}$$

$$\Delta\ell = \frac{F\ell}{AE}$$

In accordance with previous assumptions, the members of a truss have only axial force. These forces are referred to as F forces, and each member will change in length by an amount equal to its $F\ell/AE$ value.

It is desired that an expression for the deflection at a joint in the truss of Figure 11.4 be developed. A convenient means of developing such an expression is to remove the external loads from the truss, place a unit load at the joint where deflection is desired, replace the external loads, and write an expression for the internal and external work performed by the unit load and the forces it causes when the external loads are replaced.

The forces caused in the truss members by the unit load are called μ forces. They cause small deformations of the members and small external deformations of the truss. When the external loads are returned to the truss, the force in each of the

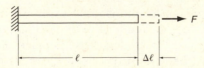

Figure 11.5

Aluminum trusses, Reynolds Hangar, Byrd Field, Richmond, Virginia. (Courtesy of Reynolds Metals Company.)

members changes by the appropriate F force, and the deformation of each member changes by its $F\ell/AE$ value. The truss deflects and the unit load is carried through a distance δ. The external work performed by the unit load when the external loads are returned to the structure may be expressed as follows:

$$W_e = 1 \times \delta$$

Internally the μ force in each member is carried through a distance $\Delta\ell = F\ell/AE$. The internal work performed by all of the μ forces as they move through these distances is

$$W_i = \Sigma\mu\frac{F\ell}{AE}$$

By equating the internal and external work, the deflection at a joint in the truss may be expressed as follows:

$$\delta = \Sigma \frac{F\mu\ell}{AE}$$

11.5 APPLICATION OF VIRTUAL WORK TO TRUSSES

Examples 11.1 and 11.2 illustrate the application of virtual work to trusses. In each case the forces due to the external loads are computed initially. Second, the external loads are removed and a unit load is placed at the point and in the direction in which deflection is desired (not necessarily horizontal or vertical). The forces due to the unit load are determined, and, finally, the value of $F\mu\ell/AE$ for each of the members is found. To simplify the numerous multiplications, a table is used. The modulus of elasticity is carried through as a constant until the summation is made for all of the members, at which time its numerical value is used. Should there be members of different Es, it is necessary that their actual or their relative values be used for the individual multiplications. A positive value of $\Sigma(F\mu\ell/AE)$ indicates a deflection in the direction of the unit load.

EXAMPLE 11.1 _____

Determine the horizontal and vertical components of deflection at joint L_4 in the truss shown in Figure 11.6. Circled figures are areas, in square inches. $E = 29 \times 10^6$ psi.

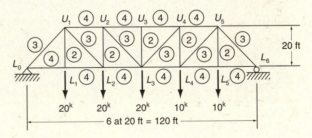

Figure 11.6

Solution.   Forces due to external loads:

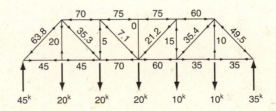

Forces due to a vertical unit load at L_4:

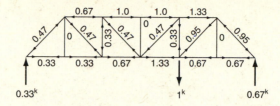

Forces due to a horizontal unit load at L_4:

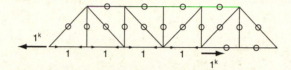

TABLE 11.1

Member	ℓ(in.)	A (in²)	$\dfrac{\ell}{A}$	F (kips)	μ_V	$\dfrac{F\mu_V\ell}{AE}$	μ_H	$\dfrac{F\mu_H\ell}{AE}$
$L_0 L_1$	240	4	60	+45	+0.33	+ 900	1.0	+2700
$L_1 L_2$	240	4	60	+45	+0.33	+ 900	1.0	+2700
$L_2 L_3$	240	4	60	+70	+0.67	+ 2800	1.0	+4200
$L_3 L_4$	240	4	60	+60	+1.33	+ 4800	1.0	+3600
$L_4 L_5$	240	4	60	+35	+0.67	+ 1400	0	0
$L_5 L_6$	240	4	60	+35	+0.67	+ 1400	0	0
$L_0 U_1$	340	3	113.3	−63.8	−0.47	+ 3400	0	0
$U_1 U_2$	240	4	60	−70	−0.67	+ 2800	0	0
$U_2 U_3$	240	4	60	−75	−1.0	+ 4500	0	0
$U_3 U_4$	240	4	60	−75	−1.0	+ 4500	0	0
$U_4 U_5$	240	4	60	−60	−1.33	+ 4800	0	0
$U_5 L_6$	340	3	113.3	−49.5	−0.95	+ 5340	0	0
$U_1 L_1$	240	2	120	+20	0	0	0	0
$U_1 L_2$	340	3	113.3	+35.3	+0.47	+ 1880	0	0
$U_2 L_2$	240	2	120	− 5	−0.33	+ 200	0	0
$U_2 L_3$	340	3	113.3	+ 7.1	+0.47	+ 378	0	0
$U_3 L_3$	240	2	120	0	0	0	0	0
$L_3 U_4$	340	3	113.3	+21.2	−0.47	− 1130	0	0
$U_4 L_4$	240	2	120	−15	+0.33	− 600	0	0
$L_4 U_5$	340	3	113.3	+35.4	+0.95	+ 3810	0	0
$U_5 L_5$	240	2	120	+10	0	0	0	0
Σ						$\dfrac{+42{,}078}{E}$		$\dfrac{+13{,}200}{E}$

Vertical deflection:

$$E\delta_{L_4} = +42,078$$

$$\delta_{L_4} = \frac{42,078 \times 1000}{29,000,000} = +1.45 \text{ in.} \downarrow$$

Horizontal deflection:

$$E\delta_{L_4} = +13,200$$

$$\delta_{L_4} = \frac{13,200 \times 1000}{29,000,000} = +0.455 \text{ in.} \rightarrow \blacksquare$$

EXAMPLE 11.2

Determine the vertical component of deflection of joint L_4 in Figure 11.7 by the virtual-work method. Circled figures are areas, in square inches. $E = 29 \times 10^6$ psi.

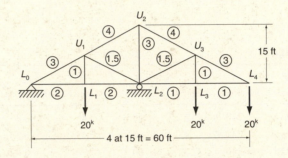

Figure 11.7

Solution.   Forces due to external loads:

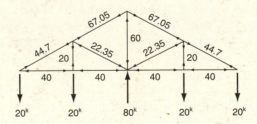

Forces due to a vertical unit load at L_4:

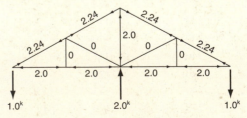

From Table 11.2, vertical deflection at $L_4 = \dfrac{(79,010)(1000)}{29 \times 10^6} = 2.72$ in. $\downarrow$ $\blacksquare$

TABLE 11.2

Member	ℓ (in.)	A (in^2)	$\dfrac{\ell}{A}$	F (kips)	μ	$\dfrac{F\mu\ell}{AE}$
L_0L_1	180	2	90	-40	-2.0	$+7,200$
L_1L_2	180	2	90	-40	-2.0	$+7,200$
L_2L_3	180	1	180	-40	-2.0	$+14,400$
L_3L_4	180	1	180	-40	-2.0	$+14,400$
L_0U_1	202	3	67.3	$+44.7$	$+2.24$	$+6,730$
U_1U_2	202	4	50.5	$+67.05$	$+2.24$	$+7,575$
U_2U_3	202	4	50.5	$+67.05$	$+2.24$	$+7,575$
U_3L_4	202	3	67.3	$+44.7$	$+2.24$	$+6,730$
U_1L_1	90	1	90	$+20$	0	0
U_1L_2	202	1.5	134.5	-22.35	0	0
U_2L_2	180	3	60	-60	-2.0	$+7,200$
L_2U_3	202	1.5	134.5	-22.35	0	0
U_3L_3	90	1	90	$+20$	0	0
Σ						$\dfrac{79,010}{E}$

11.6 DEFLECTIONS OF BEAMS AND FRAMES BY VIRTUAL WORK

The law of conservation of energy may be used to develop an expression for the deflection at any point in a beam or frame. In the following derivation each fiber of the structure is considered to be a "bar" or member similar to the members of the trusses considered in the preceding sections. The summation of the internal work performed by the force in each of the "bars" must equal the external work performed by the loads.

For the following discussion the beam of Figure 11.8(a) on page 318 is considered. Part (b) of the figure shows the beam cross section. It is desired to know the vertical deflection δ at point A in the beam caused by the external loads P_1 to P_3. If the loads were removed from the beam and a vertical unit load placed at A, small forces and deformations would be developed in the "bars," and a small deflection would occur at A. Replacing the external loads would cause increases in the "bar" forces and deformations, and the unit load at A would deflect an additional amount δ. The internal work performed by the unit load forces, as they are carried through the additional "bar" deformations, equals the external work performed by the unit load as it is carried through the additional deflection δ.

The following symbols are used in writing an expression for the internal work performed in a dx length of the beam: M is the moment at any section in the beam due to the external loads and m is the moment at any section due to the unit load. The stress in a differential area of the beam cross section due to the unit load can be found from

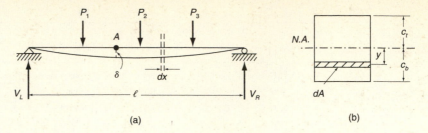

Figure 11.8

(a)   (b)

the flexure formula as follows:

$$\text{stress in } dA = \frac{my}{I}$$

$$\text{total force in } dA = \frac{my}{I} \, dA$$

The dA area has a thickness or length of dx that deforms by $\epsilon \, dx$ when the external loads are returned to the structure. The deformation is as follows:

$$\text{stress due to external loads} = f = \frac{My}{I}$$

$$\text{deformation of } dx \text{ length} = \epsilon \, dx = \frac{f}{E} \, dx = \frac{My}{EI} \, dx$$

The total force in dA due to the unit load $(my/I) \, dA$ is carried through this deformation, and the work it performs is as follows:

$$\text{work in } dA = \left(\frac{my}{I} \, dA\right)\left(\frac{My}{EI} \, dx\right) = \frac{Mmy^2}{EI^2} \, dA \, dx$$

The total work performed on the cross section equals the summation of the work in each dA area in the cross section.

$$\text{work} = \int_{C_b}^{C_t} \frac{Mmy^2}{EI^2} \, dA \, dx = \frac{Mm}{EI^2} \int_{C_b}^{C_t} y^2 \, dA \, dx$$

The expression $\int y^2 \, dA$ is a familiar one, being the moment of inertia of the section, and the equation becomes

$$\text{work} = \frac{Mm}{EI} \, dx$$

It is now possible to determine the internal work performed in the entire beam because it equals the integral from 0 to ℓ of the preceding expression.

$$W_i = \int_0^{\ell} \frac{Mm}{EI} \, dx$$

The external work performed by the unit load as it is carried through the distance δ is $1 \times \delta$. By equating the external work and the internal work, an expression for the deflection at any point in the beam is obtained.

$$W_e = W_i$$

$$1 \times \delta = \int_0^\ell \frac{Mm}{EI}\, dx$$

$$\delta = \int_0^\ell \frac{Mm}{EI}\, dx$$

11.7 EXAMPLE PROBLEMS FOR BEAMS AND FRAMES

Examples 11.3 to 11.7 illustrate the application of virtual work to beams and frames. To apply the method, a unit load is placed at the point and in the direction in which deflection is desired. Expressions are written for M and m throughout the structure and the results are integrated from 0 to ℓ. It is rarely possible to write one expression for M or one expression for m that is correct in all parts of the structure. As an illustration, consider the beam of Figure 11.9 and let the deflection under P_2 be desired. A unit load is placed at this point in the figure. The reactions V_L and V_R are due to loads P_1 and P_2, whereas v_L and v_R are due to the unit load.

The values of M and m are written with respect to the distance x from the left support. From the left support to the unit load m may be represented by one expression, $v_L x$, but the expression for M is not constant for the full distance. Its value is $V_L x$ from the left support to P_1 and $V_L x - P_1(x - a)$ from P_1 to P_2. The integration will be made from 0 to a for $V_L x$ and $v_L x$, and from a to $a + b$ for $V_L x - P_1(x - a)$ and $v_L x$. Moment expressions for all parts of the beam are shown as follows. The left support is used as the origin of x.

For $x = 0$ to a:

$$M = V_L x$$

$$m = v_L x$$

For $x = a$ to $a + b$

$$M = V_L x - P_1(x - a)$$

$$m = v_L x$$

For $x = a + b$ to ℓ

$$M = V_L x - P_1(x - a) - P_2(x - a + b)$$

$$m = v_L x - 1(x - a + b)$$

$$\delta = \int_0^a \frac{Mm}{EI}\, dx + \int_a^{a+b} \frac{Mm}{EI}\, dx + \int_{a+b}^\ell \frac{Mm}{EI}\, dx$$

A positive sign is used for a moment that causes tension in the bottom fibers of a beam. If the result of integration is positive, the direction used for the unit load is the direction of the deflection.

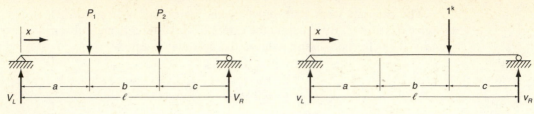

Figure 11.9

EXAMPLE 11.3 _____

Determine the deflection at point A for the beam shown in Figure 11.10 by virtual work.

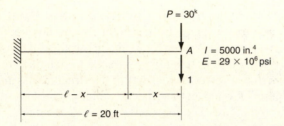

Figure 11.10

Solution.   An expression is written for M (the moment due to the 30-kip load) at any point a distance x from the free end. A unit load is placed at the free end and an expression is written for the moment m it causes at any point. The origin of x may be selected at any point as long as the same point is used for writing M and m for that portion of the beam.

$$M = -Px$$

$$m = -1x$$

$$\delta_A = \int_0^\ell \frac{Mm}{EI}\, dx = \int_0^\ell \frac{(-Px)(-1x)}{EI}\, dx = \int_0^\ell \frac{Px^2}{EI}\, dx$$

$$\delta_A = \frac{P}{EI}\left[\frac{x^3}{3}\right]_0^\ell = \frac{P\ell^3}{3EI}$$

$$\delta_A = \frac{(30{,}000)(20 \times 12)^3}{(3)(29 \times 10^6)(5000)} = 0.955 \text{ in. } \downarrow \quad \blacksquare$$

EXAMPLE 11.4 _____

Determine the deflection at point A on the beam shown in Figure 11.11.

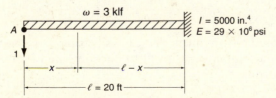

Figure 11.11

Solution.

$$M = -(w)(x)\left(\frac{x}{2}\right) = -\frac{wx^2}{2}$$

$$m = -1x$$

$$Mm = +\frac{wx^3}{2}$$

$$\delta_A = \int_0^\ell \frac{Mm}{EI}\, dx = \int_0^\ell \frac{wx^3}{2EI}\, dx$$

$$= \frac{w}{2EI}\left(\frac{x^4}{4}\right)_0^\ell = \frac{w\ell^4}{8EI}$$

$$= \frac{(3000/12)(20 \times 12)^4}{(8)(29 \times 10^6)(5000)} = 0.714 \text{ in. } \downarrow \quad \blacksquare$$

EXAMPLE 11.5

Determine the deflection at point B in the beam shown in Figure 11.12.

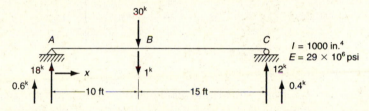

Figure 11.12

Solution. It is necessary to write one expression for M from A to B and another from B to C. The same is true for m. Frequently it is possible to simplify the mathematics by using different origins of x for different sections of the beam. The same results would have been obtained if for the M and m expressions from B to C the origin had been taken at C. The small reactions at A and C are those due to the unit load.

For $x = 0$ to 10:

$$M = 18x$$
$$m = 0.6x$$
$$Mm = 10.8x^2$$

For $x = 10$ to 25:

$$M = 18x - (30)(x - 10) = -12x + 300$$
$$m = 0.6x - 1(x - 10) = -0.4x + 10$$
$$Mm = +4.8x^2 - 240x + 3000$$
$$\delta_B = \int_0^{10} \frac{10.8x^2}{EI}\, dx + \int_{10}^{25} \frac{(4.8x^2 - 240x + 3000)}{EI}\, dx$$

$$= \frac{1}{EI}[3.6x^3]_0^{10} + \frac{1}{EI}[1.6x^3 - 120x^2 + 3000x]_{10}^{25}$$

$$= \frac{3600}{EI} + \frac{25,000 - 19,600}{EI}$$

$$= \frac{9000 \text{ ft}^3\text{-k}}{EI} = \frac{(9000)(1728)(1000)}{(29 \times 10^6)(1000)} = 0.536 \text{ in. } \downarrow \quad \blacksquare$$

EXAMPLE 11.6

Determine the deflection at point B in the beam of Example 10.12, which is reproduced in Figure 11.13.

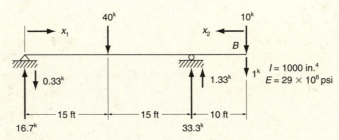

Figure 11.13

Solution.   In writing the moment expressions from the left support to the right support, the left support is used for the origin of x. For the overhanging portion of the beam, the right end is used as the origin.

For $x_1 = 0$ to 15:

$$M = 16.7x$$
$$m = -0.33x$$
$$Mm = -5.56x^2$$

For $x_1 = 15$ to 30:

$$M = 16.7x - (40)(x - 15)$$
$$M = -23.3x + 600$$
$$m = -0.33x$$
$$Mm = +7.77x^2 - 200x$$

For $x_2 = 0$ to 10 coming back from right end

$$M = -10x$$
$$m = -x$$
$$Mm = +10x^2$$

$$\delta_B = \frac{1}{EI}\int_0^{15} (-5.56x^2)\, dx + \frac{1}{EI}\int_{15}^{30} (7.77x^2 - 200x)\, dx + \frac{1}{EI}\int_0^{10} (10x^2)\, dx$$

$$= \frac{1}{EI}[-1.85x^3]_0^{15} + \frac{1}{EI}[2.59x^3 - 100x^2]_{15}^{30} + \frac{1}{EI}[3.33x^3]_0^{10}$$

$$= \frac{-6250}{EI} + \frac{-20,000 + 13,750}{EI} + \frac{3330}{EI}$$

$$= -\frac{9170 \text{ ft}^3\text{-k}}{EI} = -0.546 \text{ in. } \uparrow \quad \blacksquare$$

EXAMPLE 11.7

Find the horizontal deflection at D in the frame shown in Figure 11.14.

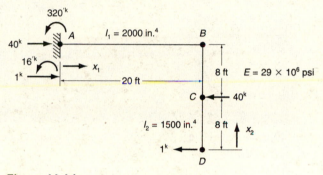

Figure 11.14

Solution.   A unit load acting horizontally to the left is placed at D. With A as the origin for x, moment expressions are written for member AB. For the vertical member, the origin is taken at D, one pair of M and m expressions is written for the portion of the member from D to C, and another pair is written for the portion of the beam from C to B. Members AB and BD do not have the same moments of inertia, and it is necessary to carry them separately as shown in the calculations.

For $x_1 = 0$ to 20:

$$M = -320$$
$$m = -16$$
$$Mm = +5120$$

For $x_2 = 0$ to 8:

$$M = 0$$
$$m = -x$$
$$Mm = 0$$

For $x_2 = 8$ to 16:

$$M = -40(x - 8) = -40x + 320$$
$$m = -x$$
$$Mm = 40x^2 - 320x$$

I-40, I-240 interchange, Oklahoma City, Oklahoma. (Courtesy of the State of Oklahoma Department of Transportation.)

$$\delta_D = \int_0^{20} \left(\frac{5120}{EI_1}\right) dx + \int_8^{16} \left(\frac{40x^2 - 320x}{EI_2}\right) dx$$

$$= \frac{1}{EI_1}[5120x]_0^{20} + \frac{1}{EI_2}\left(\frac{40}{3}x^3 - 160x^2\right)\Bigg|_8^{16}$$

$$= \frac{102,400 \text{ ft}^3\text{-k}}{EI_1} + \frac{17,066 \text{ ft}^3\text{-k}}{EI_2} = +3.73 \text{ in.} \leftarrow \blacksquare$$

Example 11.8 presents one more illustration of deflection calculations using the virtual work procedure. Its purpose is to provide the reader with a little additional practice writing the M and m expressions. For convenience, the inclined 60^k load applied to the structure is broken into its vertical and horizontal components as shown in Figure 11.15. You should note also that the moment expressions are written separately for each of the x_1, x_2, and x_3 directions shown in the figure.

EXAMPLE 11.8 _____

Determine the horizontal deflection at the 60^k load for the frame shown in Figure 11.15. Assume that each of the members has an $I = 1250$ in.[4] and an $E = 29 \times 10^6$ psi.

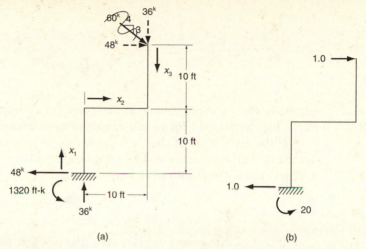

Figure 11.15

Solution.

For $x_1 = 0$ to 10

$M = 48x - 1320$

$m = 1x - 20$

$Mm = 48x^2 - 2280x + 26,400$

For $x_2 = 0$ to 10

$M = 36x + 480 - 1320 = 36x - 840$

$m = 10 - 20 = -10$

$Mm = -360x + 8400$

For $x_3 = 0$ to 10

$M = 48x$

$m = 1x$

$Mm = 48x^2$

After integration

$$EI\delta_{horiz} = (16x^3 - 1140x^2 + 26,400)_0^{10} + (-180x^2 + 8400x)_0^{10} + (16x^3)_0^{10}$$

$$= 248,000 \text{ ft}^3\text{-k}$$

$$\delta_{horiz} = \frac{(248,000)(1728)(1000)}{(29 \times 10^6)(1250)} = 11.82 \text{ in.} \rightarrow \quad \blacksquare$$

11.8 ROTATIONS OR ANGLE CHANGES BY VIRTUAL WORK

Virtual work may be used to determine the rotations or slopes at various points in a structure. On the following page, to find the slope at point A in the beam of Figure 11.16 a unit couple is applied at A, the external loads being removed from the

Figure 11.16

structure. The value of moment at any point in the beam caused by the couple is m. Replacing the external loads will cause an additional moment, at any point, of M.

If the application of the loads causes the beam to rotate through an angle θ at A, the external work performed by the couple equals $1 \times \theta$. The internal work or the internal elastic energy stored is $\int (Mm/EI)\, dx$.

$$\theta = \int \frac{Mm}{EI}\, dx$$

If a clockwise couple is assumed at the position where slope is desired and the result of integration is positive, the rotation is clockwise. Examples 11.9 and 11.10 illustrate the determination of slopes by virtual work. The slopes obtained are in radians.

EXAMPLE 11.9 _____

Find the slope at the free end A of the beam shown in Figure 11.17.

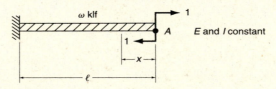

Figure 11.17

Solution.

$$M = -\frac{wx^2}{2}$$

$$m = -1$$

$$Mm = \frac{wx^2}{2}$$

$$\theta_A = \int_0^\ell \frac{wx^2}{2EI}\, dx$$

$$\theta_A = \left[\frac{wx^3}{6EI}\right]_0^\ell = +\frac{W\ell^3}{6EI} \text{ clockwise } \blacksquare$$

EXAMPLE 11.10

Find the slope at the 30-kip load in the beam of Figure 11.18.

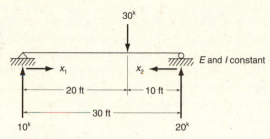

Figure 11.18

Solution.   A separate diagram is drawn for the couple and the reactions it causes. For the beam to the left of the load, the left support is used for the origin of x and the right support is used for the origin for the portion of the beam to the right of the 30-kip load.

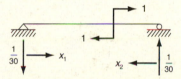

For $x_1 = 0$ to 20:

$$M = 10x$$
$$m = -\tfrac{1}{30}x$$
$$Mm = -\tfrac{1}{3}x^2$$

For $x_2 = 0$ to 10:

$$M = 20x$$
$$m = \tfrac{1}{30}x$$
$$Mm = \tfrac{2}{3}x^2$$

$$\theta = \int_0^{20}\left(-\frac{1x^2}{3EI}\right)dx + \int_0^{10}\left(\frac{2x^2}{3EI}\right)dx$$

$$= \left[-\frac{x^3}{9EI}\right]_0^{20} + \left[\frac{2x^3}{9EI}\right]_0^{10} = -\frac{888 \text{ ft}^2\text{-k}}{EI} + \frac{222 \text{ ft}^2\text{-k}}{EI}$$

$$= -\frac{666 \text{ ft}^2\text{-k}}{EI} \text{ counterclockwise} \quad \blacksquare$$

11.9 MAXWELL'S LAW OF RECIPROCAL DEFLECTIONS

The deflections of two points in a beam have a surprising relationship to each other. The reader may have noticed this relationship, which was first published by James Clerk Maxwell in 1864. Maxwell's law may be stated as follows: **The deflection at one point A in a structure due to a load applied at another point B is exactly the same as the deflection at B if the same load is applied at A.** The rule is perfectly general and applies to any type of structure, whether it is a truss, beam, or frame, which is made up of elastic materials following Hooke's law. The displacements may be caused by flexure, shear, or torsion. In preparing influence lines for continuous structures, in analyzing statically indeterminate structures, and in model-analysis problems this useful tool is frequently applied.

The law is not only applicable to the deflections in all of these types of structures but is also applicable to rotations. For instance, a unit couple at A will produce a rotation at B equal to the rotation caused at A if the same couple is applied at B.

Example 11.11 proves the law to be correct for a simple beam in which the deflections at two points are determined by the conjugate-beam method.

EXAMPLE 11.11

Compute the deflections at points A and B for a load P applied first at B and then at A in Figure 11.19.

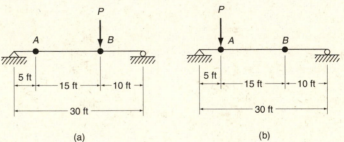

(a)   (b)

Figure 11.19

Solution.

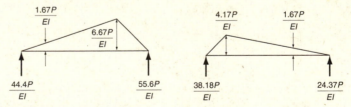

Deflection at A due to load at B:

$$\delta_A = \left(\frac{44.4P}{EI}\right)(5) - \left(\frac{1}{2}\right)\left(\frac{1.67P}{EI}\right)(5)(1.67)$$

$$= \frac{222P}{EI} - \frac{7P}{EI}$$

$$= \frac{215P}{EI}$$

Deflection at B due to load at A:

$$\delta_B = \left(\frac{24.37P}{EI}\right)(10) - \left(\frac{1}{2}\right)\left(\frac{1.67P}{EI}\right)(10)(3.33)$$

$$= \frac{243.7P}{EI} - \frac{28.7P}{EI}$$

$$= \frac{215P}{EI} \quad \blacksquare$$

11.10 INTRODUCTION TO CASTIGLIANO'S THEOREMS

Alberto Castigliano, an Italian railway engineer, published an original and elaborate book in 1879 on the study of statically indeterminate structures.[1] In this book were included the two theorems that are known today as Castigliano's first and second theorems. The first theorem, commonly known as the method of least work, is discussed in Chapter 15. It has played an important role historically in the development of the analysis of statically indeterminate structures.

Castigliano's second theorem, which provides an important method for computing deflections, is considered in this section and the next. Its application involves equating the deflection to the first partial derivative of the total internal work of the structure with respect to a load at the point where deflection is desired. In detail, the theorem can be stated as follows:

Theorem.   *For any linear elastic structure subject to a given set of loads, constant temperature, and unyielding supports, the first partial derivative of the strain energy with respect to a particular force will equal the displacement of that force in the direction of its application.*

In presenting the virtual-work method for beams and frames in Section 11.6, the following equation was derived.

$$W = \int \frac{Mm}{EI}\, dx$$

to express the total internal work accomplished by the stresses caused by a unit load placed at the point and in the direction of a desired deflection when the external loads have been replaced on the beam.

The internal work of the actual stresses caused by gradually applied external loads can be determined in a similar manner. The work in each fiber equals the average stress, $(My/2I)\, da$ (as the external loads vary from zero to full value), times the total strain in the fiber, $(My/EI)\, dx$. Integrating the product of these two expressions for the

---

[1] A. Castigliano, *Théorie de l' équilibre des systèmes élastique et ses applications,* Turin, 1879. (There was an English translation by E. S. Andrews entitled *Elastic Stresses in Structures,* published by Scott, Greenwood & Son in London in 1919. This translation was then reprinted in the United States by Dover Publications, Inc., of New York and entitled *The Theory of Equilibrium of Elastic Systems and Its Application.*)

cross section of the member and throughout the structure gives the total internal work or strain energy stored for the entire structure.

$$W = \int_0^\ell \frac{M^2\,dx}{2EI}$$

In a similar manner the internal work in the members of a truss due to a set of gradually applied loads can be shown to equal

$$W = \Sigma \frac{F^2\ell}{2AE}$$

These expressions will be used frequently in the following section, in which a detailed discussion of Castigliano's second theorem is presented.

11.11 CASTIGLIANO'S SECOND THEOREM

As a general rule the other methods of deflection computation (such as conjugate-beam or virtual-work) are a little easier to use and more popular than Castigliano's second theorem. For certain structures this method is very useful, and from the standpoint of the student its study serves as an excellent background for the extremely important second theorem.

The derivation of the second theorem (presented in this section) is very similar to that given by Kinney.[2] Figure 11.20 shows a beam that has been subjected to the gradually applied loads P_1 and P_2. These loads cause the deflections δ_1 and δ_2. It is desired to find the deflection δ_1 at load P_1.

The external work performed during the application of the loads must equal the average load times the deflection, and it also must equal the internal strain energy of the beam.

$$W = \frac{P_1\delta_1}{2} + \frac{P_2\delta_2}{2} \tag{11.1}$$

Should the load P_1 be increased by the small amount dP_1, the beam will deflect additionally. Figure 11.21 shows the additional deflections at each of the loads $d\delta_1$ and $d\delta_2$.

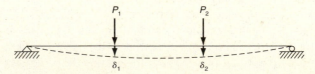

Figure 11.20

[2] J. S. Kinney, *Indeterminate Structural Analysis* (Reading, Mass.: Addison-Wesley, 1957), 84–86.

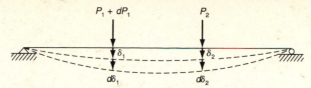

Figure 11.21

The additional work performed or strain energy stored during the application of dP_1 is as follows:

$$dW = \left(P_1 + \frac{dP_1}{2} \right) d\delta_1 + P_2\, d\delta_2$$

Performing the indicated multiplications and neglecting the product of the differentials, we find that

$$dW = P_1\, d\delta_1 + P_2\, d\delta_2 \tag{11.2}$$

The same procedure is repeated except that the loads P_1, P_2, and dP_1 are all applied at the same time and the total strain energy is represented by W'.

$$W' = \left(\frac{P_1 + dP_1}{2} \right)(\delta_1 + d\delta_1) + \left(\frac{P_2}{2} \right)(\delta_2 + d\delta_2)$$

Performing the indicated multiplications and neglecting the product of the differentials,

$$W' = \frac{P_1 \delta_1}{2} + \frac{P_1\, d\delta_1}{2} + \frac{dP_1\, \delta_1}{2} + \frac{P_2 \delta_2}{2} + \frac{P_2\, d\delta_2}{2} \tag{11.3}$$

It is obvious that dW equals $W' - W$ and can be obtained by subtracting Equation (11.1) from (11.3)

$$dW = \frac{P_1 \delta_1}{2} + \frac{P_1\, d\delta_1}{2} + \frac{dP_1\, \delta_1}{2} + \frac{P_2 \delta_2}{2} + \frac{P_2\, d\delta_2}{2} - \frac{P_1 \delta_1}{2} - \frac{P_2 \delta_2}{2}$$

$$= \frac{P_1\, d\delta_1}{2} + \frac{dP_1\, \delta_1}{2} + \frac{P_2\, d\delta_2}{2} \tag{11.4}$$

From Equation (11.2) the value of $P_2\, d\delta_2$ can be obtained as follows:

$$P_2\, d\delta_2 = dW - P_1\, d\delta_1$$

By substituting this value into Equation (11.4) and solving for the desired deflection δ_1

$$dW = \frac{P_1\, d\delta_1}{2} + \frac{dP_1\, \delta_1}{2} + \frac{dW}{2} - \frac{P_1\, d\delta_1}{2}$$

$$\delta_1 = \frac{dW}{dP_1}$$

Burro Creek Bridge, Wickieup, Mohave County, Arizona. (Courtesy of the American Institute of Steel Construction, Inc.)

Since more than one action is applied to a structure, this deflection usually is written as a partial derivative as follows:

$$\delta = \frac{\partial W}{\partial P}$$

To determine the deflection at a point in a beam or frame the second theorem would be written as

$$\delta = \frac{\partial}{\partial P} \int \frac{M^2}{2EI} \, dx$$

Should the procedure represented in the preceding expression be followed, M would be squared, integrated, and the first partial derivative would be taken. If M is rather complicated, however, as it frequently is, the process would be very tedious. For this reason usually it is simpler to differentiate under the integral sign with the following results:

$$\delta = \int M \left(\frac{\partial M}{\partial P} \right) \frac{dx}{EI}$$

For a truss a similar procedure is advisable, with the following results:

$$\delta = \frac{\partial}{\partial P} \Sigma \frac{F^2 \ell}{2AE}$$

$$\delta = \Sigma F \left(\frac{\partial F}{\partial P} \right) \frac{\ell}{AE}$$

Examples 11.12 through 11.15 illustrate the application of Castigliano's second theorem. In applying the theorem, the load at the point where the deflection is desired is referred to as P. After the operations required in the equation are completed, the numercial value of P is replaced in the expression. Should there be no load at the point or in the direction in which deflection is desired, an imaginary force P will be placed there in the direction desired. After the operation is completed, the correct value of P (zero) will be substituted in the expression (see Example 11.13).

If slope or rotation is desired in a structure, the partial derivative is taken with respect to an assumed moment P acting at the point where rotation is desired (see Example 11.15). A plus sign on the answer indicates that the rotation is in the assumed direction of the moment P.

EXAMPLE 11.12 _____

Determine the vertical deflection at the 10-kip load in the beam shown in Figure 11.22.

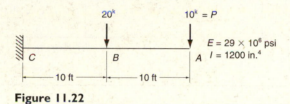

Figure 11.22

Solution.   By letting the 10-kip load be P.

TABLE 11.3

Section	M	$\dfrac{\partial M}{\partial P}$	$M\dfrac{\partial M}{\partial P}$	$\displaystyle\int M\left(\dfrac{\partial M}{\partial P}\right)\dfrac{dx}{EI}$
A to B	$-Px$	$-x$	$+Px^2$	$\displaystyle\int_0^{10}(Px^2)\dfrac{dx}{EI}=\dfrac{333P}{EI}$
B to C	$-P(x+10)-20x$	$-x-10$	$+Px^2+20Px+20x^2$ $+100P+200x$	$\displaystyle\int_0^{10}(Px^2+20Px+20x^2$ $+100P+200x)\dfrac{dx}{EI}$ $=\dfrac{2333P+16{,}667}{EI}$
Σ				$\dfrac{2666P+16{,}667}{EI}$

$$\delta=\frac{2666P+16{,}667}{EI}=\frac{43{,}333\ \text{ft}^3\text{-k}}{EI}=2.15\ \text{in.}\downarrow\quad\blacksquare$$

EXAMPLE 11.13 —————————————————————————————————————

Determine the vertical deflection at the free end of the cantilever beam shown in Figure 11.23.

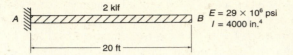

Figure 11.23

Solution.   Placing an imaginary load P at the free end.

TABLE 11.4

Section	M	$\dfrac{\partial M}{\partial P}$	$M\dfrac{\partial M}{\partial P}$	$\displaystyle\int M\left(\dfrac{\partial M}{\partial P}\right)\dfrac{dx}{EI}$
B to A	$-Px - x^2$	$-x$	$Px^2 + x^3$	$\displaystyle\int_0^{20}(Px^2 - x^3)\dfrac{dx}{EI} = \dfrac{40{,}000}{EI}$
Σ				$\dfrac{40{,}000}{EI}$

$$\delta = 0.596 \text{ in.} \downarrow \quad \blacksquare$$

EXAMPLE 11.14 —————————————————————————————————————

Determine the vertical deflection at the 30-kip load in the beam of Figure 11.24.

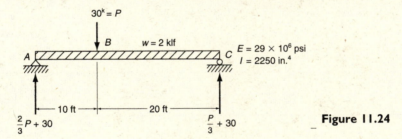

Figure 11.24

Solution.   The 30-kip load is replaced with P. Note that the beam reactions are computed with a load of P at point B and not 30 kips.

TABLE 11.5

Section	M	$\dfrac{\partial M}{\partial P}$	$M\dfrac{\partial M}{\partial P}$	$\displaystyle\int M\left(\dfrac{\partial M}{\partial P}\right)\dfrac{dx}{EI}$
A to B	$\dfrac{2}{3}Px + 30x - x^2$	$\dfrac{2}{3}x$	$\dfrac{4}{9}Px^2 + 20x^2 - \dfrac{2}{3}x^3$	$\dfrac{1}{EI}\displaystyle\int_0^{10}\left(\dfrac{4}{9}Px^2 + 20x^2 - \dfrac{2}{3}x^3\right)dx$
C to B	$\dfrac{P}{3}x + 30x - x^2$	$\dfrac{x}{3}$	$\dfrac{Px^2}{9} + 10x^2 - \dfrac{x^3}{3}$	$\dfrac{1}{EI}\displaystyle\int_0^{20}\left(\dfrac{Px^2}{9} + 10x^2 - \dfrac{x^3}{3}\right)dx$

Performing the indicated operations $\delta = 0.838$ in. $\downarrow$ ■

EXAMPLE 11.15

Find the slope at the free end of the cantilever beam of Example 11.12. The beam is drawn again in Figure 11.25.

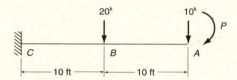

Figure 11.25

Solution.   A clockwise moment of P is assumed to be acting at the free end as shown in the figure.

TABLE 11.6

Section	M	$\dfrac{\partial M}{\partial P}$	$M\dfrac{\partial M}{\partial P}$	$\int M\left(\dfrac{\partial M}{\partial P}\right)\dfrac{dx}{EI}$
A to B	$-10x - P$	-1	$10x + P$	$\dfrac{1}{EI}\displaystyle\int_0^{10}(10x + P)\,dx$
B to C	$-30x - P - 100$	-1	$30x + P + 100$	$\dfrac{1}{EI}\displaystyle\int_0^{10}(30x + P + 100)\,dx$

Performing the indicated operations $\theta = +0.0124$ rad↺ ∎

PROBLEMS

For Problems 11.1 through 11.6 use the virtual-work method to determine the deflection of each of the joints marked on the trusses shown in the accompanying illustrations. The circled figures are areas in square inches and $E = 29 \times 10^6$ psi unless otherwise indicated.

11.1 U_2 vertical, U_2 horizontal

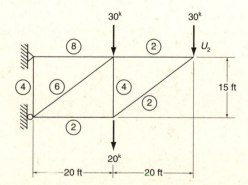

11.2 L_2 vertical, L_4 horizontal (*Ans.* $L_2 = 1.64$ in. ↓, $L_4 = 0.552$ in. →)

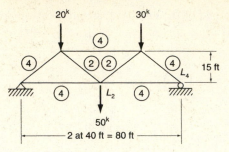

11.3 U_2 vertical, U_1 horizontal

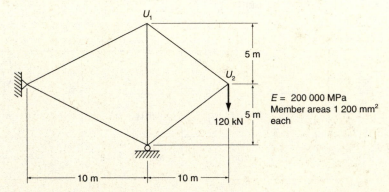

$E = 200\ 000$ MPa
Member areas 1 200 mm²
each

11.4 L_3 vertical (*Ans.* $L_3 = 5.54$ in. ↓)

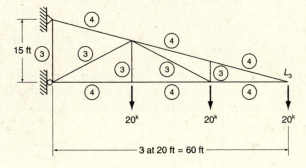

11.5 U_1 vertical, L_1 horizontal

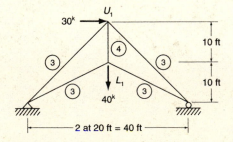

11.6 L_1 vertical, L_1 horizontal (*Ans.* L_1 = 0.143 in. ↓, 0.130 in. →)

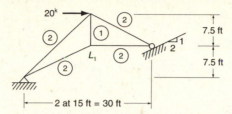

Use the virtual-work method for solving Problems 11.7 through 11.30. E = 29 × 10^6 psi for all problems unless otherwise indicated.

11.7   Determine the slope and deflection at points A and B of the structure shown in the accompanying illustration. I = 2000 in.^{4}

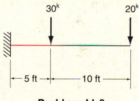

11.8   Determine the slope and deflection underneath each of the concentrated loads shown in the accompanying illustration. I = 3500 in.^{4} (*Ans.* θ_{20} = 0.00372 rad\, θ_{30} = 0.00231 rad\, δ_{20} = 0.468 in. ↓, δ_{30} = 0.078 in. ↓)

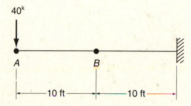

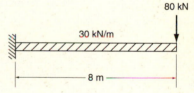

Problem 11.8   Problem 11.9

11.9   Find the slope and deflection at the free end of the beam shown in the accompanying illustration. I = 2.35 × 10^9 mm^4, E = 200 000 MPa.

11.10 Find the deflections at points A and B on the beam shown in the accompanying illustration. I = 3500 in.^{4} (*Ans.* δ_A = 1.61 in. ↓, δ_B = 0.553 in. ↓)

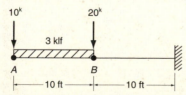

11.11 Rework Problem 10.22.

11.12 Determine the slope and deflection 10 ft from the right end of the beam of Problem 10.22 (*Ans.* θ = 0.00153 rad/, δ = 0.257 in. ↓)

11.13 Find the slope and deflection at the 20-kip load in the beam of Problem 10.24.

11.14   Calculate the slope and deflection at a point 3 m from the left support of the beam shown in the accompanying illustration. $I = 2.5 \times 10^8$ mm⁴, $E = 200\ 000$ MPa. (*Ans.* $\theta = 0.018\ 9$ rad\, $\delta = 84.7$ mm ↓)

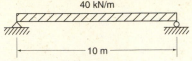

11.15   Rework Problem 10.42.

11.16   Rework Problem 10.44 (*Ans.* $\theta = 0.00234$ rad\, $\delta = 0.157$ in. ↓)

11.17   Rework Problem 10.54.

11.18   Determine the deflection at the centerline and the slope at the right end of the beam shown in the accompanying illustration. $I = 2100$ in.⁴ (*Ans.* $\delta = 0.419$ in. ↓, $\theta = 0.00244$ rad/)

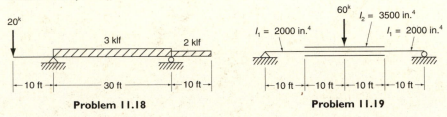

Problem 11.18   Problem 11.19

11.19   Calculate the deflection at the 60-kip load in the structure shown in the accompanying illustration.

11.20   Calculate the slope and deflection at the 60-kip load on the beam shown in the accompanying illustration. (*Ans.* $\theta_{60} = 0.00069$ rad/, $\delta_{60} = 0.0274$ in. ↓)

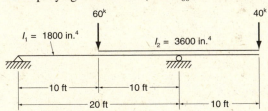

11.21   Calculate the slope and deflection at the 60-kN load on the structure shown in the accompanying illustration. $I = 1.46 \times 10^9$ mm⁴. $E = 200\ 000$ MPa.

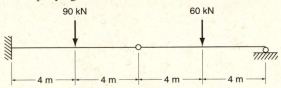

11.22   Find the slope and deflection 20 ft from the left end of the simple beam shown in the accompanying illustration. $I = 2250$ in.⁴ (*Ans.* $\theta_{20} = 0.00221$ rad/, $\delta_{20} = 0.459$ in. ↓)

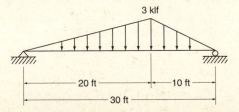

11.23 Calculate the slope and deflection at the free end of the beam shown in the accompanying illustration. $I = 3200$ in.^{4}

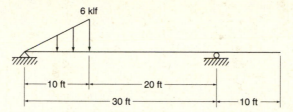

11.24 Find the vertical deflection at point A and the horizontal deflection at point B on the frame shown in the accompanying illustration. $I = 4000$ in.^{4} (*Ans.* $\delta_A = 0.92$ in. $\downarrow$, $\delta_B = 0.298$ in. $\leftarrow$)

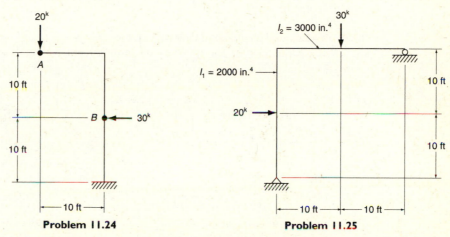

Problem 11.24   Problem 11.25

11.25 Determine the horizontal deflection at the 20-kip load and the vertical deflection at the 30-kip load on the frame shown in the accompanying illustration.

11.26 Find the vertical and horizontal components of deflection at the 30-kip load in the frame shown in the accompanying illustration. $I = 1500$ in.^{4} (*Ans.* $\delta_H = 4.37$ in. $\rightarrow$, $\delta_V = 2.38$ in. $\downarrow$)

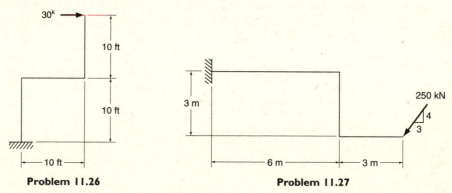

Problem 11.26   Problem 11.27

11.27 Determine the vertical and horizontal components of deflection at the free end of the frame shown in the accompanying illustration. $I = 2.60 \times 10^9$ mm^4. $E = 200\ 000$ MPa.

11.28   Find the horizontal deflection at point A and the vertical deflection at point B on the frame shown in the accompanying illustration. $I = 1200$ in.[4] (*Ans.* $\delta_B = 2.98$ in. $\leftarrow$, $\delta_n = 1.99$ in. $\uparrow$)

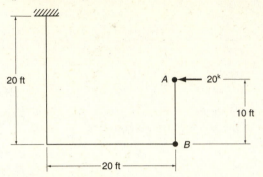

11.29   Calculate the horizontal deflection at the roller support and the vertical deflection at the 30-kip load for the frame shown in the accompanying illustration. $I = 1500$ in.[4]

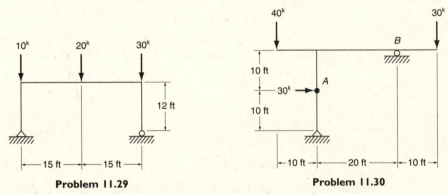

Problem 11.29   **Problem 11.30**

11.30   Find the horizontal deflections at points A and B on the frame shown in the accompanying illustration. $I = 3000$ in.[4] (*Ans.* $\delta_A = 0.463$ in. $\rightarrow$, $\delta_B = 0.430$ in. $\rightarrow$)

For Problems 11.31 through 11.46, use Castigliano's second theorem to determine the slopes and deflections indicated. $E = 29 \times 10^6$ psi for the problems having customary units and 200 000 MPa for those with SI units. Similarly circled values are cross-sectional areas in square inches or in square millimeters.

11.31   Deflection at each load; $I = 3250$ in.[4]

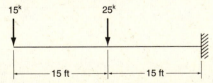

11.32   Slope and deflection at free end; $I = 1750$ in.[4] (*Ans.* $\theta = 0.00638$ rad\, $\delta = 1.53$ in. $\downarrow$)

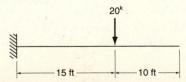

11.33   Slope and deflection at 30^k load; $I = 5500$ in.^{4}

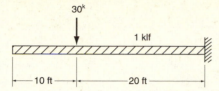

11.34   Deflection at each end of uniform load; $I = 1250$ in.^{4} (*Ans.* $\delta_{20\,ft} = 2.44$ in. $\downarrow$, $\delta_{10\,ft} = 0.83$ in. $\downarrow$)

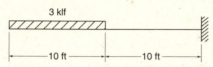

11.35   Deflection at each load; $I = 4.0 \times 10^8$ mm^4

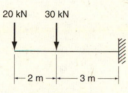

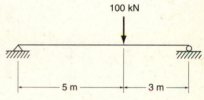

Problem 11.35   Problem 11.36

11.36   Slope and deflection at 100-kN load. $I = 1.798 \times 10^8$ mm^4 (*Ans.* $\theta = -0.003\,47$ rad/, $\delta = 26.12$ mm $\downarrow$)

11.37   Rework Problem 10.13.

11.38   Deflection at each load; $I = 1750$ in.^{4} (*Ans.* $\delta_{20k} = 2.16$ in. $\downarrow$, $\delta_{40k} = 1.65$ in. $\downarrow$)

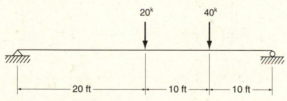

11.39   Slope and deflection at 60^k load: $I = 2700$ in.^{4}

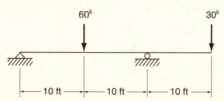

11.40   Rework Problem 10.53 (*Ans.* $\theta = 0.00144$ rad\, $\delta = 1.74$ in. $\uparrow$)

11.41   Rework Problem 11.26.

11.42   Rework Problem 11.23 (*Ans.* $\theta = -0.000772$ rad/, $\delta = 0.0868$ in. $\uparrow$)

11.43   Rework Problem 11.27.

11.44   Vertical deflection at joint L_1 (*Ans.* $\delta = 0.652$ in. ↓)

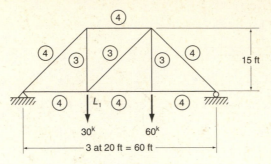

11.45   Vertical and horizontal deflection at L_0

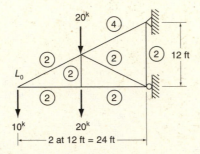

11.46   Vertical deflection at L_1 and horizontal deflection at L_2 (*Ans.* $\delta_v = 45.2$ mm ↓, $\delta_H = 21$ mm →)

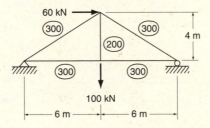

Chapter 12

Introduction to Statically Indeterminate Structures

12.1 GENERAL

When a structure has too many external reactions and/or internal forces to be determined with the equations of statics (including any equations of condition), it is statically indeterminate. A load placed on one part of a statically indeterminate or continuous structure will cause shears, moments, and deflections in the other parts of the structure. In other words, loads applied to a column affect the beams, slabs, and other columns and vice versa.

Up to this point the text has been so completely devoted to statically determinate structures that the reader may have been falsely led to believe that statically determinate beams and trusses are the rule in modern structures. The truth is that it is a difficult task to find an ideal simply supported beam. Probably the best place to look for one would be in a textbook on structures, for bolted or welded beam-to-column connections do not produce ideal simple supports with zero moments.

The same situation holds for statically determinate trusses. Bolted or welded joints are not frictionless pins, as previously assumed for analysis purposes. The other assumptions made about trusses in the earlier chapters of this book are not altogether true, and thus in a strict sense all trusses are statically indeterminate because they have some bending and secondary forces.

Almost all reinforced-concrete structures are statically indeterminate. The concrete for a whole or a large part of a concrete floor, including the support beams and girders and parts of the columns, may be placed at the same time. The reinforcing bars extend from member to member, as from one span of a beam into the next. When there are construction joints, the reinforcing bars are left protruding from the older concrete so they may be lapped or spliced to the bars in the newer concrete. In addition, the old concrete is cleaned and perhaps roughened so that the newer concrete will bond or stick to it as well as possible. The result of all these facts is that reinforced-concrete structures are generally monolithic or continuous and thus are statically indeterminate.

About the only way a statically determinate reinforced-concrete structure could be built would be with individual sections precast at a concrete plant and assembled on the job site. Even these structures could have some continuity in their joints.

Up until the early part of the 20th century statically indeterminate structures were avoided as much as possible by most American engineers. But three great developments completely changed the picture: 1. monolithic reinforced-concrete structures, 2. arc welding of steel structures, and 3. modern methods of analysis.

12.2 CONTINUOUS STRUCTURES

As the spans of simple structures become longer, their bending moments increase rapidly. If the weight of a structure per unit length remained constant, regardless of the span, the dead-load moment would vary in proportion to the square of the span length ($M = w\ell^2/8$). This proportion, however, is not correct, because the weight of structures must increase with longer spans to be strong enough to resist the increased bending moments; therefore, the dead-load moment increases at a greater rate than does the square of the span.

For economy, it pays in long spans to introduce types of structures that have smaller moments than the tremendous ones that occur in long-span, simply supported structures. Chapter 4 introduced one type of structure that considerably reduced bending moments: cantilever-type construction. Two other moment-reducing structures are discussed in the following paragraphs.

In some locations it may be possible to have a beam with fixed ends, rather than one with simple supports. A comparison of the moments developed in a uniformly loaded simple beam with those in a uniformly loaded, fixed-ended beam is made in Figure 12.1.

The maximum bending moment for the fixed-ended beam is only two thirds of that for the simply supported beam. Usually it is difficult to fix the ends, particularly in the case of bridges; for this reason, flanking spans are often used, as illustrated in Figure 12.2. These spans will partially restrain the interior supports, thus tending to reduce the moment in the center span. This figure compares the bending moments that occur in three uniformly loaded simple beams (spans 60, 100, and 60 ft) with the

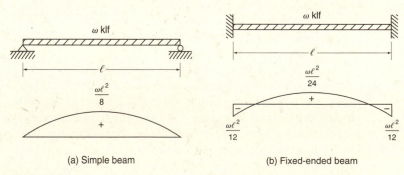

(a) Simple beam   (b) Fixed-ended beam

Figure 12.1

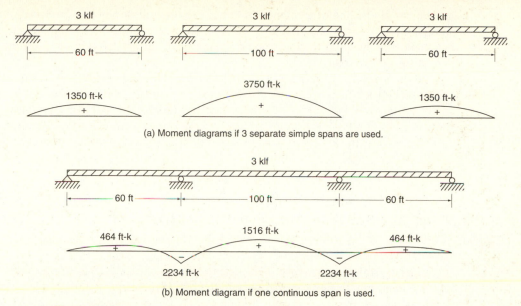

(a) Moment diagrams if 3 separate simple spans are used.

(b) Moment diagram if one continuous span is used.

Figure 12.2

moments of a uniformly loaded beam continuous over the same three spans.

The maximum bending moment for the continuous beam is approximately 40% less than that for the simple beams. Unfortunately, there will not be a corresponding 40% reduction in total cost. The cost-reduction probably is only a small percentage of the total cost of the structure because such items as foundations, connections, and floor systems are not reduced a great deal by the moment reduction. Varying

Colorado River arch bridge, Utah Route 95. (Courtesy of the Utah Department of Transportation.)

the lengths of the flanking spans will change the magnitude of the largest moment occurring in the continuous member. (For a constant uniform load over the 3 spans, the smallest moment will occur when the flanking spans are from 0.3 to 0.4 times as long as the center span.)

In the foregoing discussion, the moments developed in beams have been shown to be reduced appreciably by continuity. This reduction occurs where beams are rigidly fastened to each other or where beams and columns are rigidly connected. There is a continuity of action in resisting a load applied to any part of a continuous structure because the load is resisted by the combined efforts of all the members of the frame.

12.3 ADVANTAGES OF STATICALLY INDETERMINATE STRUCTURES

In comparing statically indeterminate structures with statically determinate ones, the first consideration, to most, would pertain to cost. However, it is impossible to make a statement favoring one type, economically, without reservation. Each structure presents a different situation, and all factors must be considered—economic or otherwise. In general, statically indeterminate structures have the advantages discussed in the following paragraphs.

Savings in Materials

The smaller moments developed permit the use of smaller members. For bridges the material saving could possibly be as high as 10 to 20%. The large number of force reversals occurring in railroad bridges keeps their maximum saving nearer 10%.

A structural member of a given size can support more load if it is part of a continuous structure than if it is simply supported. The continuity permits the use of smaller members for the same loads and spans or increased spacing of supports for the same size members. The possibility of fewer columns in buildings or fewer piers in bridges may permit a reduction in overall costs.

Continuous structures of concrete or steel are cheaper without the joints, pins, and so on required to make them statically determinate, as was frequently the practice in past years. Monolithic reinforced-concrete structures are erected so that they are naturally continuous and statically indeterminate. To install the hinges and other devices necessary to make them statically determinate would not only be a difficult construction problem but also be very expensive. Furthermore, if a building frame consisted of columns and simple beams, it would be necessary to have objectionable diagonal bracing between the joints to make the frame stable and rigid.

Larger Safety Factors

Statically indeterminate structures often have higher safety factors than statically determinate ones. The student, in his or her studies of statically indeterminate structural steel or reinforced-concrete structures, will learn that when portions of those structures are overstressed they will often have the ability to redistribute portions of those stresses to less-stressed areas. Statically determinate structures generally do not have this ability.[1]

Should the moment in a statically indeterminate beam or frame reach the ultimate moment capacity of the structure at a particular point, the structure will fail. This is not the case for statically indeterminate structures.

It can be clearly shown that a statically indeterminate beam or frame normally will not collapse when its ultimate moment capacity is reached at just one section. Instead, there is a redistribution of the moments in the structure. Its behavior is rather similar to the case in which three men are walking along with a log on their shoulders and one of the men gets tired and lowers his shoulder just a little. The result is a redistribution of loads to the other men and thus changes in the shears and moments along the log.

Greater Rigidity and Smaller Deflections

Statically indeterminate structures are more rigid than statically determinate ones and have smaller deflections. Because of their continuity, they are stiffer and have greater stability against all types of loads (horizontal, vertical, moving, etc.).

More Attractive Structures

It is difficult to imagine statically determinate structures having the gracefulness and beauty of many statically indeterminate arches and rigid frames being erected today.

Broadway Street Bridge showing cantilever erection, Kansas City, Missouri. (Courtesy of USX Corporation.)

[1] J. C. McCormac, *Structural Steel Design LRFD Method* 2nd ed. (New York: HarperCollins, 1995), 225–235.

Adaptation to Cantilever Erection

The cantilever method of erecting bridges is of particular value where conditions underneath (probably marine traffic or deep water) hinder the erection of falsework. Continuous statically indeterminate bridges and cantilever-type bridges are conveniently erected by the cantilever method.

12.4 DISADVANTAGES OF STATICALLY INDETERMINATE STRUCTURES

A comparison of statically determinate and statically indeterminate structures shows the latter have several disadvantages that make their use undesirable on many occasions. They are discussed in the following paragraphs.

Support Settlement

Statically indeterminate structures are not desirable where foundation conditions are poor, because seemingly minor support settlements or rotations may cause major changes in the moments, shears, reactions, and bar forces. Where statically indeterminate bridges are used despite the presence of poor foundation conditions, it is occasionally felt necessary to physically measure the dead-load reactions. The supports of the bridge are jacked up or down until the calculated reaction is obtained, after which the support is built to that elevation.

Development of Other Stresses

Support settlement is not the only condition that causes stress variations in statically indeterminate structures. Variation in the relative positions of members caused by temperature changes, poor fabrication, or internal deformation of members of the structure under load may cause serious force changes throughout the structure.

Difficulty of Analysis and Design

The forces in statically indeterminate structures depend not only on their dimensions but also on their elastic properties (moduli of elasticity, moments of inertia, and cross-sectional areas). This situation presents a design difficulty: The forces cannot be determined until the member sizes are known, and the member sizes cannot be determined until their forces are known. The problem is handled by assuming member sizes and computing the forces, designing the members for these forces and computing the forces for the new sizes, and so on, until the final design is obtained. Design by this method—*the method of successive approximations*—takes more time than the design of a comparable statically determinate structure, but the extra cost is only a small part of the total cost of the structure. Such a design is best done by interaction between the designer and the computer. Interactive computing is now used extensively in the aircraft and automobile industries.

Force Reversals

Generally, more force reversals occur in statically indeterminate structures than in statically determinate ones. Additional material may be required at certain sections to resist the different force conditions and to prevent fatigue failures.

12.5 LOOKING AHEAD

In Part II of this book (Chapters 13–16) several classical methods of analyzing statically indeterminate structures are presented. These include the method of consistent distortions, the three-moment theorem, the least-work method of Castigliano, and slope deflection. These methods are primarily of historical interest and are almost never used in practice.

The authors have placed in Part III (Chapters 17–21) methods for analyzing statically indeterminate structures that are frequently used today by the structural engineering profession. First, several approximate methods of analysis are introduced, and then moment distribution and the matrix methods are discussed at length.

PART TWO

STATICALLY INDETERMINATE STRUCTURES—CLASSICAL METHODS

Chapter 13

Force Methods of Analyzing Statically Indeterminate Structures

13.1 METHODS OF ANALYZING STATICALLY INDETERMINATE STRUCTURES

There are two general approaches used for analysis of statically indeterminate structures: *force* methods and *displacement* methods. These are discussed in this section.

Force Methods   With these methods a sufficient number of redundants (reactions and/or member forces) are assumed to be removed from a statically indeterminate structure so that a stable and statically determinate structure remains. The deformations at and in the direction of the removed redundants are computed. The redundants must be sufficiently large to force the redundant points back to zero deformations. One equation is written for the deformation condition at each redundant and the resulting equations solved for the redundants. The remaining sections of this chapter are devoted to the force method (which also is called the *flexibility* method or the *compatibility* method). Statically indeterminate beams, frames, and trusses are considered, as well.

James Clerk Maxwell published in 1864 the first consistent force method for analyzing statically indeterminate structures. His method was based on a consideration of deflections, but the presentation (which included the reciprocal deflection theorem) was rather brief and attracted little attention. Ten years later Otto Mohr independently extended the theory to almost its present stage of development. Analysis of redundant structures with the use of deflection computations often is referred to as the *Maxwell-Mohr method* or the *method of consistent distortions.*[1,2]

[1] J. I. Parcel and R. B. B. Moorman, *Analysis of Statically Indeterminate Structures* (New York: Wiley, 1955), 48.

[2] J. S. Kinney, *Indeterminate Structural Analysis* (Reading, Mass.: Addison-Wesley, 1957), 12–13.

The force methods of structural analysis are somewhat useful for analyzing beams, frames, and trusses that are statically indeterminate to the first or second degree. They also are convenient for some one-story frames with unusual dimensions. For structures that are highly statically indeterminate, such as multistory buildings, the methods such as moment distribution and the matrix methods that are introduced in later chapters are clearly more satisfactory.

The force methods have been almost completely superseded by the modern methods of analysis described in Part III of this text. Perhaps, however, a study of the force methods will provide a clear understanding of the behavior of statically indeterminate structures that might not be obtained if only the modern methods are studied.

Displacement or Stiffness Methods   In the displacement methods of analysis the displacement of the joints (rotations and translations) necessary to describe fully the deformed shape of the structure are used in the equations instead of the redundant actions as used in the force methods. When the simultaneous equations are solved, these displacements are determined and substituted into the original equations to determine the various internal forces. Displacement methods are introduced in Chapters 20 and 21.

13.2 BEAMS AND FRAMES WITH ONE REDUNDANT

The two-span beam of Figure 13.1(a) is assumed to consist of a material following Hooke's law. This statically indeterminate structure supports the loads P_1 and P_2 and is in turn supported by reaction components at points A, B, and C. Removal of support B would leave a statically determinate beam, proving the structure to be statically indeterminate to the first degree. It is a simple matter to find the deflection at B, δ_B in Figure 13.1(b), caused by the external loads.

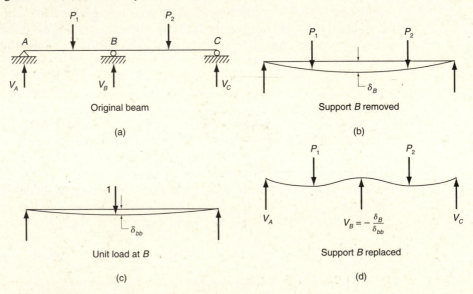

Original beam

(a)

Support B removed

(b)

Unit load at B

(c)

Support B replaced

(d)

Figure 13.1

I-180 viaduct, Cheyenne, Wyoming. (Courtesy of the Wyoming State Highway Department.)

If the external loads are removed from the beam and a unit load is placed at B, a deflection at B equal to δ_{bb} will be developed, as indicated in Figure 13.1(c). Deflections due to external loads are denoted with capital letters herein. The deflection at point C on a beam due to external loads would be δ_C. Deflections due to the imaginary unit load are denoted with two small letters. The first letter indicates the location of the deflection and the second letter indicates the location of the unit load. The deflection at E caused by a unit load at B would be δ_{eb}. Displacements due to unit loads often are called *flexibility coefficients*.

Support B is unyielding, and its removal is merely a convenient assumption. An upward force is present at B and is sufficient to prevent any deflection, or, continuing with the fictitious line of reasoning, there is a force at B that is large enough to push point B back to its original nondeflected position. The distance the support must be pushed is δ_B.

A unit load at B causes a deflection at B equal to δ_{bb}, and a 10-kip load at B will cause a deflection of $10\delta_{bb}$. Similarly, an upward reaction at B of V_B will push B up an amount $V_B\delta_{bb}$. The total deflection at B due to the external loads and the reaction V_B is zero and may be expressed as follows:

$$\delta_B + V_B\delta_{bb} = 0$$

$$V_B = -\frac{\delta_B}{\delta_{bb}}$$

The minus sign in this expression indicates V_B is in the opposite direction from the downward unit load. If the solution of the expression yields a positive value, the reaction is in the same direction as the unit load. Examples 13.1 to 13.4 illustrate the force method of computing the reactions for statically indeterminate beams having one redundant reaction component. Example 13.5 shows that the method may be extended to include statically indeterminate frames as well. The necessary deflections for the first four examples are determined with the conjugate-beam procedure, while those for Example 13.5 are determined by virtual work. After the value of the redundant reaction in each problem is found, the other reactions are determined by statics, and shear and moment diagrams are drawn.

EXAMPLE 13.1

Determine the reactions and draw shear and moment diagrams for the two-span beam of Figure 13.2; assume V_B to be the redundant; E and I are constant.

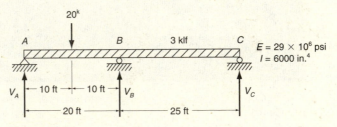

Figure 13.2

Solution.   Drawing the M/EI diagrams shown on page 357.

$$EI\delta_B = (20)(11{,}400) - (\tfrac{1}{2})(20)(750)(6.67) - \frac{(3)(20)^3}{12}(10)$$

$$+ (1430)(25) - (\tfrac{1}{2})(25)(111)(8.33)$$

$$EI\delta_B = 182{,}100 \text{ ft}^3\text{-k}$$

$$EI\delta_{bb} = (20)(130) - (\tfrac{1}{2})(20)(11.1)(6.67)$$

$$EI\delta_{bb} = 1860 \text{ ft}^3\text{-k}$$

$$V_B = -\frac{\delta_B}{\delta_{bb}} = -\frac{182{,}100}{1860} = -98^k \uparrow$$

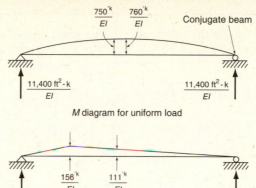

$\dfrac{750^{'k}}{EI}$ $\dfrac{760^{'k}}{EI}$ Conjugate beam

$\dfrac{11,400\ ft^2\text{-}k}{EI}$ $\dfrac{11,400\ ft^2\text{-}k}{EI}$

M diagram for uniform load

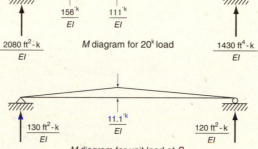

$\dfrac{156^{'k}}{EI}$ $\dfrac{111^{'k}}{EI}$

$\dfrac{2080\ ft^2\text{-}k}{EI}$ *M* diagram for 20^k load $\dfrac{1430\ ft^4\text{-}k}{EI}$

$\dfrac{130\ ft^2\text{-}k}{EI}$ $\dfrac{11.1^{'k}}{EI}$ $\dfrac{120\ ft^2\text{-}k}{EI}$

M diagram for unit load at *B*

Computing reactions at *A* and *C* by statics,

$$\Sigma M_A = 0$$
$$(20)(10) + (45)(3)(22.5) - (20)(98) - 45V_C = 0$$
$$V_C = 28.5^k \uparrow$$
$$\Sigma V = 0$$
$$20 + (3)(45) - 98 - 28.5 - V_A = 0$$
$$V_A = 28.5^k \uparrow$$

Shear and moment diagrams:

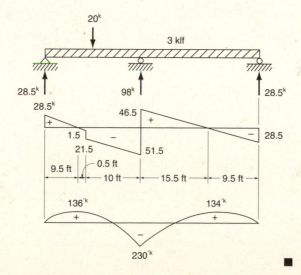

20^k 3 klf

28.5^k 98^k 28.5^k

28.5^k 46.5

1.5   +

21.5   51.5

9.5 ft   0.5 ft   10 ft   15.5 ft   9.5 ft

$136^{'k}$ $134^{'k}$

$230^{'k}$

■

EXAMPLE 13.2

Determine the reactions and draw shear and moment diagrams for the propped beam shown in Figure 13.3. Consider V_B to be the redundant; E and I are constant.

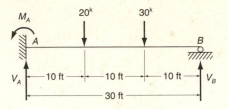

Figure 13.3

Solution.

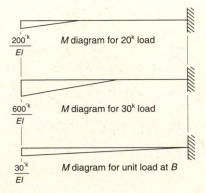

$EI\delta_B = (\frac{1}{2})(200)(10)(26.67) + (\frac{1}{2})(600)(20)(23.33) = 166{,}670 \text{ ft}^3\text{-k}$

$EI\delta_{bb} = (\frac{1}{2})(30)(30)(20) = 9000 \text{ ft}^3\text{-k}$

$$= -\frac{166{,}670}{9000} = -18.5^k \uparrow$$

By statics

$$V_A = 31.5^k \uparrow \text{ and } M_A = 245'^k$$

Shear and moment diagrams:

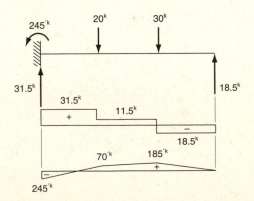

■

EXAMPLE 13.3

Rework Example 13.2 by using the resisting moment at the fixed end as the redundant.

Solution.   Any one of the reactions may be considered to be the redundant and taken out, provided a stable structure remains. If the resisting moment at *A* is removed, a simple support remains and the beam loads cause the tangent to the elastic curve to rotate an amount θ_A. A brief discussion of this condition will reveal a method of determining the moment.

The value of θ_A equals the shear at *A* in the conjugate beam loaded with the *M/EI* diagram. If a unit moment is applied at *A*, the tangent to the elastic curve will rotate an amount θ_{aa}, which can also be found from the conjugate beam. The tangent to the elastic curve at *A* actually does not rotate; therefore, when M_A is replaced, it must be of sufficient magnitude to rotate the tangent back to its original horizontal position. The following expression equating θ_A to zero may be written and solved for the redundant M_A.

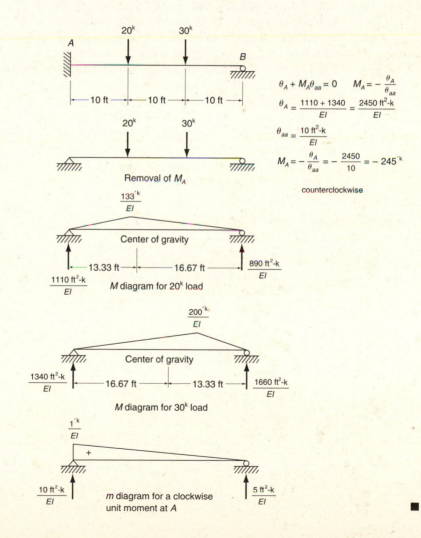

$$\theta_A + M_A\theta_{aa} = 0 \qquad M_A = -\frac{\theta_A}{\theta_{aa}}$$

$$\theta_A = \frac{1110 + 1340}{EI} = \frac{2450 \text{ ft}^2\text{-k}}{EI}$$

$$\theta_{aa} = \frac{10 \text{ ft}^2\text{-k}}{EI}$$

$$M_A = -\frac{\theta_A}{\theta_{aa}} = -\frac{2450}{10} = -245^{\text{'k}}$$

counterclockwise

Removal of M_A

$\frac{133^{\text{'k}}}{EI}$

Center of gravity

13.33 ft — 16.67 ft

$\frac{1110 \text{ ft}^2\text{-k}}{EI}$ $\frac{890 \text{ ft}^2\text{-k}}{EI}$

M diagram for 20^k load

$\frac{200^{\text{'k}}}{EI}$

Center of gravity

16.67 ft — 13.33 ft

$\frac{1340 \text{ ft}^2\text{-k}}{EI}$ $\frac{1660 \text{ ft}^2\text{-k}}{EI}$

M diagram for 30^k load

$\frac{1^{\text{'k}}}{EI}$

+

$\frac{10 \text{ ft}^2\text{-k}}{EI}$ $\frac{5 \text{ ft}^2\text{-k}}{EI}$

m diagram for a clockwise unit moment at *A*

EXAMPLE 13.4

Find the reactions and draw the shear and moment diagrams for the two-span beam shown in Figure 13.4. Assume the moment at the interior support to be the redundant.

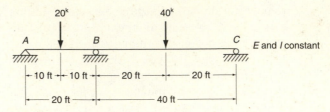

Figure 13.4

Solution.   Removal of moment from the interior support changes the support into a hinge, and the beams on each side are free to rotate independently as indicated by the angles θ_{b_1} and θ_{b_2} in the deflection curve shown. The numerical values of the angles can be found by placing the M/EI diagram on the conjugate structure and computing the shear on each side of the support. In the actual beam there is no change of slope of the tangent to the elastic curve from a small distance to the left of B to a small distance to the right of B.

The angle represented in the diagram by θ_B is the angle between the tangents to the elastic curve on each side of the support (that is, $\theta_{b_1} + \theta_{b_2}$). The actual moment M_B, when replaced, must be of sufficient magnitude to bring the tangents back together or reduce θ_B to zero. A unit moment applied on each side of the hinge produces a change of slope of θ_{bb}; therefore, the expression at the top of page 361 is applicable:

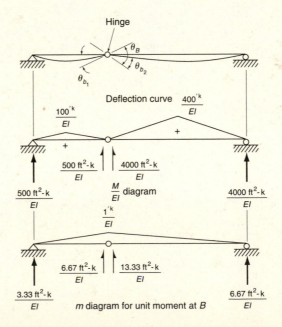

$$M_B = -\frac{\theta_B}{\theta_{bb}}$$

$$\theta_{b_1} = 500$$

$$\theta_{b_2} = 4000$$

$$\theta_B = \theta_{b_1} + \theta_{b_2} = 4500$$

$$\theta_{bb} = 6.67 + 13.33 = 20$$

$$M_B = -\frac{\theta_B}{\theta_{bb}} = -\frac{4500}{20} = -225'^k$$

By statics the following reactions are found:

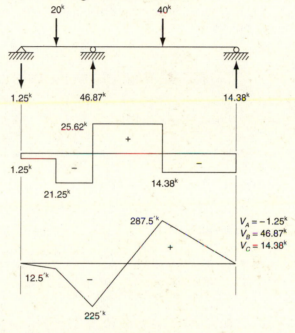

EXAMPLE 13.5

Compute the reactions and draw the moment diagram for the structure shown in Figure 13.5.

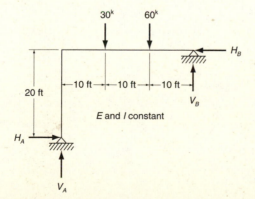

Figure 13.5

Solution.   Remove H_A as the redundant. This will change A to a roller type of support.

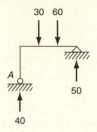

Compute the horizontal deflection at A by virtual work. The result is

$$\delta_A = \frac{86,600 \text{ ft}^{3\text{-k}}}{EI} \leftarrow$$

Apply a unit horizontal load at A and determine the horizontal deflection δ_{aa}.

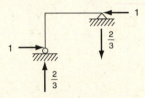

The result is

$$\delta_{aa} = + \frac{6667 \text{ ft}^{3\text{-k}}}{EI} \rightarrow$$

$$\delta_A + H_A \delta_{aa} = 0$$

$$H_A = - \frac{\delta_A}{\delta_{aa}} = - \frac{-86,660}{+6667} = +13^k \rightarrow$$

Compute the remaining reactions by statics.

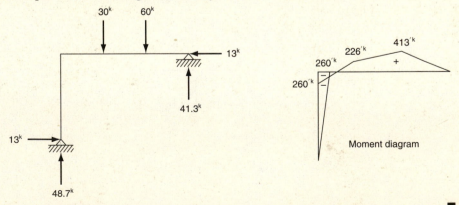

Moment diagram

Raritan River Bridge in New Jersey. (Courtesy of Steinman, Boynton, Gronquist & Birdsall Consulting Engineers.)

13.3 BEAMS AND FRAMES WITH TWO OR MORE REDUNDANTS

The force method of analyzing beams and frames with one redundant may be extended to beams and frames having two or more redundants. The continuous beam of Figure 13.6, which has two redundant reactions, is considered here.

To make the beam statically determinate, it is necessary to remove two supports. Supports B and C are assumed to be removed, and their deflections δ_B and δ_C due to the external loads are computed. The external loads are theoretically removed from the beam; a unit load is placed at B; and the deflections at B and C—δ_{bb} and δ_{cb}—are found. The unit load is moved to C, and the deflections at the two points δ_{bc} and δ_{cc} are again determined.

The reactions at supports B and C push these points up until they are in their original positions of zero deflection. The reaction V_B will raise B an amount $V_b\delta_{bb}$ and C an amount $V_B\delta_{cb}$. The reaction V_C raises C by $V_C\delta_{cc}$ and B by $V_C\delta_{bc}$.

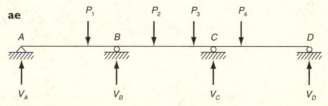

Figure 13.6

I-81 river relief route interchange, Harrisburg, Pennsylvania. (Courtesy of Gannett Fleming.)

An equation may be written for the deflection at each of the supports. Both equations contain the two unknowns V_B and V_C, and their values may be obtained by solving the equations simultaneously.

$$\delta_B + V_B\delta_{bb} + V_C\delta_{bc} = 0$$
$$\delta_C + V_B\delta_{cb} + V_C\delta_{cc} = 0$$

The force method of computing redundant reactions may be extended indefinitely for structures with any number of redundants. The calculations become quite lengthy, however, if there are more than two or three redundants. Considering the beam of Figure 13.7 the following expressions may be written:

$$\delta_B + V_B\delta_{bb} + V_C\delta_{bc} + V_D\delta_{bd} = 0$$
$$\delta_C + V_B\delta_{cb} + V_C\delta_{cc} + V_D\delta_{cd} = 0$$
$$\delta_D + V_B\delta_{db} + V_C\delta_{dc} + V_D\delta_{dd} = 0$$

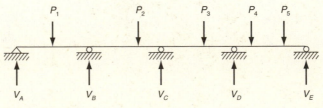

Figure 13.7

13.4 SUPPORT SETTLEMENT

Continuous beams with unyielding supports have been considered in the preceding sections. Should the supports settle or deflect from their theoretical positions major changes may occur in the reactions, shears, moments, and stresses. Whatever the factors causing displacement (weak foundations, temperature changes, poor erection or fabrication, and so on), analysis may be made with the deflection expressions previously developed for continuous beams.

An expression for deflection at point B in the two-span beam of Figure 13.1 was written in Section 13.2. The expression was developed on the assumption that support B was temporarily removed from the structure, allowing point B to deflect, after which the support was replaced. The reaction at B, V_B, was assumed to be of a sufficient magnitude to push B up to its original position of zero deflection. Should B actually settle 1.0 in., V_B will be smaller because it will only have to push B up an amount $\delta_B - 1.0$ in., and the deflection expression may be written as

$$\delta_B - 1.0 + V_B \delta_{bb} = 0$$

If three men are walking along with a log on their shoulders (a statically indeterminate situation) and one of them lowers his shoulder slightly, he will not have to support as much of the total weight as before. He has, in effect, backed out from under the log and thrown more of its weight to the other men. The settlement of a support in a statically indeterminate continuous beam has the same effect.

The values of δ_b and δ_{bb} must be calculated in inches if the support movement is given in inches; they are calculated in feet if the support movement is given in feet, and so on. Example 13.6 illustrates the analysis of the two-span beam of Example 13.1 with the assumption of a $\frac{3}{4}$-in. settlement of the interior support. The moment diagram is drawn after settlement occurs and is compared with the diagram before settlement. The seemingly small displacement has completely changed the moment picture.

EXAMPLE 13.6

Determine the reactions and draw shear and moment diagrams for the beam of Example 13.1, which is reproduced in Figure 13.8, if support B settles $\frac{3}{4}$ in.

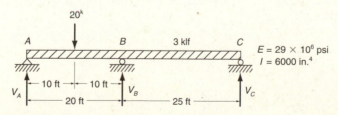

Figure 13.8

Solution.   The values of δ_B and δ_{bb} previously found are computed in inches, and the effect of the support settlement on V_B is determined. By statics the new values of V_A and V_C are found and the shear and moment diagrams are drawn. The moment diagram before settlement is repeated to illustrate the striking changes.

$$\delta_B = \frac{182{,}100 \text{ ft}^3\text{-k}}{EI} = 1.81 \text{ in.}$$

$$\delta_{bb} = \frac{1860 \text{ ft}^3\text{-k}}{EI} = 0.0185 \text{ in.}$$

$$\delta_B - 0.750 + V_B \delta_{bb} = 0$$

$$V_B = -\frac{1.81 - 0.750}{0.0185} = -57.3^k \uparrow$$

$$V_A = 51.2^k \uparrow$$

$$V_C = 46.5^k \uparrow$$

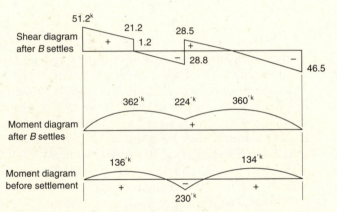

When several or all of the supports are displaced, the analysis may be conducted on the basis of relative settlement values. For example, if all of the supports of the beam of Figure 13.9(a) were to settle 1.5 in., the stress

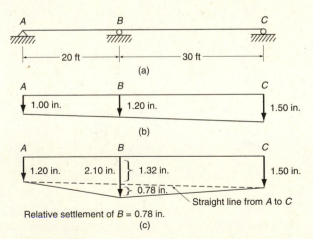

Figure 13.9

conditions would be unchanged. If the supports settle different amounts but remain in a straight line, as illustrated in Figure 13.9(b), the situation theoretically is the same as before settlement.

Where inconsistent settlements occur and the supports no longer lie on a straight line, the stress conditions change because the beam is distorted. The situation may be handled by drawing a line through the displaced positions of two of the supports, usually the end ones. The distances of the other supports from this line are determined and used in the calculations, as illustrated in Figure 13.9(c).

It is assumed that the supports of the three-span beam of Figure 13.10(a) settle as follows: A is 1.25 in., B is 2.40 in., C is 2.75 in., and D is 1.10 in. A diagram of these settlements is plotted in Figure 13.10(b) and the relative settlements of supports B and C are determined.

The solution of the two simultaneous equations that follow will yield the values of V_B and V_C. By statics the values of V_A and V_D may then be computed.

$$\delta_B - 1.19 + V_B\delta_{bb} + V_C\delta_{bc} = 0$$
$$\delta_C - 1.59 + V_B\delta_{cb} + V_C\delta_{cc} = 0$$

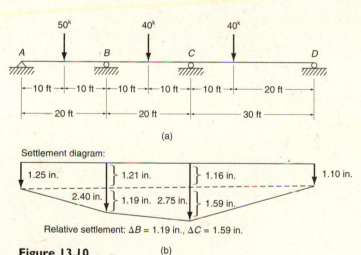

(a)

Settlement diagram:

Relative settlement: $\Delta B = 1.19$ in., $\Delta C = 1.59$ in.

Figure 13.10   (b)

13.5 ANALYSIS OF EXTERNALLY REDUNDANT TRUSSES

Trusses may be statically indeterminate because of redundant reactions, redundant members, or a combination of redundant reactions and members. Externally redundant trusses will be considered initially, and they will be analyzed on the basis of deflection computations in a manner closely related to the procedure used for statically indeterminate beams.

The two-span continuous truss of Figure 13.11 on page 369 is considered for the following discussion. One reaction component, for example, V_B, is removed, and the deflection at that point caused by the external loads is determined. Next the external loads are removed from the truss, and the deflection at the support point B due to a unit load at that point is determined. The reaction is replaced, and it supplies the force necessary to push the support back to its original position. The familiar deflection expression is then written as follows:

$$\delta_B + V_B \delta_{bb} = 0$$

$$V_B = -\frac{\delta_B}{\delta_{bb}}$$

The forces in the truss members due to the external loads, when the redundant is removed, are not the correct final forces and are referred to as F' forces. The deflection at the removed support due to the external loads can be computed by $\Sigma(F'\mu\ell/AE)$. The deflection caused at the support by placing a unit load there can be found by applying the same virtual work expression, except the unit load is now the external load and the forces caused are the same as the μ forces. The deflection at the support due to the unit load is $\Sigma(\mu^2\ell/AE)$, and the redundant reaction may be expressed as follows:

$$V_B = -\frac{\Sigma(F'\mu_B\ell/AE)}{\Sigma(\mu_B^2\ell/AE)}$$

Example 13.7 illustrates the complete analysis of a two-span truss by the method just described. After the redundant reaction is found, the other reactions and the final member forces may be determined by statics. However, another method is available for finding the final forces and should be used as a mathematics check. When the redundant reaction V_B is returned to the truss, it causes the force in each member to change by V_B times its μ force value. The final force in a member becomes

$$F = F' + V_B\mu$$

Rio Grande Gorge Bridge, Taos County, New Mexico. (Courtesy of the American Institute of Steel Construction, Inc.)

EXAMPLE 13.7 _____

Compute the reactions and member forces for the two-span continuous truss shown in Figure 13.11. Circled figures are member areas, in square inches.

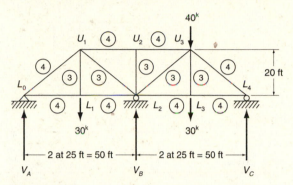

Figure 13.11

Solution.   Remove center support as the redundant and compute F' forces.

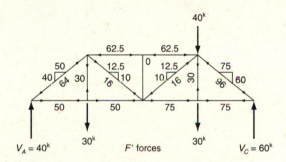

Remove the external loads and place a unit load at the center support. Then compute the μ forces.

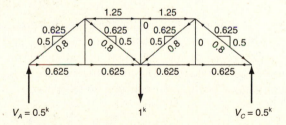

Computing the values of δ_B and δ_{bb} in Table 13.1 on page 370,

$$V_B = -\frac{35,690}{637.6} = -56.0^k \uparrow$$

The following reactions and member forces are found by statics in order to check the final values in Table 13.1.

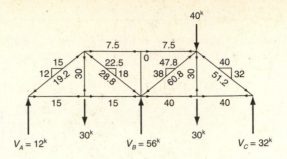

TABLE 13.1

Member	ℓ(in.)	A (in.^{2})	$\dfrac{\ell}{A}$	F' (kips)	μ	$\delta_B = \dfrac{F'\mu\ell}{AE}$	$\delta_{bb} = \dfrac{\mu^2\ell}{AE}$	$F = F' + V_B\mu$
L_0L_1	300	4	75	+50	+0.625	+2340	+29.2	+15.0
L_1L_2	300	4	75	+50	+0.625	+2340	+29.2	+15.0
L_2L_3	300	4	75	+75	+0.625	+3510	+29.2	+40.0
L_3L_4	300	4	75	+75	+0.625	+3510	+29.2	+40.0
L_0U_1	384	4	96	−64	−0.800	+4920	+61.4	−19.2
U_1U_2	300	4	75	−62.5	−1.25	+5850	+117.0	+ 7.5
U_2U_3	300	4	75	−62.5	−1.25	+5850	+117.0	+ 7.5
U_3L_4	384	4	96	−96	−0.800	+7370	+ 61.4	−51.2
U_1L_1	240	3	80	+30	0	0	0	+30.0
U_1L_2	384	3	128	+16	+0.800	+1640	+ 82.0	−28.8
U_2L_2	240	3	80	0	0	0	0	0
L_2U_3	384	3	128	−16	+0.800	−1640	+ 82.0	−60.8
U_3L_3	240	3	80	+30	0	0	0	+30.0
Σ						35,690	637.6	

It should be evident that the deflection procedure may be used to analyze trusses that have two or more redundant reactions. The truss of Figure 13.12 is continuous over three spans, and the reactions at the interior supports V_B and V_C are considered to be the redundants. The following expressions, previously written for a three-span continuous beam, are applicable to the truss:

$$\delta_B + V_B\delta_{bb} + V_C\delta_{bc} = 0$$
$$\delta_C + V_B\delta_{cb} + V_C\delta_{cc} = 0$$

The forces due to a unit load at B are called the μ_B forces; those due to a unit load at C are called the μ_C forces. A unit load at B will cause a deflection at C equal to $\Sigma(\mu_B\mu_C\ell/AE)$, and a unit load at C causes the same deflection at B, $\Sigma(\mu_C\mu_B\ell/AE)$, which is another illustration of Maxwell's law. The deflection expressions become

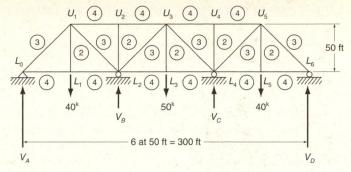

Figure 13.12

$$\Sigma \frac{F'\mu_B \ell}{AE} + V_B \Sigma \frac{\mu_B^2 \ell}{AE} + V_C \Sigma \frac{\mu_B \mu_C \ell}{AE} = 0$$

$$\Sigma \frac{F'\mu_C \ell}{AE} + V_B \Sigma \frac{\mu_C \mu_B \ell}{AE} + V_C \Sigma \frac{\mu_C^2 \ell}{AE} = 0$$

A simultaneous solution of these equations will yield the values of the redundants. Should support settlement occur, the deflections would have to be worked out numerically in the same units given for the settlements.

Tennessee River Bridge, Stevenson, Alabama. (Courtesy of USX Corporation.)

13.6 ANALYSIS OF INTERNALLY REDUNDANT TRUSSES

The truss of Figure 13.13 has one more member than necessary for stability and is therefore statically indeterminate internally to the first degree, as can be proved by applying the equation

$$m = 2j - 3$$

Internally redundant trusses may be analyzed in a manner closely related to the one used for externally redundant trusses. One member is assumed to be the redundant and is theoretically cut or removed from the structure. The remaining members must form a statically determinate and stable truss. The F' forces in these members are assumed to be of a nature causing the joints at the ends of the removed member to pull apart, the distance being $\Sigma(F'\mu\ell/AE)$.

The redundant member is replaced in the truss and is assumed to have a unit tensile force. The μ forces in each of the members caused by the redundant's force of $+1$ are computed and they will cause the joints to be pulled together an amount equal to $\Sigma(\mu^2\ell/AE)$. If the redundant has an actual force of X, the joints will be pulled together an amount equal to $X\Sigma(\mu^2\ell/AE)$.

If the member had been sawed in half, the F' forces would have opened a gap of $\Sigma(F'\mu\ell/AE)$; therefore, X must be sufficient to close the gap, and the following expressions may be written:

$$X\Sigma\frac{\mu^2\ell}{AE} + \Sigma\frac{F'\mu\ell}{AE} = 0$$

$$X = -\frac{\Sigma(F'\mu\ell/AE)}{\Sigma(\mu^2\ell/AE)}$$

The application of this method of analyzing internally redundant trusses is illustrated by Example 13.8. After the truss force in the redundant member is found, the force in any other member equals its F' force plus X times its μ force. Final forces may also be calculated by statics as a check on the mathematics.

EXAMPLE 13.8

Determine the forces in the members of the internally redundant truss shown in Figure 13.13. Members U_1L_1, U_1L_2, L_1U_2, and U_2L_2 have areas of 1 in.^{2} The areas are 2 in.^{2} for each of the other members. $E = 29 \times 10^6$ psi.

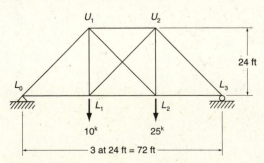

Figure 13.13

Solution.   Assume $L_1 U_2$ to be the redundant, remove it, and compute the F' forces.

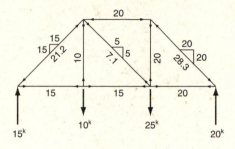

Replace $L_1 U_2$ with a force of $+1$ and compute the μ forces.

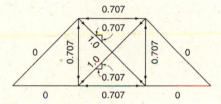

After going through the calculations of Table 13.2 on page 374 and computing X, we can determine the following final forces.

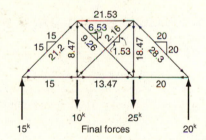

Final forces

$$X = -\frac{\Sigma(F'\mu\ell/AE)}{\Sigma(\mu^2\ell/AE)} = -\frac{-2700}{+1248} = +2.16^k \qquad \blacksquare$$

For trusses that have more than one redundant internally, simultaneous equations are necessary in the solution. Two members, with forces of X_A and X_B, assumed to be the redundants are theoretically cut. The F' forces in the remaining truss members pull the cut places apart by $\Sigma(F'\mu_A\ell/AE)$ and $\Sigma(F'\mu_B\ell/AE)$, respectively. Replacing the first redundant member with a force of $+1$ causes μ_A forces in the truss members and causes the gaps to close by $\Sigma(\mu_A^2\ell/AE)$ and $\Sigma\mu_A\mu_B\ell/AE$. Repeating the process with the other redundant causes μ_B forces and additional gap closings of $\Sigma(\mu_B\mu_A\ell/AE)$ and $\Sigma(\mu_B^2\ell/AE)$. The redundant

TABLE 13.2

Member	ℓ(in.)	A (in.²)	$\dfrac{\ell}{A}$	F'(kips)	μ	$\dfrac{F'\mu\ell}{AE}$	$\dfrac{\mu^2\ell}{AE}$	$F = F'' + X\mu$ (kips)
L_0L_1	288	2	144	+15	0	0	0	+15.00
L_1L_2	288	2	144	+15	−0.707	−1530	+ 72	+13.47
L_2L_3	288	2	144	+20	0	0	0	+20.00
L_0U_1	408	2	204	−21.2	0	0	0	−21.20
U_1U_2	288	2	144	−20	−0.707	+2040	+ 72	−21.53
U_2L_3	408	2	204	−28.3	0	0	0	−28.30
U_1L_1	288	1	288	+10	−0.707	−2040	+ 144	+ 8.47
U_1L_2	408	1	408	+ 7.1	+1.0	+2900	+ 408	+ 9.26
L_1U_2	408	1	408	0	+1.0	0	+ 408	+ 2.16
U_2L_2	288	1	288	+20	−0.707	−4070	+ 144	+18.47
Σ						−2700	+1248	

forces must be sufficient to close the gaps, permitting the writing of the following equations:

$$\Sigma\frac{F'\mu_A\ell}{AE} + X_A\Sigma\frac{\mu_A^2\ell}{AE} + X_B\Sigma\frac{\mu_A\mu_B\ell}{AE} = 0$$

$$\Sigma\frac{F'\mu_B\ell}{AE} + X_A\Sigma\frac{\mu_B\mu_A\ell}{AE} + X_B\Sigma\frac{\mu_B^2\ell}{AE} = 0$$

13.7 ANALYSIS OF TRUSSES REDUNDANT INTERNALLY AND EXTERNALLY

Deflection equations have been written so frequently in the past few sections that the reader probably is able to set up his or her own equations for types of statically indeterminate beams and trusses not previously encountered. Nevertheless, one more group of equations is developed here—those necessary for the analysis of a truss that is statically indeterminate internally and externally. For the following discussion the truss of Figure 13.14, which has two redundant members and one redundant reaction component, will be considered.

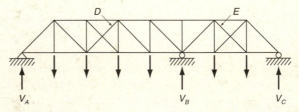

Figure 13.14

Delaware River Turnpike Bridge. (Courtesy of USX Corporation.)

The diagonals lettered D and E and the interior reaction V_B are removed from the truss, which leaves a statically determinate structure. The openings of the gaps in the cut members and the deflections at the interior support due to the external loads may be computed from the following:

$$\delta_B = \Sigma \frac{F'\mu_B \ell}{AE} \qquad \delta_D = \Sigma \frac{F'\mu_D \ell}{AE} \qquad \delta_E = \Sigma \frac{F'\mu_E \ell}{AE}$$

Placing a unit load at the interior support will cause deflections at the gaps in the cut members as well as at the point of application.

$$\delta_{bb} = \Sigma \frac{\mu_B^2 \ell}{AE} \qquad \delta_{db} = \Sigma \frac{\mu_B \mu_D \ell}{AE} \qquad \delta_{cb} = \Sigma \frac{\mu_B \mu_E \ell}{AE}$$

Replacing member D and assuming it to have a positive unit tensile force will cause the following deflections:

$$\delta_{bd} = \Sigma \frac{\mu_B \mu_D \ell}{AE} \qquad \delta_{dd} = \Sigma \frac{\mu_D^2 \ell}{AE} \qquad \delta_{ed} = \Sigma \frac{\mu_E \mu_D \ell}{AE}$$

Similarly, replacement of member E with a force of $+1$ will cause these deflections:

$$\delta_{be} = \Sigma \frac{\mu_B \mu_E \ell}{AE} \qquad \delta_{de} = \Sigma \frac{\mu_D \mu_E \ell}{AE} \qquad \delta_{ee} = \Sigma \frac{\mu_E^2 \ell}{AE}$$

Computation of these sets of deflections permits the calculation of the numerical values of the redundants, because the total deflection at each may be equated to zero.

$$\delta_B + V_B\delta_{bb} + X_D\delta_{bd} + X_E\delta_{be} = 0$$
$$\delta_D + V_B\delta_{db} + X_D\delta_{dd} + X_E\delta_{de} = 0$$
$$\delta_E + V_B\delta_{eb} + X_D\delta_{ed} + X_E\delta_{ee} = 0$$

Nothing new is involved in the solution of this type of problem, and space is not taken for the lengthy calculations necessary for an illustrative example.

13.8 TEMPERATURE CHANGES, SHRINKAGE, FABRICATION ERRORS, AND SO ON

Structures are subject to deformations due not only to external loads but also to temperature changes, support settlements, inaccuracies in fabrication dimensions, shrinkage in reinforced concrete members caused by drying and plastic flow, and so forth. Such deformations in statically indeterminate structures can cause the development of large additional forces in the members. For an illustration it is assumed that the top-chord members of the truss of Figure 13.15 may be exposed to the sun much more than are the other members. As a result, on a hot sunny day they may have much higher temperatures than the other members and the member forces may undergo some appreciable changes.

Problems such as these may be handled exactly as were the previous problems of this chapter. The changes in the length of each of the members due to temperature are computed. (These values, which correspond to the $F'\ell/AE$ values, each equal the temperature change times the coefficient of expansion of the material times the member length.) The redundant is removed from the structure, a unit load is placed at the support in the direction of the redundant, and the μ forces are computed. Then the values $\Sigma(F'\mu\ell/AE)$ and $\Sigma(\mu^2\ell/AE)$ are determined in the same units and the usual deflection expression is finally written. Such a problem is illustrated in Example 13.9.

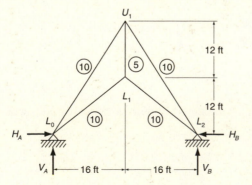

Figure 13.15

EXAMPLE 13.9

The top-chord members of the statically indeterminate truss of Figure 13.15 are assumed to increase in temperature by 60°F. If $E = 29 \times 10^6$ psi and the coefficient of linear temperature expansion is 0.0000065/°F, determine the forces induced in each of the truss members. The circled figures are areas in square inches.

Solution.   Assume H_B is the redundant and compute the μ forces.

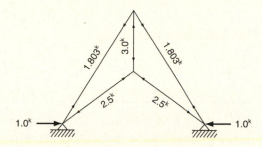

TABLE 13.3

Member	ℓ(in.)	A (in.^{2})	$\dfrac{\ell}{A}$	μ	$\dfrac{\mu^2 \ell}{AE}$	$\Delta\ell = \Delta t \cdot \text{coeff} \cdot \ell$ $\left(\text{equivalent to } \dfrac{F'L}{AE}\right)$	$\mu(\Delta\ell) = \dfrac{F'\mu\ell}{AE}$
$L_0 L_1$	240	10	24	-2.5	$+150$	—	—
$L_1 L_2$	240	10	24	-2.5	$+150$	—	—
$L_0 U_1$	346	10	34.6	$+1.803$	$+112.48$	$(60)(0.0000065)(346) =$ 0.1349	0.2433
$U_1 L_2$	346	10	34.6	$+1.803$	$+112.48$	$(60)(0.0000065)(346) =$ 0.1349	0.2433
$U_0 L_1$	144	5	28.8	-3.0	$+259.2$	—	—
				$\Sigma = +\dfrac{784.16}{E}$			$\Sigma = 0.4866$

$$\delta_{bb} = + \frac{(784.16)(1000)}{29 \times 10^6} = +0.02704 \text{ in.}$$

$$\delta_B = +0.4866 \text{ in.}$$

$$\delta_B + H_B\delta_{bb} = 0$$

$$H_B = -\frac{0.4866}{0.02704} = -18.02^k \rightarrow$$

The final forces in the truss members due to the temperature change are as follows:

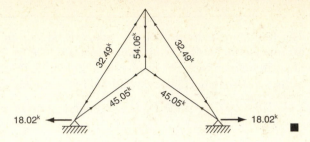

13.9 COMPUTER EXAMPLE

In Example 13.10 the computer program SABLE is used to repeat Example 13.8.

EXAMPLE 13.10 ───

Determine the forces in the truss of Figure 13.13, repeated as Figure 13.16. Assume the moments of inertia of each of the members are equal to 20 in.^{4} Members $U_1 L_1$, $U_1 L_2$, $L_1 U_2$, and $U_2 L_2$ each have areas of 1 in.^{2} while the other members each have areas equal to 2 in.^{2} $E = 29 \times 10^6$ psi. The joints and members are numbered as shown in the figure.

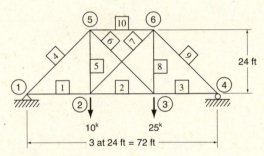

Figure 13.16

Solution.

Structural Data

Data: Nodal Location and Restraint Data

	Coordinates			Restraints		
Node	X	Y	Z	X	Y	Rot
-----	----------	----------	----------	---	---	---
1	0.000*E*+00	0.000*E*+00	0.000*E*+00	Y	Y	Y
2	2.880*E*+02	0.000*E*+00	0.000*E*+00	N	N	Y
3	5.760*E*+02	0.000*E*+00	0.000*E*+00	N	N	Y
4	8.640*E*+02	0.000*E*+00	0.000*E*+00	N	Y	Y
5	2.880*E*+02	2.880*E*+02	0.000*E*+00	N	N	Y
6	5.760*E*+02	2.880*E*+02	0.000*E*+00	N	N	Y

Data: Beam Location and Property Data

Beam	i	j	Type	Area	Beam Properties Izz	E
1	1	2	P-P	2.000E+00	2.000E+01	2.900E+04
2	2	3	P-P	2.000E+00	2.000E+01	2.900E+04
3	3	4	P-P	2.000E+00	2.000E+01	2.900E+04
4	1	5	P-P	2.000E+00	2.000E+01	2.900E+04
5	2	5	P-P	1.000E+00	2.000E+01	2.900E+04
6	5	3	P-P	1.000E+00	2.000E+01	2.900E+04
7	2	6	P-P	1.000E+00	2.000E+01	2.900E+04
8	3	6	P-P	1.000E+00	2.000E+01	2.900E+04
9	6	4	P-P	2.000E+00	2.000E+01	2.900E+04
10	5	6	P-P	2.000E+00	2.000E+01	2.900E+04

Data: Applied Joint Loads

Node	Case	Force-X	Force-Y	Moment-Z
1	1	0.000E+00	0.000E+00	0.000E+00
2	1	0.000E+00	−1.000E+01	0.000E+00
3	1	0.000E+00	−2.500E+01	0.000E+00
4	1	0.000E+00	0.000E+00	0.000E+00
5	1	0.000E+00	0.000E+00	0.000E+00
6	1	0.000E+00	0.000E+00	0.000E+00

Results

Data: Calculated Beam End Forces

Beam	Case	End	Axial	Shear-Y	Moment-Z
1	1	i	−1.500E+01	0.000E+00	0.000E+00
		j	1.500E+01	0.000E+00	0.000E+00
2	1	i	−1.346E+01	0.000E+00	0.000E+00
		j	1.346E+01	0.000E+00	0.000E+00
3	1	i	−2.000E+01	0.000E+00	0.000E+00
		j	2.000E+01	0.000E+00	0.000E+00
4	1	i	2.121E+01	0.000E+00	0.000E+00
		j	−2.121E+01	0.000E+00	0.000E+00
5	1	i	−8.457E+00	0.000E+00	0.000E+00
		j	8.457E+00	0.000E+00	0.000E+00
6	1	i	−9.253E+00	0.000E+00	0.000E+00
		j	9.253E+00	0.000E+00	0.000E+00
7	1	i	−2.182E+00	0.000E+00	0.000E+00
		j	2.182E+00	0.000E+00	0.000E+00
8	1	i	−1.846E+01	0.000E+00	0.000E+00
		j	1.846E+01	0.000E+00	0.000E+00
9	1	i	2.828E+01	0.000E+00	0.000E+00
		j	−2.828E+01	0.000E+00	0.000E+00
10	1	i	2.154E+01	0.000E+00	0.000E+00
		j	−2.154E+01	0.000E+00	0.000E+00

PROBLEMS

For Problems 13.1 to 13.23 compute the reactions and draw shear and moment diagrams for the continuous beams or frames; *E* and *I* are constant unless noted otherwise. The method of consistent distortions is to be used.

13.1

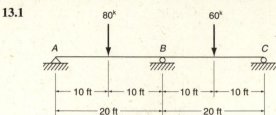

13.2   (*Ans.* $V_A = 95^k \uparrow$, $M_A = -400$ ft-k)

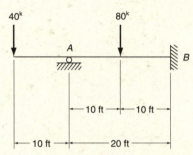

13.3

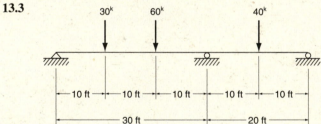

13.4   (*Ans.* $V_A = 45.00^k \uparrow$, $M_A = -270$ ft-k)

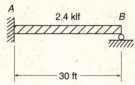

13.5

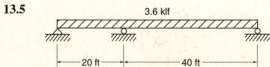

13.6   (*Ans.* $M_B = 204.2$ ft-k, $V_C = 29.19$ k)

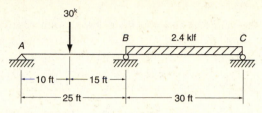

13.7

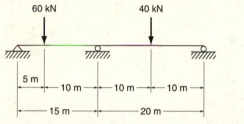

13.8   (*Ans.* $V_A = 9.86^k \uparrow$, $M_B = -154.2'^k$, $V_C = 83.92^k \uparrow$, $M_C = -276.2'^k$)

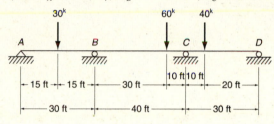

13.9

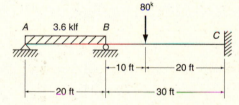

13.10   (*Ans.* $V_A = 6.00^k \uparrow$, $V_B = 91.60^k$, $M_B = -480$ ft-k)

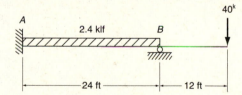

13.11

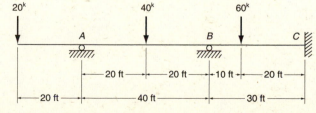

13.12 The beam of Problem 13.1 assuming support B settles 2.50 in. $E = 29 \times 10^6$ psi. $I = 1200$ in.^{4} (*Ans.* $V_B = 58.51^k \uparrow$, $M_B = +115$ ft-k)

13.13 The beam of Problem 13.1 assuming the following support settlements: $A = 4.00$ in., $B = 2.00$ in., and $C = 3.50$ in. $E = 29 \times 10^6$ psi. $I = 1200$ in.^{4}

13.14 The beam of Problem 13.8 assuming the following support settlements: $A = 1.00$ in., $B = 3.00$ in., $C = 1.50$ in., and $D = 2.00$ in. $E = 29 \times 10^6$ psi. $I = 3200$ in.^{4} (*Ans.* $M_B = 156$ ft-k, $M_C = -508.8$ ft-k)

13.15

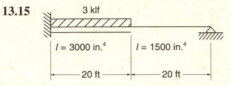

13.16 (*Ans.* $M_A = -129.6$ ft-k, $V_B = 8.52^k \uparrow$)

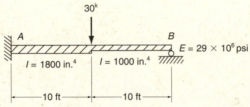

13.17 $E = 29 \times 10^6$ psi

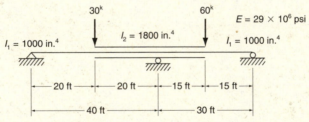

13.18 (*Ans.* $V_A = 25.62^k \uparrow$, $M_B = -112.4'^k$, $H_C = 15.62^k \leftarrow$)

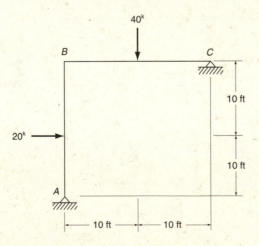

13.19

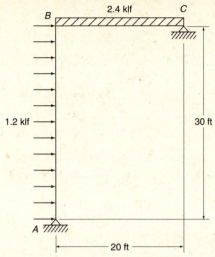

13.20   (*Ans.* $V_A = 3^k \downarrow$, $V_D = 27^k \uparrow$, $M_B = 123.4$ ft-k)

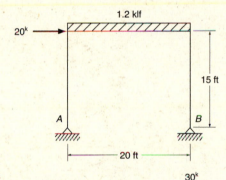

13.21

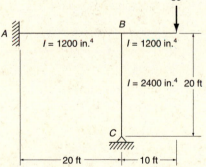

13.22   Repeat Problem 13.21 if the column is fixed at its base, (*Ans.* $M_A = 50$ ft-k, $V_C = 37.50^k \uparrow$)

13.23

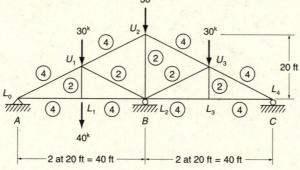

For Problems 13.24 to 13.39 determine the reactions and member forces for the trusses. Circled figures are member areas, in square inches unless shown otherwise. E is constant.

13.24   (Ans. $V_B = 19.8^k \uparrow$, $L_0 L_1 = +15.1^k$, $U_1 L_2 = -21.35^k$)

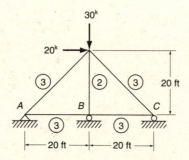

13.25

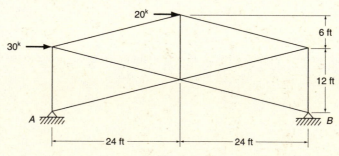

13.26   All areas are equal (Ans. $H_B = 23.2^k \rightarrow$, $U_0 L_1 = -32.55^k$, $U_1 L_1 = +4.2^k$)

13.27   All areas are equal

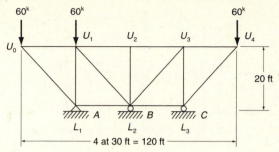

13.28   (*Ans.* $V_B = 80.67$ kN $\uparrow$, $L_1L_2 = +42.86^k$, $U_1L_1 = -80.67^k$)

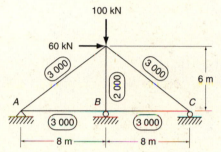

13.29

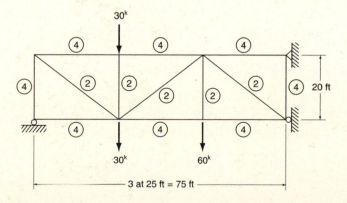

13.30   All areas are equal (*Ans.* $L_1L_2 = +47.5^k$, $U_2U_3 = +41.5^k$, $U_2L_2 = +60^k$)

13.31

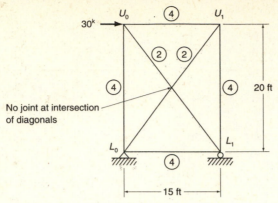

13.32   (*Ans.* $U_0 L_1 = +8.88^k$, $L_1 U_2 = +7.15^k$)

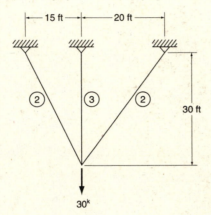

13.33   All areas are equal

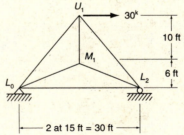

13.34   All areas are equal (*Ans.* $L_1 U_2 = -180.3$ kN, $U_0 L_0 = -40.2$ kN, $U_1 L_1 = -140.2$ kN)

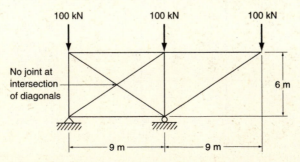

13.35

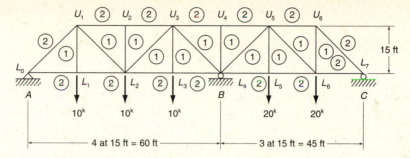

13.36 (*Ans.* $L_3 L_4 = -192^k$, $U_3 U_4 = +192^k$)

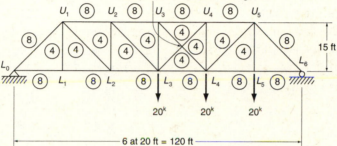

No joint at intersection of diagonals

13.37

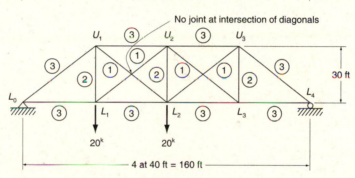

No joint at intersection of diagonals

13.38 (*Ans.* $V_A = 1.1^k \downarrow$, $U_1 U_2 = +2.2^k$, $U_2 L_3 = +22.0^k$)

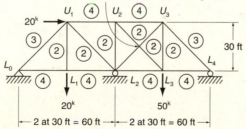

No joint at intersection of diagonals

13.39 Determine the forces in all the members of the truss shown in the accompanying illustration if the top-chord members, $U_0 U_1$ and $U_1 U_2$, have an increase in temperature of 75°F and no change in temperature in other members. Coefficient of linear expansion $\epsilon = 0.0000065$.

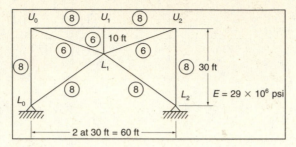

For Problems 13.40 through 13.43 use the computer program SABLE and repeat the following problems.

13.40   Problem 13.1 ($V_B = 96.27^k$, $M_B = -262.6$ ft-k)

13.41   Problem 13.2

13.42   Problem 13.24 ($U_1 L_1 = -19.8^k$, $L_1 L_2 = +15.1^k$)

13.43   Problem 13.38

Chapter 14

Influence Lines for Statically Indeterminate Structures

14.1 INFLUENCE LINES FOR STATICALLY INDETERMINATE BEAMS

The uses of influence lines for statically indeterminate structures are the same as those for statically determinate structures. They enable the designer to locate the critical positions for live loads and to compute forces for various positions of the loads. Influence lines for statically indeterminate structures are not as simple to draw as they are for statically determinate structures. For the latter case it is possible to compute the ordinates for a few controlling points and connect those values with a set of straight lines. Unfortunately, influence lines for continuous structures require the computation of ordinates at a large number of points because the diagrams are either curved or made up of a series of chords. The chord-shaped diagram occurs where loads can only be transferred to the structure at intervals, as at the panel points of a truss or at joists framing into a girder.

The problem of preparing the diagrams is not as difficult as the preceding paragraph seems to indicate because a large percentage of the work may be eliminated by applying Maxwell's law of reciprocal deflections. The preparation of an influence line for the interior reaction of the two-span beam of Figure 14.1 is considered in the following paragraphs.

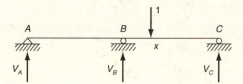

Figure 14.1

389

The procedure for calculating V_B has been to remove it from the beam and then compute δ_B and δ_{bb} and substitute their values in the usual formula. The same procedure may be used in drawing an influence line for V_B. A unit load is placed at some point x causing δ_B to equal δ_{bx}, from which the following expression is written:

$$V_B = -\frac{\delta_B}{\delta_{bb}} = -\frac{\delta_{bx}}{\delta_{bb}}$$

At first glance it appears that the unit load will have to be placed at numerous points on the beam and the value of δ_{bx} laboriously computed for each. A study of the deflections caused by a unit load at point x, however, proves these computations to be unnecessary. By Maxwell's law the deflection at B due to a unit load at $x(\delta_{bx})$ is identical with the deflection at x due to a unit load at $B(\delta_{xb})$. The expression for V_B becomes

$$V_B = -\frac{\delta_{xb}}{\delta_{bb}}$$

It is now evident that the unit load need only be placed at B, and the deflections at various points across the beam computed. Dividing each of these values by δ_{bb} gives the ordinates for the influence line. If a deflection curve is plotted for the beam for a unit load at B (support B being removed), an influence line for V_B may be obtained by dividing each of the deflection ordinates by δ_{bb}. Another way of expressing this principle is as follows: If a unit deflection is caused at a support for which the influence line is desired, the beam will draw its own influence line because the deflection at any point in the beam is the ordinate of the influence line at that point for the reaction in question.

Maxwell's presentation of his theorem in 1864 was so brief that its value was not fully appreciated until 1886 when Heinrich Müller-Breslau clearly showed its true worth as described in the preceding paragraph.[1] Müller-Breslau's principle may be stated in detail as follows: **The deflected shape of a structure represents to some scale the influence line for a function such as stress, shear, moment, or reaction component if the function is allowed to act through a unit displacement.** This principle, which is applicable to statically determinate and indeterminate beams, frames, and trusses, is proved in the next section of this chapter.

The influence line for the reaction at the interior support of two-span beam is presented in Example 14.1. Influence lines also are shown for the end reactions, the values for ordinates having been obtained by statics from those computed for the interior reaction. The conjugate-beam procedure is an excellent method of determining the beam deflections necessary for preparing the diagrams.

EXAMPLE 14.1

Draw influence lines for reactions at each support of the structure shown in Figure 14.2.

---

[1] J. S. Kinney, *Indeterminate Structural Analysis* (Reading, Mass.: Addison-Wesley, 1957), 14.

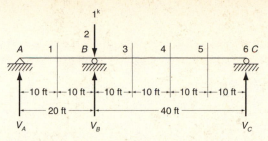

Figure 14.2

Solution. Remove V_B, place a unit load at B, and compute the deflections caused at 10-ft intervals by the conjugate-beam method:

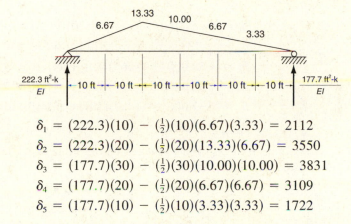

$$\delta_1 = (222.3)(10) - (\tfrac{1}{2})(10)(6.67)(3.33) = 2112$$
$$\delta_2 = (222.3)(20) - (\tfrac{1}{2})(20)(13.33)(6.67) = 3550$$
$$\delta_3 = (177.7)(30) - (\tfrac{1}{2})(30)(10.00)(10.00) = 3831$$
$$\delta_4 = (177.7)(20) - (\tfrac{1}{2})(20)(6.67)(6.67) = 3109$$
$$\delta_5 = (177.7)(10) - (\tfrac{1}{2})(10)(3.33)(3.33) = 1722$$

Noting $\delta_{bb} = \delta_2$, the values of the influence-line ordinates for V_B are found by dividing each deflection by δ_2.

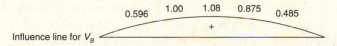

Having the values of V_B for various positions of the unit load, the values of V_A and V_C for each load position can be determined by statics with the following results:

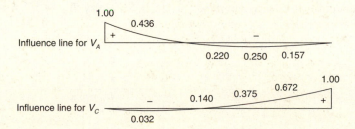

For checking results, support C is removed, a unit load is placed there, and the resulting deflections are computed at 10-ft intervals.

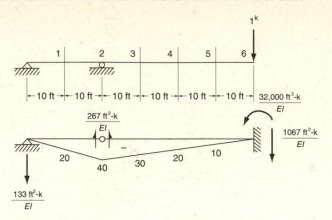

$$\delta_1 = -(133)(10) + (\tfrac{1}{2})(10)(20)(3.33) = -1000$$

$$\delta_2 = -(133)(20) + (\tfrac{1}{2})(20)(40)(6.67) = 0$$

$$\delta_3 = +32,000 - (1067)(30) + (\tfrac{1}{2})(30)(30)(10) = +4500$$

$$\delta_4 = +32,000 - (1067)(20) + (\tfrac{1}{2})(20)(20)(6.67) = +12,000$$

$$\delta_5 = +32,000 - (1067)(10) + (\tfrac{1}{2})(10)(10)(3.33) = +21,500$$

$$\delta_6 = +32,000$$

Since $\delta_{cc} = \delta_6$, the ordinates of the influence line for V_C are determined by dividing each deflection by δ_6.

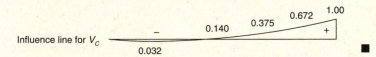

The next problem is to draw the influence lines for beams continuous over three spans, which have two redundants. For this discussion the beam of Figure 14.3 is considered, and the reactions V_B and V_C are assumed to be the redundants.

It will be necessary to remove the redundants and compute the deflections at various sections in the beam for a unit load at B and also for a unit load at C. By Maxwell's law a unit load at any point x causes a deflection at $B(\delta_{bx})$ equal to the deflection at x due to a unit load at $B(\delta_{xb})$. Similarly, $\delta_{cx} = \delta_{xc}$. After computing δ_{xb} and

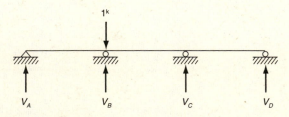

Figure 14.3

δ_{xc} at the several sections, their values at each section may be substituted into the following simultaneous equations, whose solution will yield the values of V_B and V_C.

$$\delta_{xb} + V_B\delta_{bb} + V_C\delta_{bc} = 0$$
$$\delta_{xc} + V_B\delta_{cb} + V_C\delta_{cc} = 0$$

The simultaneous equations are solved quickly, even though a large number of ordinates are being computed, because the only variables in the equations are δ_{xb} and δ_{xc}. After the influence lines are prepared for the redundant reactions of a beam, the ordinates for any other function (moment, shear, and so on) can be determined by statics. Example 14.2 illustrates the calculations necessary for preparing influence lines for several functions of a three-span continuous beam.

EXAMPLE 14.2

Draw influence lines for V_B, V_C, V_D, M_7, and shear at section 6 in Figure 14.4.

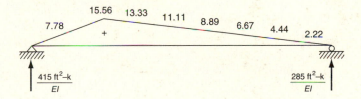

Figure 14.4

Solution. Remove V_B and V_C, place a unit load at B, and load the conjugate beam with the M/EI diagram.

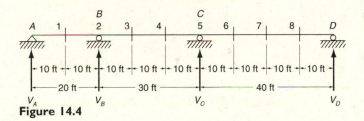

Place a unit load at C and load the conjugate beam with the M/EI diagram.

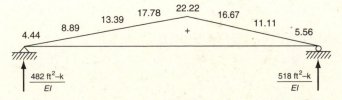

Compute the values of δ_{xb} and δ_{xc}, from which V_B and V_C are obtained by solving the simultaneous equations. Ordinates for M_7, V_D, and shear at section 6 are obtained by statics and shown in Table 14.1 on the following page. ■

TABLE 14.1

Section	δ_{xb}	δ_{xc}	V_B	V_C	V_D	M_7	V_6
1	4020	4746	+0.646	−0.0735	+0.008	+0.160	−0.008
2	$7255 = \delta_{bb}$	$9040 = \delta_{bc}$	+1.00	0	0	0	0
3	9100	12.460	+0.855	+0.320	−0.0344	−0.688	+0.0344
4	9630	14.540	+0.432	+0.720	−0.0513	−1.026	+0.0513
5	$9040 = \delta_{cb}$	$14.825 = \delta_{cc}$	0	+1.00	0	0	0
6	7550	13.040	−0.227	+1.02	+0.1504	+3.008	−0.1504
							+0.8496
7	5404	9618	−0.257	+0.805	+0.388	+7.76	+0.612
8	2813	5088	−0.159	+0.440	+0.688	+3.68	+0.316

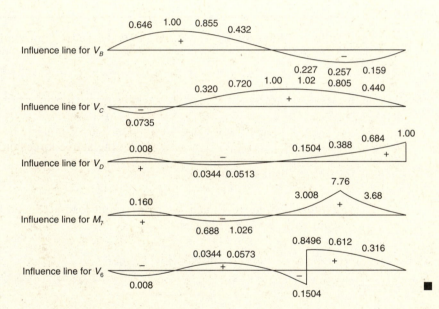

Influence lines for the reactions, shears, moments, and so on for frames can be prepared exactly as they are for the statically indeterminate beams just considered. Space is not taken to present such calculations. The next section of this chapter, which is concerned with the preparation of qualitative influence lines, will show the reader how to place live loads in frames so as to cause maximum values.

14.2 QUALITATIVE INFLUENCE LINES

Müller-Breslau's principle is based upon Castigliano's theorem of least work, which is presented in Chapter 15 of this text. This theorem can be expressed as follows: **when a displacement is induced into a structure, the total virtual work done by all the active forces equals zero.**

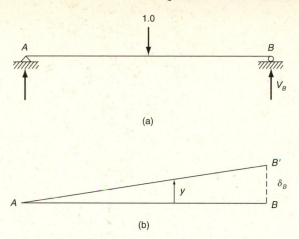

Figure 14.5

Proof of Müller-Breslau's principle can be made by considering the beam of Figure 14.5(a) when it is subjected to a moving unit load. To determine the magnitude of the reaction V_B, we can remove the support at B and allow V_B to move through a small distance δ_B, as shown in part (b) of the figure. Notice the beam's position is now represented by the line AB' in the figure and the unit load has been moved through the distance y.

Writing the virtual work expression for the active forces on the beam

$$(V_B)(\delta_B) = (1.0)(y)$$

$$V_B = \frac{y}{\delta_B}$$

Should δ_B be given a unit value, V_B will equal y.

$$V_B = \frac{y}{1.0} = y$$

You can now see that y is the ordinate of the deflected beam at the unit load and also is the value of the right-hand reaction V_B due to that moving unit load. Therefore, the deflected beam position AB' represents the influence line for V_B. Similar proofs can be developed for the influence lines for other functions of a structure, such as shear and moments.

Müller-Breslau's principle is of such importance that space is taken to emphasize its value. The shape of the usual influence line needed for continuous structures is so simple to obtain from his principle that in many situations it is unnecessary to perform the labor needed to compute the numerical values of the ordinates. It is possible to sketch the diagram roughly with sufficient accuracy to locate the critical positions for live load for various functions of the structure. This possibility is of particular importance for building frames, as will be illustrated in subsequent paragraphs.

If the influence line is desired for the left reaction of the continuous beam of Figure 14.6(a), its general shape can be determined by letting the reaction act upward

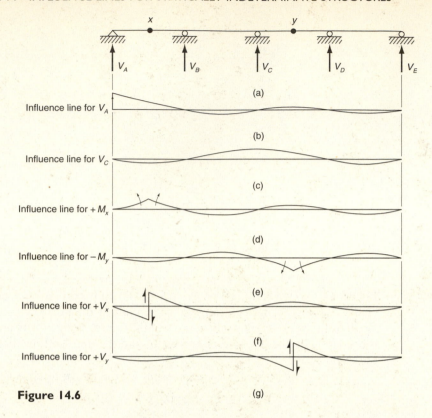

Figure 14.6

through a unit distance as shown in Figure 14.6(b) of the figure. If the left end of the beam were pushed up, the beam would take the shape shown. This distorted shape can be easily sketched, remembering the other supports are unyielding. Influence lines obtained by sketching are said to be *qualitative influence lines,* whereas the exact ones are said to be *quantitative influence lines.* The influence line for V_C in Figure 14.6(c) is another example of qualitative sketching for reaction components.

Figure 14.6(d) shows the influence line for positive moment at point x near the center of the left-hand span. The beam is assumed to have a pin or hinge inserted at x and a couple applied adjacent to each side of the pin that will cause compression on the top fibers (plus moment). Twisting the beam on each side of the pin causes the left span to take the shape indicated, and the deflected shape of the remainder of the beam may be roughly sketched. A similar procedure is used to draw the influence line for negative moment at point y in the third span, except that a moment couple is applied at the assumed pin, which will tend to cause compression on the bottom beam fibers, corresponding with negative moment.

Finally, qualitative influence lines are drawn for positive shear at points x and y. At point x the beam is assumed to be cut, and two vertical forces of the nature required to give positive shear are applied to the beam on the sides of the cut section. The beam will take the shape shown in Figure 14.6(f). The same procedure is used to draw the diagram for positive shear at point y.

From these diagrams considerable information is available concerning critical live-loading conditions. If a maximum positive value of V_A was desired for a uniform live load, the load would be placed in spans 1 and 3, where the diagram has positive ordinates; if maximum negative moment was required at point x, spans 2 and 4 would be loaded, and so on.

Qualitative influence lines are particularly valuable for determining critical load positions for buildings, as shown by the moment influence line for the building frame of Figure 14.7. In drawing the diagrams for an entire frame, the joints are assumed to be free to rotate, but the members at each joint are assumed to be rigidly connected to each other so that the angles between them do not change during rotation. The diagram of this figure is sketched for positive moment at the center of beam AB.

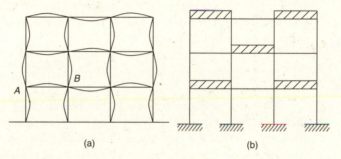

(a)   (b)

Figure 14.7 Qualitative influence line for positive moment at center of span AB. To obtain maximum positive moment at ℄ span AB, place live load as shown.

The spans that should be loaded to cause maximum positive moment are obvious from the diagram. It should be noted that loads on a beam more than approximately three spans away have little effect on the function under consideration. This fact can be seen in the influence lines of Example 14.2, where the ordinates even two spans away are quite small.

A warning should be given regarding qualitative influence lines. They should be drawn for functions near the center of spans or at the supports, but for sections near one-fourth points they should not be sketched without a good deal of study. Near the one-fourth point of a span is a point called the *fixed point*, at which the influence line changes in type. The subject of fixed points is discussed at some length in the book *Continuous Frames of Reinforced Concrete* by H. Cross and N. D. Morgan.[2]

14.3 INFLUENCE LINES FOR STATICALLY INDETERMINATE TRUSSES

For analyzing statically indeterminate trusses, influence lines are necessary to determine the critical positions for live loads, as they were for statically determinate trusses.

[2] H. Cross and N. D. Morgan, *Continuous Frames of Reinforced Concrete* (New York: Wiley, 1932).

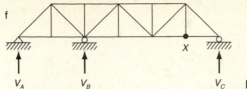

V_A V_B V_C **Figure 14.8**

The discussion of the details of construction of these diagrams for statically indeterminate trusses is quite similar to the one presented for statically indeterminate beams in Sections 14.1 and 14.2. To prepare the influence line for a reaction of a continuous truss, the support is removed and a unit load is placed at the support point. For this position of the unit load, the deflection at each of the truss joints is determined. For example, the preparation of an influence line for the interior reaction of the truss of Figure 14.8 is considered. The value of the reaction when the unit load is at joint x may be expressed as follows:

$$V_B = -\frac{\delta_{xb}}{\delta_{bb}} = -\frac{\Sigma(\mu_x \mu_B \ell / AE)}{\Sigma(\mu_B^2 \ell / AE)}$$

After the influence line for V_B has been plotted, the influence line for another reaction may be prepared by repeating the process of removing it as the redundant, introducing a unit load there, and computing the necessary deflections. A simpler procedure is to compute the other reactions, or any other functions for which influence lines are desired, by statics after the diagram for V_B is prepared. This method is employed in Example 14.3 for a two-span truss for which influence lines are desired for the reactions and several member forces.

EXAMPLE 14.3

Draw influence lines for the three vertical reactions and for forces in members $U_1 U_2$, $L_0 U_1$, and $U_1 L_2$ (Figure 14.9). Circled figures are member areas, in square inches.

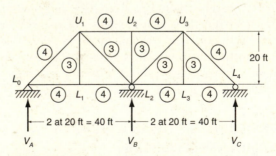

Figure 14.9

Solution.   Remove the interior support and compute the forces for a unit load at L_1 and at L_2. (Note that the deflection at L_1 caused by a unit load at L_2 is the same as the deflection caused at L_3 due to symmetry.)

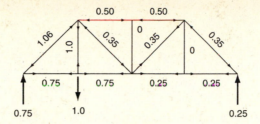

Unit load at L_1

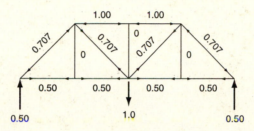

Unit load at L_2

TABLE 14.2

Member	ℓ(in.)	A (in.²)	$\dfrac{\ell}{A}$	μ_B	μ_A	$\dfrac{\mu_B^2 \ell}{AE}$	$\dfrac{\mu_B \mu_A \ell}{AE}$
L_0L_1	240	4	60	+0.50	+0.75	+15	+22.5
L_1L_2	240	4	60	+0.50	+0.75	+15	+22.5
L_2L_3	240	4	60	+0.50	+0.25	+15	+ 7.5
L_3L_4	240	4	60	+0.50	+0.25	+15	+ 7.5
L_0U_1	340	4	85	−0.707	−1.06	+42.5	+63.6
U_1U_2	240	4	60	−1.00	−0.50	+60	+30.0
U_2U_3	240	4	60	−1.00	−0.50	+60	+30.0
U_3L_4	340	4	85	−0.707	−0.35	+42.5	+21.0
U_1L_1	240	3	80	0	+1.00	0	0
U_1L_2	340	3	113	+0.707	−0.35	+56.5	−28.0
U_2L_2	240	3	80	0	0	0	0
L_2U_3	340	3	113	+0.707	+0.35	+56.5	+28.0
U_3L_3	240	3	80	0	0	0	0
Σ						$\dfrac{378}{E}$	$\dfrac{204.6}{E}$

Divide each of the values by δ_{bb} to obtain the influence-line ordinates for V_B and compute the ordinates for the other influence diagrams by statics.

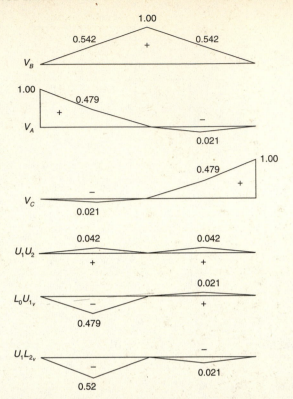

Example 14.4 shows that influence lines for members of an internally redundant truss may be prepared by an almost identical procedure. The member assumed to be the redundant is given a unit force, and deflections caused thereby at each of the joints are calculated. The ordinates of the diagram for the member are obtained by dividing each of these deflections by the deflection at the member. All other influence lines are prepared by statics.

EXAMPLE 14.4 _____

Prepare influence lines for force in members L_1U_2, U_1U_2, and U_2L_2 of the truss of Figure 13.13, which is reproduced in Figure 14.10. Circled figures are member areas, in square inches.

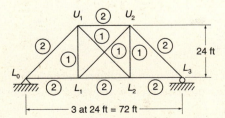

Figure 14.10

Solution.   Remove $L_1 U_2$ as the redundant and compute the forces caused by unit loads at L_1 and L_2; replace $L_1 U_2$ with a force of $+1$ and compute the forces in the remaining truss members.

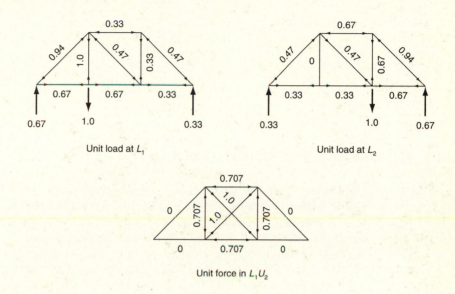

Unit load at L_1

Unit load at L_2

Unit force in $L_1 U_2$

TABLE 14.3

Member	ℓ(in.)	A(in.²)	$\dfrac{\ell}{A}$	μ_{L_1}	μ_{L_2}	μ_A	$\delta_{AL_1} = \dfrac{\mu_{L_1}\mu_A\ell}{AE}$	$\delta_{AL_2} = \dfrac{\mu_{L_2}\mu_A\ell}{AE}$	$\delta_{aa} = \dfrac{\mu_A^2\ell}{AE}$
$L_0 L_1$	288	2	144	+0.67	+0.33	0	0	0	0
$L_1 L_2$	288	2	144	+0.67	+0.33	−0.707	− 68	− 34	+ 72
$L_2 L_3$	288	2	144	+0.33	+0.67	0	0	0	0
$L_0 U_1$	408	2	204	−0.94	−0.47	0	0	0	0
$U_1 U_2$	288	2	144	−0.33	−0.67	−0.707	+ 34	+ 68	+ 72
$U_2 L_3$	408	2	204	−0.47	−0.94	0	0	0	0
$U_1 L_1$	288	1	288	+1.00	0	−0.707	−204	0	+ 144
$U_1 L_2$	408	1	408	−0.47	+0.47	+1.00	−192	+192	+ 408
$L_1 U_2$	408	1	408	0	0	+1.00	0	0	+ 408
$U_2 L_2$	288	1	288	+0.33	+0.67	−0.707	− 34	−136	+ 144
Σ							$\dfrac{-498}{E}$	$\dfrac{+ 90}{E}$	$\dfrac{+1248}{E}$

Draw the influence line for the redundant, $L_1 U_2$, and note that the force in any other member equals $F' + X\mu$.

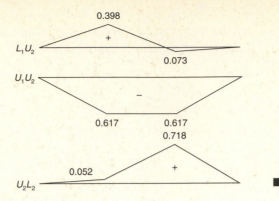

14.4 INFLUENCE LINES USING SABLE

As we have seen in this chapter, the hand calculations necessary for preparing quantitative influence lines for statically indeterminate beams, frames, and trusses can be extremely tedious. For applications such as this, computers can be very helpful, saving time and avoiding math mistakes. (By the way, the preparation of qualitative influence lines will provide rather good checks on the computer-generated diagrams.)

With SABLE, the unit load is placed at one point on a structure and the analysis is automatically performed. Then the unit load is placed at another point and the analysis is made, and so on. The resulting solutions then can be easily used to draw quantitative influence lines.

PROBLEMS

Draw quantitative lines for the situations listed in Problems 14.1 to 14.6.

14.1 Reactions for all supports of the beam shown. Place unit load at 10-ft intervals.

Problem 14.1	**Problem 14.2**

14.2 The left vertical reaction and the moment reaction at the fixed end for the beam shown in the accompanying illustration. Place unit load at 5-ft intervals. (*Ans.* Load at ₵: $V_A = 0.687 \uparrow$, $M_A = -3.75$)

14.3 Shear immediately to the left of support B and moment at support B for the beam shown. Place unit load at 10-ft intervals.

14.4 Shear and moment at a point 20 ft to the left of the fixed-end support *C* of the beam shown. Place unit load at 10-ft intervals. (*Ans.* Load at left end: $V = -0.32$, $M = +3.2$; load halfway from *B* to *C*: $V = +0.24$, $M = -2.4$)

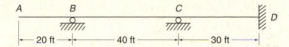

14.5 Vertical reactions and moments at *A* and *B*. Place unit load at 10-ft intervals.

Problem 14.5
Problem 14.6

14.6 Vertical reactions at *A* and *B*, moment at *A* and *x*, and shear just to left of *x*. Place unit load at 10-ft intervals. (*Ans.* Load at *x*: $V_A = 0.722 \uparrow$, $V_B = 0.278 \uparrow$, $M_A = -8.88$, $M_x = 5.56$)

Using Müller-Breslau's principle, sketch influence lines qualitatively for the functions indicated in the structures of Problems 14.7 through 14.14.

14.7 With reference to the accompanying illustration: (a) reactions at *A* and *C*, (b) positive moment at *x* and *y*, and (c) positive shear at *x*.

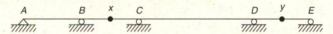

14.8 With reference to the accompanying illustration: (a) reaction at *A*, (b) positive and negative moments at *x*, and (c) negative moment at *B*.

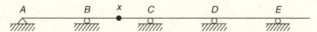

14.9 With reference to the accompanying illustration: (a) vertical reactions at *A* and *C*, (b) negative moment at *A* and positive moment at *x*, and (c) shear just to left of *C*.

14.10 With reference to the accompanying illustration: (a) vertical reactions at *A* and *C*, (b) negative moments at *x* and *C*, and (c) shear just to right of *B*.

14.11 With reference to the accompanying illustration: (a) reaction at *A*, (b) positive moment at *x*, (c) positive shear at *y*, and (d) positive moment at *z*, assuming right side of column is bottom side.

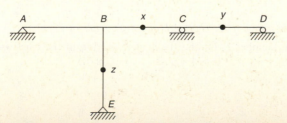

14.12   With reference to the accompanying illustration: (a) positive moment at x, (b) positive shear at x, and (c) negative moment just to the right of y.

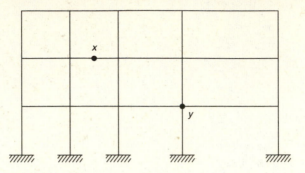

14.13   With reference to the accompanying illustration: (a) positive moment and shear at x, and (b) negative moment just to the right of y.

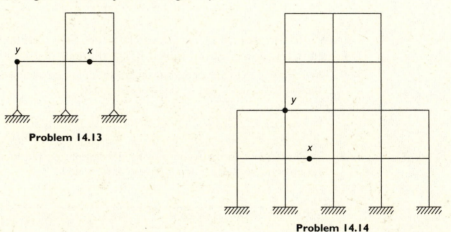

Problem 14.13

Problem 14.14

14.14   With reference to the accompanying illustration: (a) positive and negative moment at x and (b) positive shear just to right of y.

Draw quantitative influence lines for the situations listed in Problems 14.15 through 14.20.

14.15   Reactions at all supports for the truss of Problem 13.35

14.16   Forces in members U_1L_2, U_3U_4, and L_4L_5 of the truss of Problem 13.35 (*Ans.* Load at L_2: $U_1L_2 = +0.634$, $U_3U_4 = +0.208$, $L_4L_5 = -0.140$)

14.17   Reactions for all supports for the truss of Figure 13.27. Assume the loads are moving across the top of the structure.

14.18   Forces in members U_3L_4, L_3L_4, U_1U_2, and U_2L_2 of the truss of Problem 13.36 (*Ans.* Load at L_3: $U_3L_4 = -0.417$, $L_3L_4 = +1.67$; Load at L_2: $U_1U_2 = -1.78$, $U_2L_2 = +0.33$)

14.19   Forces in members L_1U_2 and U_2L_3 of the truss of Problem 13.37

14.20   Force in member L_2U_3 and the center reaction of the truss of Problem 13.38 (*Ans.* Load at L_1: $V_B = +0.504 \uparrow$, $L_2U_3 = -0.0078$)

Chapter 15

Castigliano's Theorems and the Three-Moment Theorem

15.1 CASTIGLIANO'S SECOND THEOREM

Castigliano developed two theorems that are useful in structural analysis. In Chapter 11 we saw that the first partial derivative of the total internal work, or strain energy, with respect to a load P (real or imaginary) applied at a point in a structure is equal to the deflection in the direction of P. This is a statement of Castigliano's second theorem, and we used it to compute deflections in statically determinate structures. This theorem also is applicable to the analysis of statically indeterminate structures. For this discussion the continuous beam of Figure 15.1 and the vertical reaction at support B, V_B, are considered.

If the first partial derivative of the strain energy in this beam, which is equal to the complementary strain energy in a linearly elastic structure, is taken with respect to the reaction V_B, the deflection at B will be obtained. In this problem that deflection is zero.

$$\frac{\partial U}{\partial P} = \delta_B = 0 \tag{15.1}$$

Equations of this type can be written for each point of constraint of a statically indeterminate structure. A structure will deform in a manner consistent with its physical limitations, or so that the internal work of deformation will be at a minimum.

To analyze an internally redundant statically indeterminate structure with Castigliano's second theorem, certain members are assumed to be the redundants and are considered removed from the structure. The removal of the members must be sufficient to leave a statically determinate and stable base structure. The F' forces in

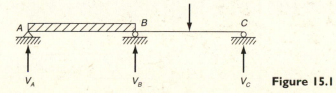

Figure 15.1

the structure are determined by means of the external loads; the redundant members are replaced as loads X_1, X_2, and so on; and the forces caused by these loads are determined.

The total internal work of deformation may be set up in terms of the F' forces and the forces caused by the redundant loads. The result is differentiated successively with respect to the redundants. The derivatives are made equal to zero in order to determine the values of the redundants.

Examples 15.1 to 15.6 illustrate the analysis of statically indeterminate structures using Castigliano's second theorem. Although consistent distortion methods are the most general methods for analyzing various types of statically indeterminate structures, they are not frequently used today because other methods previously discussed are more easily applied.

EXAMPLE 15.1 ──────────────────────────────

Determine the reaction at support C in the beam of Figure 15.2 by Castigliano's second theorem. E and I are constant.

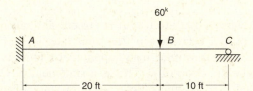

Figure 15.2

Solution. The reaction at C is assumed to be V_C and the other reactions are determined as follows:

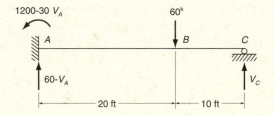

TABLE 15.1

Section	M	$\dfrac{\partial M}{\partial V_c}$	$M\dfrac{\partial M}{\partial V_c}$	$\displaystyle\int M\left(\dfrac{\partial M}{\partial V_c}\right)\dfrac{dx}{EI} = 0$
C to B	$V_c x$	x	$V_c x^2$	$\dfrac{I}{EI}\displaystyle\int_0^{10}(V_c x^2)\,dx$
B to A	$V_c x + 10V_c - 60x$	$x + 10$	$V_c x^2 + 20V_c x - 60x^2 + 100V_c - 600x$	$\dfrac{I}{EI}\displaystyle\int_0^{20}(V_c x^2 + 20V_c x - 60x^2 + 100V_c - 600x)\,dx$
Σ				$9000V_c - 280{,}000 = 0$

$$V_c = 31.1^k \uparrow$$

■

EXAMPLE 15.2

Determine the value of the reaction at support C, Figure 15.3, with Castigliano's second theorem.

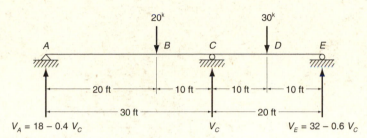

Figure 15.3

$V_A = 18 - 0.4 V_C$ V_C $V_E = 32 - 0.6 V_C$

TABLE 15.2

Section	M	$\dfrac{\partial M}{\partial V_c}$	$\displaystyle\int M\left(\dfrac{\partial M}{\partial V_B}\right)\dfrac{dx}{EI} = 0$
A to B	$18x - 0.4V_c x$	$-0.4x$	$\displaystyle\int_0^{20} (-7.2x^2 + 0.16V_c x^2)\, dx$
B to C	$-2x - 0.4V_c x$ $- 8V_c + 360$	$-0.4x - 8$	$\displaystyle\int_0^{10} (+0.8x^2 + 0.16V_c x^2 + 6.4V_c x$ $- 128x + 64V_c - 2880)\, dx$
E to D	$32x - 0.6V_c x$	$-0.6x$	$\displaystyle\int_0^{10} (-19.2x^2 + 0.36V_c x^2)\, dx$
D to C	$2x - 0.6V_c x$ $- 6V_c + 320$	$-0.6x - 6$	$\displaystyle\int_0^{10} (-1.2x^2 + 0.36V_c x^2 + 7.2V_c x$ $- 204x + 36V_c - 1920)\, dx$

Solution. By integrating the $\int M(\partial M/\partial V_B)(dx/EI)$ expressions and substituting the values of the proper limits, the result for the entire beam is

$$-90{,}333 + 2400.3V_C = 0$$
$$V_C = +37.6^k \uparrow \quad \blacksquare$$

EXAMPLE 15.3

Determine the force in member CD of the truss shown in Figure 15.4 on the following page. Circled values are the bar areas in square inches. E is constant.

Solution. Member CD is assumed to be the redundant and is assigned a force of T. Using the equation below, the results shown on the next page are obtained.

$$\frac{\partial W}{\partial T} = \frac{\partial}{\partial T} \Sigma \frac{F^2 \ell}{2AE} = \Sigma F \frac{\partial F}{\partial T} \frac{\ell}{AE}$$

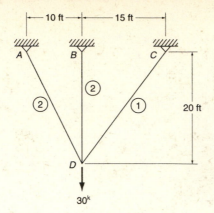

Figure 15.4

TABLE 15.3

Member	ℓ(in.)	A(in.²)	$\dfrac{\ell}{A}$	F	$\dfrac{\partial F}{\partial T}$	$F\left(\dfrac{\partial F}{\partial T}\right)\dfrac{\ell}{AE}$
AD	268	2	134	$+1.34T$	$+1.34$	$+240T$
BD	240	2	120	$-2T + 30$	-2	$+480T - 7200$
CD	300	1	300	$+T$	$+1$	$+300T$
Σ						$1020T - 7200 = 0$

$$T = 7.60^k$$

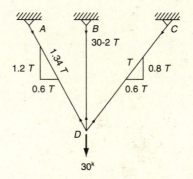

EXAMPLE 15.4 ──

Analyze the truss of Figure 15.5 using Castigliano's second theorem. Circled numbers are member areas, in square inches.

Solution.   Remove the center support and compute the F' forces.

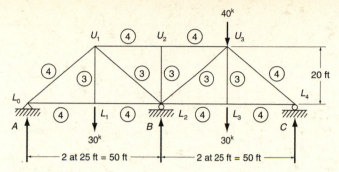

Figure 15.5

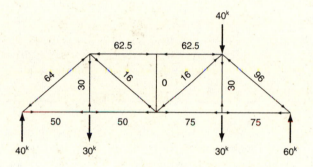

Replace the center support and determine its effect on member forces in terms of V_B.

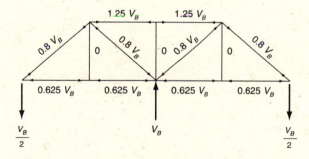

Compute the values in Table 15.4 on the next page and then:

$$E\frac{\partial W}{\partial V_B} = \frac{F\ell}{A}\frac{\partial F}{\partial V_B}$$

$$-35{,}738 + 638.2V_B = 0$$

$$V_B = +56^k \uparrow$$

TABLE 15.4

Member	ℓ (in.)	A (in²)	$\dfrac{\ell}{A}$	F (kips)		$\dfrac{F\ell}{A}$		$\dfrac{\partial F}{\partial V_B}$	$\dfrac{F\ell}{A}\dfrac{\partial F}{\partial V_B}$	
$L_0 L_1$	300	4	75	$+50$	$-0.625V_B$	$+3750-$	$46.9V_B$	-0.625	$-2345+$	$29.3V_B$
$L_1 L_2$	300	4	75	$+50$	$-0.625V_B$	$+3750-$	$46.9V_B$	-0.625	$-2345+$	$29.3V_B$
$L_2 L_3$	300	4	75	$+75$	$-0.625V_B$	$+5625-$	$46.9V_B$	-0.625	$-3520+$	$29.3V_B$
$L_3 L_4$	300	4	75	$+75$	$-0.625V_B$	$+5625-$	$46.9V_B$	-0.625	$-3520+$	$29.3V_B$
$L_0 U_1$	384	4	96	-64	$+0.8V_B$	$-6144+$	$76.8V_B$	$+0.8$	$-4915+$	$61.4V_B$
$U_1 U_2$	300	4	75	-62.5	$+1.25V_B$	$-4687+$	$93.8V_B$	$+1.25$	$-5860+$	$117.2V_B$
$U_2 U_3$	300	4	75	-62.5	$+1.25V_B$	$-4687+$	$93.8V_B$	$+1.25$	$-5860+$	$117.2V_B$
$U_3 L_4$	384	4	96	-96	$+0.8V_B$	$-9216+$	$76.8V_B$	$+0.8$	$-7373+$	$61.4V_B$
$U_1 L_1$	240	3	80	$+30$	$+0$	$+2400+$	0	0	$0+$	0
$U_1 L_2$	384	3	128	$+16$	$-0.8V_B$	$+2048-$	$102.4V_B$	-0.8	$-1638+$	$81.9V_B$
$U_2 L_2$	240	3	80	0	$+0$	$0+$	0	0	$0+$	0
$L_2 U_3$	384	3	128	-16	$-0.8V_B$	$-2048-$	$102.4V_B$	-0.8	$+1638+$	$81.9V_B$
$U_3 L_3$	240	3	80	$+30$	$+0$	$+2400+$	0	0	$0+$	0
Σ									$-35,738+$	$638.2V_B$

■

It should be obvious from the preceding example problems that the amount of work involved in the analysis of statically indeterminate trusses by Castigliano's second theorem or by consistent distortions is about equal. For statically indeterminate beams and frames, other methods such as the moment distribution method presented in Chapters 18 and 19 are much simpler to apply.

Castigliano's second theorem, however, is particularly useful for analyzing composite structures, such as those considered in Examples 15.5 and 15.6. In these types of structures both bending and truss action take place. The reader will be convinced of the advantage of Castigliano's second theorem for the analysis of composite structures if he or she attempts to solve the following two problems by consistent distortions.

EXAMPLE 15.5 _____

Using Castigliano's second theorem, calculate the force in the steel rod of the composite structure shown in Figure 15.6.

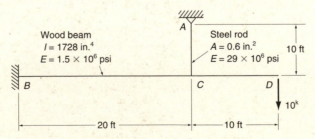

Figure 15.6

Solution. The rod is assumed to be the redundant with a force of T and the values of $\Sigma F(\partial F/\partial T)(\ell/AE)$ and $\int M(\partial M/\partial T)(dx/EI)$ are computed. Relative values of E of 29 and 1.5 are used for the steel and wood, respectively.

TABLE 15.5

Member	ℓ(in.)	A(in.²)	$\dfrac{\ell}{A}$	$\dfrac{\ell}{AE}$	F(kip)	$\dfrac{\partial F}{\partial T}$	$F\left(\dfrac{\partial F}{\partial T}\right)\dfrac{\ell}{AE}$
AC	120	0.6	200	6.90	$+T$	-1	$+6.90T$
Σ							$+6.90T$

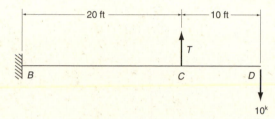

TABLE 15.6

Section	M	$\dfrac{\partial M}{\partial T}$	$M\dfrac{\partial M}{\partial T}$	$\int M\left(\dfrac{\partial M}{\partial T}\right)\dfrac{dx}{EI}$
D to C	$-10x$	0	0	0
C to B	$Tx - 10x - 100$	x	$Tx^2 - 10x^2 - 100x$	$\displaystyle\int_0^{20}(Tx^2 - 10x - 100x)\dfrac{dx}{(1.5)(1728)}$
Σ				$1.03T - 18$

To change the value of $1.03T - 18$ to inches it is necessary to multiply it by 1728×1000 and divide by 1×10^6, as 29 was used for E instead of 29×10^6. To change the value $6.90T$ to inches it is necessary to multiply it by 1000 and divide by 1×10^6. The only difference in the two conversions is the 1728; therefore, the expression can be written as follows to change them to the same units.

$$\int M\left(\frac{\partial M}{\partial T}\right)\frac{dx}{EI} + \Sigma F\left(\frac{\partial F}{\partial T}\right)\frac{\ell}{AE} = (1.03T - 18)(1728) + 6.90T = 0$$

$$T = +17.4^k \quad \blacksquare$$

EXAMPLE 15.6 ────────────

Find the forces in all members of the king post truss shown in Figure 15.7 on page 412.

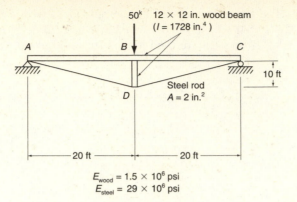

$E_{wood} = 1.5 \times 10^6$ psi
$E_{steel} = 29 \times 10^6$ psi

Figure 15.7

Solution.   By letting BD be the redundant with a force of F, the deflection of the beam at B is found in terms of F.

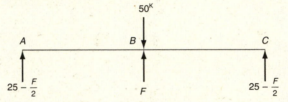

From A to B and from C to B:

$$M = 25x - \frac{F}{2}x \qquad \frac{\partial M}{\partial F} = -\frac{x}{2}$$

$$\int M\frac{\partial M}{\partial F}\frac{dx}{EI} = 2\int_0^{20} \frac{(-12.5x^2 + 0.25Fx^2)\,dx}{EI}$$

$$\frac{2(-12.5x^3/3 + 0.25Fx^3/3)_0^{20}}{EI} = \frac{-66{,}600 + 1333F}{EI}$$

$$\frac{-66{,}600 + 1333F}{1.5} = -44{,}400 + 889F$$

Determine the forces in the various members in terms of the unknown force F.

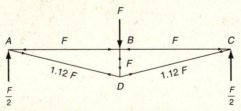

By changing these values to equivalent units and solving for F,

$$F = 49.2^k$$

Final forces:

$$AC = -(1)(49.2) = -49.2^k \qquad AD = +(1.12)(49.2) = +55.1^k$$

$$BD = -(1)(49.2) = -49.2^k \qquad DC = +(1.12)(49.2) = +55.1^k$$

TABLE 15.7

Member	ℓ(in.)	A(in^2)	E	$\dfrac{\ell}{AE}$	F'(kips)	$\dfrac{F'\ell}{AE}$	$\dfrac{\partial F'}{\partial F}$	$\dfrac{\partial F'}{\partial F} \times \dfrac{F'\ell}{AE}$
AC	480	144	1.5×10^6	2.22	$-F$	$-2.22F$	-1	$+\ 2.22F$
BD	120	144	1.5×10^6	0.555	$-F$	$-0.555F$	-1	$+\ 0.555F$
AD	268	2	29×10^6	4.62	$+1.12F$	$+5.19F$	$+1.12$	$+\ 5.80F$
DC	268	2	29×10^6	4.62	$+1.12F$	$+5.19F$	$+1.12$	$+\ 5.80F$
Σ								$+14.375F$

$$-44.400 + 889F + 14.375F = 0 \qquad \blacksquare$$

EXAMPLE 15.7 _____

Using SABLE, determine the total tension in the cable and the shears and moment in the beam of the structure shown in Figure 15.8. $E = 29,000$ ksi.

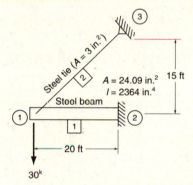

Figure 15.8

Solution.

Structural Data

Data: Nodal Location and Restraint Data

	Coordinates			Restraints		
Node	X	Y	Z	X	Y	Rot
1	0.000E+00	0.000E+00	0.000E+00	N	N	N
2	2.400E+02	0.000E+00	0.000E+00	Y	Y	Y
3	2.400E+02	1.800E+02	0.000E+00	Y	Y	Y

Data: Beam Location and Property Data

				Beam Properties		
Beam	i	j	Type	Area	Izz	E
1	1	2	F-F	2.409E+01	2.364E+03	2.900E+04
2	1	3	P-P	3.000E+00	1.000E+02	2.900E+04

Data: Applied Joint Loads

Node	Case	Force-X	Force-Y	Moment-Z
1	1	0.000E+00	-3.000E+01	0.000E+00
2	1	0.000E+00	0.000E+00	0.000E+00
3	1	0.000E+00	0.000E+00	0.000E+00

Results

Data: Calculated Beam End Forces

Beam	Case	End	Axial	Shear-Y	Moment-Z
1	1	i	3.473E+01	-3.949E+00	5.317E-06
		j	-3.473E+01	3.949E+00	-9.478E+02
2	1	i	-4.342E+01	0.000E+00	0.000E+00
		J	4.342E+01	0.000E+00	0.000E+00

Data: Calculated Nodal Forces

Node	Case	Force-X	Force-Y	Moment-Z
1	1	0.000E+00	0.000E+00	0.000E+00
2	1	-3.473E+01	3.949E+00	-9.478E+02
3	1	3.473E+01	2.605E+01	0.000E+00 ■

15.2 CASTIGLIANO'S FIRST THEOREM—THE METHOD OF LEAST WORK

As we discussed briefly in Chapter 11, Castigliano developed two theorems. We have already discussed in detail his second theorem for evaluating statically determinate and statically indeterminate structural systems. As you recall from that discussion, Castigliano's second theorem is based on complementary strain energy, which, conveniently, for a linearly elastic structure is equal to strain energy. Castigliano's first theorem, which is really the converse of the second theorem, states that the partial derivative of strain energy with respect to a displacement yields the magnitude of the force at that point moving through that displacement.

Castigliano's first theorem, commonly known as the *method of least work,* has played an important role in the development of structural analysis through the years and is occasionally used today. Conceptually, the columns and girders meeting at a joint in a building will all deflect the same amount: the smallest possible value. Each member does no more work than is necessary, and the total work performed by all of the members and loads is the least possible. The method of least work has the disadvantage that it is not applicable in its usual form to forces caused by displacements due to temperature changes, support settlements, and fabrication errors.

The method of least work states that from all the admissible configurations of an elastic structure that satisfy the conditions of equilibrium, the configuration with the minimum potential energy is the stable equilibrium configuration. An admissible configuration in general is one that does not violate the boundary conditions, the equilibrium, and the continuity of the system. An important phrase in the definition of least work is *admissible system configuration.* Let us look at the beam shown in Figure 15.9 to better understand this concept.

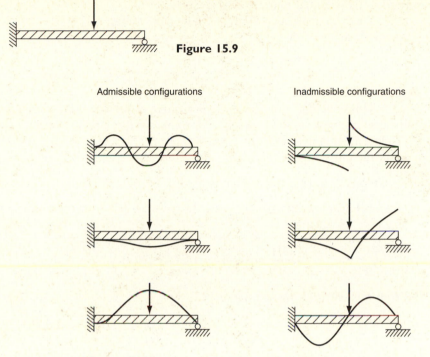

Figure 15.9

Figure 15.10

Possible admissible deformed configurations for this beam are shown on the left side of Figure 15.10. Notice that although we know from our experience that the probable configuration is the middle configuration, the other configurations are possible because they have not violated the continuity or the boundary conditions. The other configurations shown on the right side of Figure 15.10 are inadmissible because continuity of the beam or the boundary conditions has been violated.[1]

The potential energy in a structure is

$$\Pi = U + \Omega \tag{15.2}$$

The first term on the right side of the equation is the strain energy stored in the structure as a result of the applied loads. The second term is the potential energy of the applied loads. From this equation we observe that if we can minimize the term Ω, the entire expression takes on a minimum value because, as we have already seen, the strain energy is a function of the distortion caused by the applied loads.

The work done by the applied loads, which is a loss of potential to do work, is equal to the sum of all the loads times the distances through which they move. Substituting this into Equation 15.2 it becomes

$$\Pi = U - \Sigma P_i D_i \tag{15.3}$$

[1] Robert D. Cook and Warren C. Young, *Advanced Mechanics of Materials* (New York: Macmillan Publishing Company, 1985), 231–235.

The minus sign in Equation 15.3 exists because a force loses potential energy as it does work—its potential to do work decreases.

The potential energy in the system is minimum for the configuration that satisfies the equation

$$d\Pi = 0 = \frac{\partial \Pi}{\partial D_1} dD_1 + \frac{\partial \Pi}{\partial D_2} dD_2 + \cdots \tag{15.4}$$

This equation is the complete derivative of the potential energy with respect to the displacements in the structure. For a nontrivial solution to exist, at least one of the displacement derivatives, dD_i, must be nonzero. In fact, they may all be nonzero. As such, the only way for this expression always to be true is for all of the coefficients, $\frac{\partial \Pi}{\partial D_i}$, modifying the displacement derivatives to be zero. This leads to as many simultaneous linear equations as there are degrees of freedom in the problem being solved. The coefficients are:

$$\frac{\partial \Pi}{\partial D_i} = \frac{\partial}{\partial D_i} [U - \Sigma P_i D_i] = 0 = \frac{\partial U}{\partial D_i} - P_i \tag{15.5}$$

From this equation we see that:

$$P_i = \frac{\partial U}{\partial D_i} \tag{15.6}$$

This last equation is a statement of Castigliano's first theorem. That theorem is:

Theorem. *The derivative of strain energy with respect to a displacement is equal to the force at and in the direction of that displacement.*

This method is not as easy to use as the second theorem, or, for that matter, other energy methods such as virtual work, because strain energy is more easily formulated in terms of load than in terms of displacement. However, for some structural responses, and for the development of system matrices when using matrix methods (see Chapter 20), Castigliano's first theorem is the easiest method to use. This method is particularly useful for the evaluation of nonlinear structures.

15.3 THE THREE-MOMENT THEOREM

Another subject in the study of the "classical" methods of analyzing statically indeterminate structures is the three-moment theorem, initiated in 1855 by the Frenchman, Bertot. Extensions of his original theorem were presented in 1857 by Clapeyron and in 1862 by Bresse, both Frenchmen.[2]

The theorem, which presents a relationship between the moments at the supports in a continuous beam, usually is developed for beams of constant cross section

[2] H. Sutherland and H. L. Bowman, *Structural Theory* (New York: Wiley, 1954), Chapter 8.

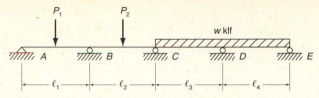

Figure 15.11

between each pair of supports. Theoretically, beams of varying sections may be considered, but the results are so complicated as to be of little practical value. The conjugate-beam method is convenient for developing the relationship between support moments. The resulting equation applies to the moments at any three consecutive supports, as long as the beam is continuous between those supports (that is, there are no internal hinges or other breaks in the continuity of the beam).

Considering the continuous beam of Figure 15.11, the three-moment theorem can be written as follows: for the moments at supports A, B, and C; for the moments at supports B, C, and D; and, finally, for the moments at supports C, D, and E. Supports A and E are simple ends and must have zero moments. Three unknown moments (M_B, M_C, and M_D) remain, but three simultaneous equations are available from which their values may be determined. The theorem applies equally well to beams with fixed ends, as will be seen in the following sections.

15.4 DEVELOPMENT OF THE THEOREM

The elastic curve of a continuous beam has the same numerical value of slope (in radians), an infinitesimal distance on each side of an interior support, although the signs of slope are different. Considering three consecutive interior supports of a continuous beam and loading the beam with the M/EI diagrams, expressions may easily be written for the slope on each side of the center support. The two expressions, which are numerically equal, are equated to each other, the result being a statement of a relationship between the moments at the three supports.

On the following page Figure 15.12 shows a section of a uniformly loaded beam, the deflected shape of the elastic curve of the beam, and the M/EI diagrams for the section. The first M/EI diagram is for the negative support moments, and the second M/EI diagram is for the positive simple beam moments caused by the uniform loads.

Span 1

The slope θ_{BL} is equal to the shear or the upward force at the center support. By elastic weights the slope can be expressed as follows:

$$\theta_{BL} = \frac{\frac{1}{2}(M_A/EI_1)(\ell_1)(\frac{1}{3}\ell_1) + \frac{1}{2}(M_B/EI_1)(\ell_1)(\frac{2}{3}\ell_1) + \frac{2}{3}(w_1\ell_1^2/8EI_1)(\ell_1)(\ell_1/2)}{\ell_1}$$

$$= \frac{M_A\ell_1}{6EI_1} + \frac{M_B\ell_1}{3EI_1} + \frac{w_1\ell_1^3}{24EI_1}$$

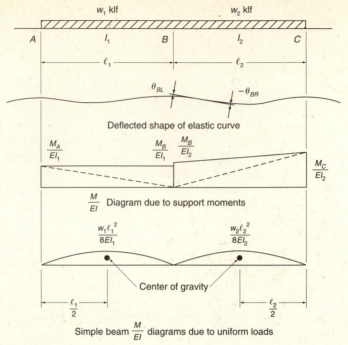

Figure 15.12

Span 2

A similar expression is written for the slope θ_{BR} to the right of support B.

$$\theta_{BR} = \frac{\frac{1}{2}(M_C/EI_2)(\ell_2)(\frac{1}{3}\ell_2) + \frac{1}{2}(M_B/EI_2)(\ell_2)(\frac{2}{3}\ell_2) + \frac{2}{3}(w_2\ell_2^2/8EI_2)(\ell_2)(\ell_2/2)}{\ell_2}$$

$$= \frac{M_C\ell_2}{6EI_2} + \frac{M_B\ell_2}{3EI_2} + \frac{w_2\ell_2^3}{24EI_2}$$

The slope on the left of the center support is equal to minus the slope on the right, or $\theta_{BL} = -\theta_{BR}$. Equating the two expressions results in an equation for the three support moments.

$$\frac{M_A\ell_1}{6EI_1} + \frac{M_B\ell_1}{3EI_1} + \frac{w_1\ell_1^3}{24EI_1} = -\frac{M_C\ell_2}{6EI_2} - \frac{M_B\ell_2}{3EI_2} - \frac{w_2\ell_2^3}{24EI_2}$$

$$\frac{M_A\ell_1}{6EI_1} + \frac{M_B}{3E}\left(\frac{\ell_1}{I_1} + \frac{\ell_2}{I_2}\right) + \frac{M_C\ell_2}{6EI_2} = -\frac{w_1\ell_1^3}{24EI_1} - \frac{w_2\ell_2^3}{24EI_2}$$

Assuming the beam to consist of the same material throughout results in the elimination of E. Multiplying through by six changes the expression into a more convenient form.

$$\frac{M_A\ell_1}{I_1} + 2M_B\left(\frac{\ell_1}{I_1} + \frac{\ell_2}{I_2}\right) + \frac{M_C\ell_2}{I_2} = -\frac{w_1\ell_1^3}{4I_1} - \frac{w_2\ell_2^3}{4I_2}$$

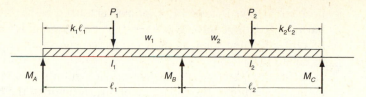

Figure 15.13

The derivation has considered the spans to be loaded with uniform loads only, but concentrated loads may be easily included by taking into account their triangular-shaped simple beam M/EI diagrams. The general equation for both concentrated and uniform loads developed using Figure 15.13 is as follows.

General Equation

$$\frac{M_A \ell_1}{I_1} + 2M_B\left(\frac{\ell_1}{I_1} + \frac{\ell_2}{I_2}\right) + \frac{M_C \ell_2}{I_2}$$

$$= -\Sigma \frac{P_1 \ell_1^2}{I_1}(k_1 - k_1^3) - \Sigma \frac{P_2 \ell_2^2}{I_2}(k_2 - k_2^3) - \frac{w_1 \ell_1^3}{4I_1} - \frac{w_2 \ell_2^3}{4I_2}$$

If the moments of inertia are the same for both spans, the expression becomes

$$M_A \ell_1 + 2M_B(\ell_1 + \ell_2) + M_C \ell_2$$

$$= -\Sigma P_1 \ell_1^2(k_1 - k_1^3) - \Sigma P_2 \ell_2^2(k_2 - k_2^3) - \frac{w_1 \ell_1^3}{4} - \frac{w_2 \ell_2^3}{4}$$

If both span lengths and moments of inertia should be equal, the equation becomes

$$M_A + 4M_B + M_C = -\Sigma P_1 \ell(k_1 - k_1^3) - \Sigma P_2 \ell(k_2 - k_2^3) - \frac{w_1 \ell^2}{4} - \frac{w_2 \ell^2}{4}$$

15.5 APPLICATION OF THE THREE-MOMENT THEOREM

The determination of support moments by the three-moment theorem is illustrated by Examples 15.8 to 15.12. Little explanation is necessary for the analysis of the beams in the first three examples. These beams are simply supported on each end, indicating zero moments. It will be seen that application of the theorem to a simply end-supported continuous beam presents two fewer equations than supports, but the end-support moments are zero, and the equations may be solved simultaneously for the unknowns.

A slightly different approach is needed for the solution of Examples 15.11 and 15.12. The beam of Example 15.11 is fixed on the left end, and the beam of Example 15.12 is fixed on both ends. It would appear that there are not going to be enough equations for the unknowns, because the fixed-end supports have moments. The problem is solved, however, by assuming a fixed end to create another span beyond it having a length of zero. The three-moment theorem is written for the assumed span and the adjoining span, as shown in these two examples, giving the required number of equations.

Shear and moment diagrams are drawn for several of the examples to present a complete picture of the stress conditions throughout the beam. When the solution of an equation yields a negative support moment, it indicates the beam is bending over the support, which causes tension in the top fibers $\left(\underset{\uparrow}{\overset{-M}{\frown}} \right)$.

EXAMPLE 15.8

Determine the support moments of the structure in Figure 15.14 by the three-moment theorem and draw shear and moment diagrams.

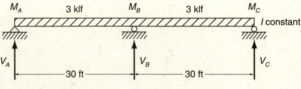

Figure 15.14

Solution.   For equal spans and constant I:

$$M_A + 4M_B + M_C = -\Sigma P_1 \ell(k_1 - k_1^3) - \Sigma P_2 \ell(k_2 - k_2^3) - \frac{w_1 \ell^2}{4} - \frac{w_2 \ell^2}{4}$$

$$M_A = M_C = 0$$

$$4M_B = -\frac{(3)(30)^2}{4} - \frac{(3)(30)^2}{4} = -675 - 675 = -1350$$

$$M_B = -337.5'^k$$

$$\Sigma M_B \text{ to left} = -337.5$$

$$-337.5 = (V_A)(30) - (3 \times 30)(15)$$

$$-337.5 = 30V_A - 1350$$

$$V_A = \frac{1012.5}{30} = +33.75^k$$

$$\Sigma M_B \text{ to right} = +337.5$$

$$+337.5 = -(V_C)(30) + (3)(30)(15)$$

$$+337.5 = -30V_C + 1350$$

$$V_C = \frac{1012.5}{30} = +33.75^k$$

$$\Sigma V = 0$$

$$33.75 + V_B + 33.75 = (3)(60) = 180$$

$$V_B = 112.5^k$$

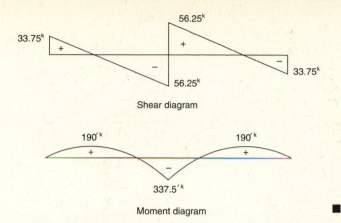

Shear diagram

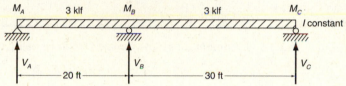

Moment diagram

EXAMPLE 15.9

By using the three-moment theorem, determine all support moments for the structure in Figure 15.15 and draw shear and moment diagrams.

Figure 15.15

Solution.   For constant I:

$$M_A\ell_1 + 2M_B(\ell_1 + \ell_2) + M_C\ell_2$$

$$= -\Sigma P_1\ell_1^2(k_1 - k_1^3) - \Sigma P_2\ell_2^2(k_2 - k_2^3) - \frac{w_1\ell_1^3}{4} - \frac{w_2\ell_2^3}{4}$$

$$M_A = M_C = 0$$

$$2M_B(20 + 30) = -\frac{(3)(20)^3}{4} - \frac{(3)(30)^3}{4} = -6000 - 20{,}250$$

$$100M_B = -26{,}250$$

$$M_B = -262.5'^{k}$$

$$\Sigma M_B \text{ to left} = -262.5$$

$$-262.5 = 20V_A - (3)(20)(10)$$

$$-262.5 = 20V_A - 600$$

$$V_A = \frac{337.5}{20} = 16.9^{k}$$

$$\Sigma M_B \text{ to right} = +262.5$$

$$+262.5 = -(30)(V_C) + (3 \times 30)(15)$$

$$+262.5 = -30V_C + 1350$$

$$V_C = \frac{1087.5}{30} = 36.2^{k}$$

$$\Sigma V = 0$$

$$16.9 + V_B + 36.2 - 150 = 0$$

$$V_B = 96.9^k$$

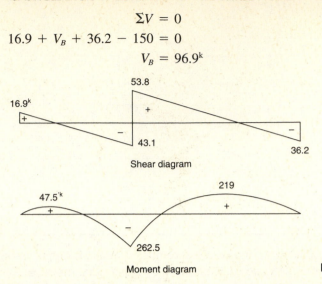

Shear diagram

Moment diagram

EXAMPLE 15.10

Draw shear and moment diagrams for the structure of Figure 15.16.

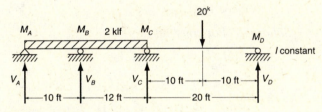

Figure 15.16

Solution.   For constant I:

$$M_A \ell_1 + 2M_B(\ell_1 + \ell_2) + M_C \ell_2$$

$$= -\Sigma P_1 \ell_1^2 (k_1 - k_1^3) - \Sigma P_2 \ell_2^2 (k_2 - k_2^3) - \frac{w_1 \ell_1^3}{4} - \frac{w_2 \ell_2^3}{4}$$

$$M_A = M_D = 0$$

$$2M_B(10 + 12) + M_C(12) = -\frac{(2)(10)^3}{4} - \frac{(2)(12)^3}{4}$$

$$= -500 - 864 = -1364$$

$$44M_B + 12M_C = -1364$$

$$M_B \ell_2 + 2M_C(\ell_2 + \ell_3) + M_D \ell_3 \tag{1}$$

$$= -\Sigma P_2 \ell_2^2 (k_2 - k_2^3) - \Sigma P_3 \ell_3^2 (k_3 - k_3^3) - \frac{w_2 \ell_2^3}{4} - \frac{w_3 \ell_3^3}{4}$$

$$M_B(12) + 2M_C(12 + 20) = -(20)(20)^2\left[\frac{1}{2} - \left(\frac{1}{2}\right)^3\right] - \frac{(2)(12)^3}{4}$$

$$12M_B + 64M_C = -3864 \qquad\qquad (2)$$

By solving Equations (1) and (2) simultaneously for M_B and M_C,

$$M_B = -15.3'^k \qquad M_C = -57.5'^k$$

By determining reactions by statics,

$$V_A = 8.5^k \qquad V_C = 28.4^k$$
$$V_B = 20.0^k \qquad V_D = 7.1^k$$

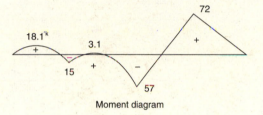

Shear diagram

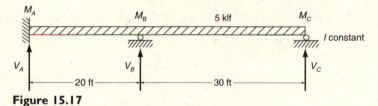

Moment diagram ■

EXAMPLE 15.11

Draw shear and moment diagrams for the structure shown in Figure 15.17.

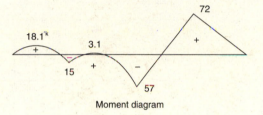

Figure 15.17

Solution. For constant I:

$$M_A\ell_1 + 2M_B(\ell_1 + \ell_2) + M_C\ell_2$$

$$= -\Sigma P_1\ell_1^2(k_1 - k_1^3) - \Sigma P_2\ell_2^2(k_2 - k_2^3) - \frac{w_1\ell_1^3}{4} - \frac{w_2\ell_2^3}{4}$$

By assuming a span of zero length to left of fixed end,

$$M_0\ell_0 + 2M_A(\ell_0 + \ell_1) + M_B\ell_1 = -\frac{w_0\ell_0^3}{4} - \frac{w_1\ell_1^3}{4}$$

$$M_0 = 0$$

$$2M_A(0 + 20) + M_B(20) = -\frac{(5)(20)^3}{4}$$

$$40M_A + 20M_B = -10,000 \tag{1}$$

$$M_A(20) + 2M_B(20 + 30) = -\frac{(5)(20)^3}{4} - \frac{(5)(30)^3}{4}$$

$$20M_A + 100M_B = -43,750 \tag{2}$$

By solving Equations (1) and (2) simultaneously,

$$M_A = -34.75'^k$$
$$M_B = -430.5'^k$$

By computing reactions by statics,

$$V_A = +30.3^k \qquad V_C = +60.7^k$$
$$V_B = +159.0^k$$

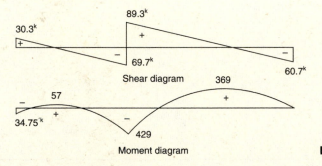

Shear diagram

Moment diagram ∎

EXAMPLE 15.12

Determine the moments at the ends of the uniformly loaded fixed-ended beam shown in Figure 15.18.

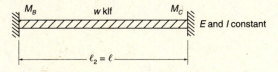

Figure 15.18

Solution.   By assuming spans of zero length (ℓ_1 and ℓ_3) outside each end with zero moments (M_A and M_D),

$$M_A\ell_1 + 2M_B(\ell_1 + \ell_2) + M_C\ell_2 = -\frac{w_1\ell_1^3}{4} - \frac{w_2\ell_2^3}{4}$$

$$M_A = 0 = M_D$$

$$2M_B(\ell) + M_C(\ell) = -\frac{w\ell^3}{4}$$

$$2M_B\ell + M_C\ell = -\frac{w\ell^3}{4} \tag{1}$$

$$M_B\ell_2 + 2M_C(\ell_2 + \ell_3) + M_D\ell_3 = -\frac{w\ell^3}{4}$$

$$M_B\ell + 2M_C\ell = -\frac{w\ell^3}{4} \qquad (2)$$

By solving Equations (1) and (2) simultaneously

$$M_B = -\frac{w\ell^2}{12}$$

$$M_C = -\frac{w\ell^2}{12}$$

The moments at the ends of a fixed-ended beam loaded with a concentrated load can be determined in a similar manner, with the following results:

$$M_A = -\frac{Pab^2}{\ell^2} \qquad M_B = -\frac{Pa^2b}{\ell^2}$$

These expressions will be needed for solution of later problems. See Figure 15.19.

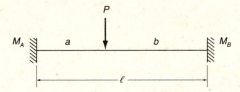

Figure 15.19 ∎

PROBLEMS

For Problems 15.1 through 15.17 analyze the structures by least work; E and I are constant unless otherwise indicated. Circled values are areas, in square inches.

15.1

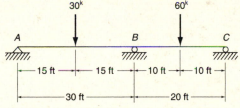

15.2   (*Ans.* $V_A = 18.25^k \uparrow$, $V_C = 18.12^k \uparrow$, $M_B = -236'^k$

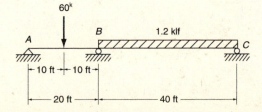

15.3

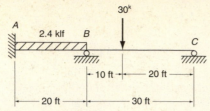

15.4   Problem 13.32 (*Ans.* Member forces left to right $+8.88^k$, 16.10^k, 7.15^k)

15.5

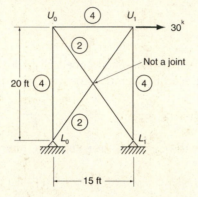

15.6   (*Ans.* $L_0 L_1 = -209.0$ kN, $U_0 L_1 = +161.3$ kN)

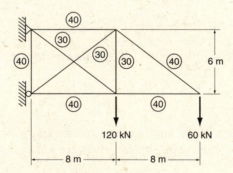

15.7   Problem 13.33

15.8   Find force in tie (*Ans.* Cable tension $+43.4^k$)

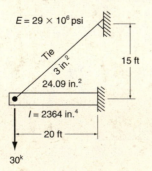

15.9

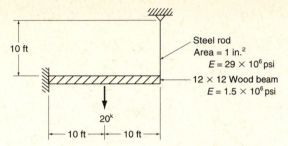

15.10   (*Ans.* $L_0 U_1 = -39.2^k$, $L_1 L_2 = -4.68^k$, cable tension $= +6.48^k$)

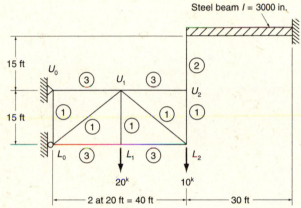

15.11   Problem 13.24

15.12   Problem 13.30 (*Ans.* $L_1 L_2 = +47.5^k$, $U_2 U_3 = +41.5^k$, $U_2 L_2 = +60^k$)

15.13   Problem 13.34

15.14   All member areas $= 10$ in.^{2} (*Ans.* $L_1 L_2 = -18.1^k$, $U_1 U_2 = +31.9^k$, $U_1 L_2 = -16.8^k$)

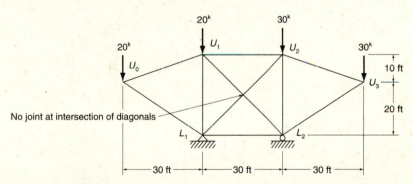

15.15

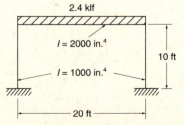

15.16   (*Ans.* $H_A = 18.16^k \rightarrow$, $M_B = 363.2$ ft-k, $M_C = 18.4$ ft-k)

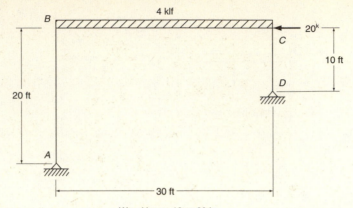

15.17

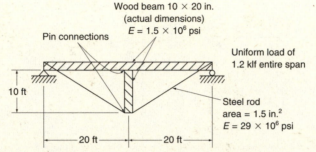

For Problems 15.18 through 15.26 compute all moments for the continuous beams by using the three-moment theorem. Draw shear and moment diagrams.

15.18   (*Ans.* $V_A = 44.0^k \uparrow$, $V_B = 87.57^k \uparrow$, $M_B = -300'^k$, $M_C = -157.2'^k$)

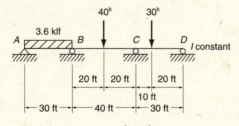

15.19

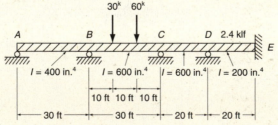

15.20. (*Ans.* $V_A = 148.5^k \uparrow$, $V_B = 31.5^k \uparrow$, $M_A = -720'^k$, $M_B = -45'^k$)

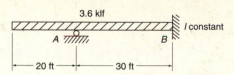

15.21

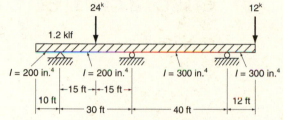

15.22 (*Ans.* $V_A = 131.2$ kN $\uparrow$, $V_C = 233.3$ kN, $M_B = -437.5$ kN · m)

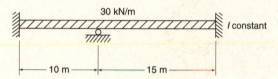

15.23

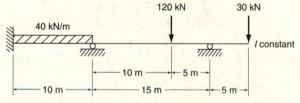

15.24 Problem 13.6 (*Ans.* $M_B = -204.5$ ft-k, $V_C = 29.18^k$)

15.25 Problem 13.8

15.26 Problem 13.11 (*Ans.* $V_A = 46^k \uparrow$, $M_B = -160$ ft-k, $M_C = -183.4$ ft-k)

For Problems 15.27 through 15.31 rework the problems given using SABLE.

15.27 Problem 15.5.

15.28 Problem 15.8 (*Ans.* Cable tension $= +43.4^k$)

15.29 Problem 15.9.

15.30 Problem 15.10 (*Ans.* $U_1 U_2 = 0$, $U_1 L_2 = +6.49^k$)

15.31 Problem 15.17

Chapter 16

Slope Deflection—
A Displacement Method
of Analysis

16.1 INTRODUCTION

George A. Maney introduced slope deflection in a 1915 University of Minnesota engineering publication.[1] His work was an extension of earlier studies of secondary stresses by Manderla[2] and Mohr.[3] For nearly 15 years, until the introduction of moment distribution, slope deflection was the popular "exact" method used for the analysis of continuous beams and frames in the United States.

Slope deflection is a method that takes into account the flexural deformations of beams and frames (that is rotations, settlements, etc.), but that neglects shear and axial deformations. Although this classical method is generally considered to be obsolete, its study can be useful for several reasons. These include:

1. Slope deflection is convenient for hand analysis of some small structures.
2. A knowledge of the method provides an excellent background for understanding the moment distribution method of Chapters 18 and 19.
3. It is a special case of the displacement or stiffness method of analysis previously defined in Section 13.1 and provides a very effective introduction for the matrix formulation of structures described in Chapters 20 to 21.
4. The slopes and deflections determined by slope deflection enable the analyst to sketch easily the deformed shape of a particular structure. The result is that he or she has a better "feel" for the behavior of structures.

[1] G. A. Maney, *Studies in Engineering,* No. 1 (Minneapolis: University of Minnesota, 1915).

[2] H. Manderla, "Die Berechnung der Sekundarspannungen," *Allg. Bautz* 45 (1880): 34.

[3] O. Mohr, "Die Berechnung der Fachwerke mit starren knotenverbingungen," *Zivilinginieur* (1892).

Hospital, Stamford, Connecticut. (Courtesy of the American Concrete Institute.)

16.2 DERIVATION OF SLOPE-DEFLECTION EQUATIONS

The name "slope deflection" comes from the fact that the moments at the ends of the members in statically indeterminate structures are expressed in terms of the rotations (or slopes) and deflections of the joints. For developing the equations, members are assumed to be of constant cross section between each pair of supports. Although it is possible to derive expressions for members of varying section, the results are so complex as to be of little practical value. It is further assumed that the joints in a structure may rotate or deflect, but the angles between the members meeting at a joint remain unchanged.

Span AB of the continuous beam of Figure 16.1(a), as shown on page 432, is considered for the following discussion. If the span is completely fixed at each end, the slope of the elastic curve of the beam at the ends is zero. External loads produce fixed-end moments, and these moments cause the span to take the shape shown in Figure 16.1(b). Joints A and B are actually not fixed and will rotate slightly under load to some position such as the one shown in Figure 16.1(c). In addition to the rotation of the joints there may possibly be some settlement of one or both of the supports, which will cause a chord rotation of the member as shown in part (d), where support B is assumed to have settled an amount Δ.

From the study of Figure 16.1 the values of the final end moments at A and B (M_{AB} and M_{BA}) are seen to be equal to the sum of the moments caused by the following:

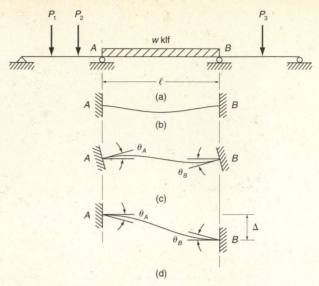

Figure 16.1

1. The fixed-end moments (FEM_{AB} and FEM_{BA}), which can be determined with the moment-area theorems as illustrated in Example 10.6. A detailed list of fixed-end moment expressions is presented in Figure 18.5.
2. The moments caused by the rotations of joints A and B (θ_A and θ_B).
3. The moments caused by chord rotation ($\psi = \Delta/\ell$) if one or both of the joints settles or deflects.

Rotations of the joints in a structure cause changes in the slopes of the tangents to the elastic curves at those points. For a particular beam the change in slope equals the end shear of the beam when it is loaded with the M/EI diagram. The beam is assumed to have the end moments FEM_{AB} and FEM_{BA}, shown in Figure 16.2.

The end reactions or end slopes are as follows:

$$\theta_A = \frac{(\tfrac{1}{2})(M_{AB}/EI)(\ell)(\tfrac{2}{3}\ell) - (\tfrac{1}{2})(M_{BA}/EI)(\ell)(\tfrac{1}{3}\ell)}{\ell}$$

$$= \frac{\ell}{6EI}(2M_{AB} - M_{BA})$$

$$\theta_B = \frac{(\tfrac{1}{2})(M_{BA}/EI)(\ell)(\tfrac{2}{3}\ell) - (\tfrac{1}{2})(M_{AB}/EI)(\ell)(\tfrac{1}{3}\ell)}{\ell}$$

$$= \frac{\ell}{6EI}(2M_{BA} - M_{AB})$$

If one of the supports of the beam settled or deflected an amount Δ, the angles θ_A and θ_B caused by joint rotation would be changed by Δ/ℓ (or ψ), as illustrated in

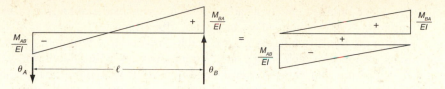

Figure 16.2

Figure 16.1(d). Adding chord rotation to the expressions results in the following total values for the slopes of the tangents to the elastic curves at the ends of the beams.

$$\theta_A = \frac{\ell}{6EI}(2M_{AB} - M_{BA}) + \psi$$

$$\theta_B = \frac{\ell}{6EI}(2M_{BA} - M_{AB}) + \psi$$

Solving the equations simultaneously for M_{AB} and M_{BA} gives the values of the end moments due to slopes and deflections. In these expressions I/ℓ has been replaced with K, the so-called *stiffness factor* (see Chapter 18).

$$M_{AB} = 2EK(2\theta_A + \theta_B - 3\psi)$$
$$M_{BA} = 2EK(\theta_A + 2\theta_B - 3\psi)$$

The final end moments are equal to the moments due to slopes and deflections plus the fixed-end moments. The slope-deflection equations are as follows:

$$M_{AB} = 2EK(2\theta_A + \theta_B - 3\psi) + FEM_{AB}$$
$$M_{BA} = 2EK(\theta_A + 2\theta_B - 3\psi) + FEM_{BA}$$

With these equations it is possible to express the end moments in a structure in terms of joint rotations and settlements. By previous methods it has been necessary to write one equation for each redundant in the structure. The number of unknowns in each equation totaled the number of redundants. The work of solving those equations was considerable for highly redundant structures. Slope deflection appreciably reduces the amount of work involved in analyzing multiredundant structures because the unknown moments are expressed in terms of only a few unknown joint rotations and settlements. Even for multistory frames, the number of unknown θ and ψ values appearing in any one equation is rarely more than five or six, whereas the degree of indeterminacy of the structure is many times that figure.

16.3 APPLICATION OF SLOPE-DEFLECTION EQUATIONS TO CONTINUOUS BEAMS

Examples 16.1 to 16.4 illustrate the analysis of statically indeterminate beams with the slope-deflection equations. Each member in the beams is considered individually, its fixed-end moments are computed, and one equation is written for the moment at each end of the member. For span AB of Example 16.1, equations for M_{AB} and M_{BA} are written; for span BC, equations for M_{BC} and M_{CB} are written, and so on.

The moment equations are written in terms of the unknown values of θ at the supports. The two moments at an interior support must total zero, as $M_{BA} + M_{BC} = 0$ at support B in Example 16.1. Expressions are therefore written for the total moment at each support, which gives a set of simultaneous equations from which the unknown θ values may be obtained. Two conditions that will simplify the solution of the equations may exist. These are fixed ends for which the θ values must be zero and simple ends for which the moment is zero. A special expression is derived at the end of Example 16.2 for end spans that are simply supported.

The beams of Examples 16.1 and 16.2 have unyielding supports, and ψ is zero for all of the equations. Some support settlement occurs for the beams of Examples 16.3 and 16.4, and ψ is included in the equations. Chord rotation is considered positive when the chord of a beam is rotated clockwise by the settlements, meaning that the sign is the same no matter which end is being considered as the entire beam rotates in that direction.

When span lengths, moduli of elasticity, and moments of inertia are constant for the spans of a continuous beam, the $2EK$ values are constant and may be canceled from the equations. Should the values of K vary from span to span, as they often do, it is convenient to express them in terms of relative values, as is done in Example 16.6.

The major difficulty experienced in applying the slope-deflection equations lies in the use of correct signs. It is essential to understand these signs before attempting to apply the equations.

For previous work in this book a plus sign for the moment has indicated tension in the bottom fibers, whereas a negative sign has indicated tension in the top fibers. This sign convention was necessary for drawing moment diagrams.

In applying the slope-deflection equations it is simpler to use a convention in which signs are given to clockwise and counterclockwise moments at the member ends. Once these moments are determined, their signs can be easily converted to the usual convention for drawing moment diagrams (known as beam notation).

The following convention is used in this chapter for slope deflection and in Chapters 18 and 19 for moment distribution: Should a member cause a moment that tends to rotate a joint clockwise, the joint moment is considered negative; if counterclockwise, it is positive. In other words, a clockwise resisting moment on the member herein is considered positive, and a counterclockwise resisting moment is considered negative.

Figure 16.3 is presented to demonstrate this convention. The moment at the left end, M_{AB}, tends to rotate the joint clockwise and is considered negative. Note that the

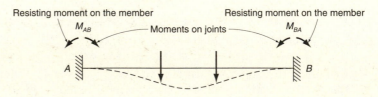

Figure 16.3

resisting moment would be counterclockwise. At the right end of the beam the loads have caused a moment that tends to rotate the joint counterclockwise, and M_{BA} is considered positive.

EXAMPLE 16.1

Determine all support moments of the structure in Figure 16.4 by the method of slope deflection.

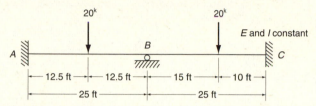

Figure 16.4

Solution.   By computing fixed-end moments (with reference to Figure 18.5),

$$\text{FEM}_{AB} = -\frac{(20)(12.5)(12.5)^2}{(25)^2} = -62.5^{'k}$$

$$\text{FEM}_{BA} = +\frac{(20)(12.5)(12.5)^2}{(25)^2} = +62.5^{'k}$$

$$\text{FEM}_{BC} = -\frac{(20)(15)(10)^2}{(25)^2} = -48^{'k}$$

$$\text{FEM}_{CB} = +\frac{(20)(10)(15)^2}{(25)^2} = +72^{'k}$$

By writing the equations, and noting $\theta_A = \theta_C = \psi = 0$,

$$M_{AB} = 2EK\theta_B - 62.5$$
$$M_{BA} = 4EK\theta_B + 62.5$$
$$M_{BC} = 4EK\theta_B - 48$$
$$M_{CB} = 2EK\theta_B + 72$$
$$\Sigma M_B = 0 = M_{BA} + M_{BC}$$
$$4EK\theta_B + 62.5 + 4EK\theta_B - 48 = 0$$
$$EK\theta_B = -1.8125$$

Final moments:

$$M_{AB} = (2)(-1.8125) - 62.5 = -66.125^{'k}$$
$$M_{BA} = (4)(-1.8125) + 62.5 = +55.25^{'k}$$
$$M_{BC} = (4)(-1.8125) - 48 = -55.25^{'k}$$
$$M_{CB} = (2)(-1.8125) + 72 = +68.375^{'k} \quad \blacksquare$$

EXAMPLE 16.2

Find all moments in the structure of Figure 16.5 by slope deflection. Use the modified equation for simple ends developed in the discussion at the end of the problem.

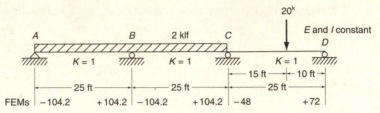

Figure 16.5

Solution.

$$M_{BA} = 3EK(\theta_B - \psi) + \text{FEM}_{BA} - \tfrac{1}{2}\text{FEM}_{AB}$$
$$M_{AB} = 2EK(2\theta_A + \theta_B - 3\psi) + \text{FEM}_{AB}$$

By writing the equations, and canceling *EK* because it is constant for all spans,

$$M_{AB} = M_{DC} = 0$$
$$M_{BA} = (3)(\theta_B) + 104.2 - (\tfrac{1}{2})(-104.2) = 3\theta_B + 156.3$$
$$M_{BC} = (2)(2\theta_B + \theta_C) - 104.2 = 4\theta_B + 2\theta_C - 104.2$$
$$M_{CB} = (2)(\theta_B + 2\theta_C) + 104.2 = 2\theta_B + 4\theta_C + 104.2$$
$$M_{CD} = (3)(\theta_C) - 48 - (\tfrac{1}{2})(+72) = 3\theta_C - 84$$
$$\Sigma M_B = 0 = M_{BA} + M_{BC}$$
$$3\theta_B + 156.3 + 4\theta_B + 2\theta_C - 104.2 = 0$$
$$7\theta_B + 2\theta_C = -52.1 \tag{1}$$
$$\Sigma M_C = 0 = M_{CB} + M_{CD}$$
$$2\theta_B + 4\theta_C + 104.2 + 3\theta_C - 84 = 0$$
$$2\theta_B + 7\theta_C = -20.2 \tag{2}$$

By solving Equations (1) and (2) simultaneously,

$$\theta_B = -7.2$$
$$\theta_C = -0.83$$

Final end moments:

$$M_{BA} = (3)(-7.2) + 156.3 = +134.7^{'k}$$
$$M_{BC} = (4)(-7.2) + (2)(-0.83) - 104.2 = -134.7^{'k}$$
$$M_{CB} = (2)(-7.2) + (4)(-0.83) + 104.2 = +86.5^{'k}$$
$$M_{CD} = (3)(-0.83) - 84 = -86.5^{'k} \quad \blacksquare$$

Discussion

For a beam with simply supported ends, it is obvious that the moments at those ends must be zero for equilibrium ($M_{AB} = M_{DC} = 0$ for the beam of Figure 16.5). Application of the usual slope-deflection equations will yield zero moments, but it seems a waste of time to go through a process to determine the value of all of the moments in the beam when by inspection two of them obviously are zero. The usual slope-deflection equations are as follows:

$$M_{AB} = 2EK(2\theta_A + \theta_B - 3\psi) + \text{FEM}_{AB} \tag{1}$$

$$M_{BA} = 2EK(\theta_A + 2\theta_B - 3\psi) + \text{FEM}_{BA} \tag{2}$$

Assuming end A to be simply end supported, the value of M_{AB} is zero. Solving the two equations simultaneously by eliminating θ_A gives a simplified expression for M_{BA}, which has only one unknown, θ_B. The resulting simplified equation will expedite considerably the solution of continuous beams with simple ends.

Twice Equation (2)

$$2M_{BA} = 2EK(2\theta_A + 4\theta_B - 6\psi) + 2\text{FEM}_{BA}$$

minus Equation (1)

$$0 = 2EK(2\theta_A + \theta_B - 3\psi) + \text{FEM}_{AB}$$

gives

$$2M_{AB} = 2EK(3\theta_B - 3\psi) + 2\text{FEM}_{BA} - \text{FEM}_{AB}$$

$$M_{BA} = 3EK(\theta_B - \psi) + \text{FEM}_{BA} - \tfrac{1}{2}\text{FEM}_{AB}$$

EXAMPLE 16.3 ——

Find the moment at support B in the beam of Figure 16.6, assuming B settles 0.25 in., or 0.0208 ft.

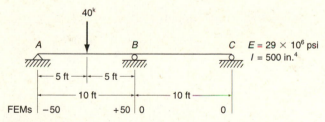

FEMs | -50 $+50 \mid 0$ $0 \mid$

Figure 16.6

Solution.   By writing the equations, and noting that $M_{AB} = M_{CB} = 0$,

$$M_{BA} = 3EK\left(\theta_B - \frac{+0.0208}{10}\right) + 50 - (\tfrac{1}{2})(-50)$$

$$M_{BA} = 3EK\theta_B - 0.00624EK + 75$$

$$M_{BC} = 3EK\left(\theta_B - \frac{-0.0208}{10}\right)$$

$$M_{BC} = 3EK\theta_B + 0.00624EK$$

$$\Sigma M_B = 0 = M_{BA} + M_{BC}$$

$$3EK\theta_B - 0.0624EK + 75 + 3EK\theta_B + 0.0624EK = 0$$

$$6EK\theta_B + 75 = 0$$

$$EK\theta_B = -12.5$$

$$M_{BA} = (3)(-12.5) - 0.00624EK + 75$$

$$EK = \frac{(29 \times 10^6)(500)}{(12 \times 1000)(12 \times 10)} = 10,070$$

$$M_{BA} = -37.5 - 62.8 + 75 = -25.3^{'k}$$

$$M_{BC} = (3)(-12.5) + 62.8 = +25.3^{'k} \quad \blacksquare$$

EXAMPLE 16.4 ──

Determine all moments for the beam of Figure 16.7, which is assumed to have the following support settlements: $A = 1.25$ in. $= 0.104$ ft, $B = 2.40$ in. $= 0.200$ ft $C = 2.75$ in. $= 0.229$ ft, and $D = 1.10$ in. $= 0.0917$ ft.

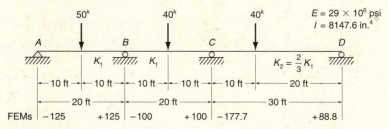

Figure 16.7

Solution.   For support B:

$$M_{BA} = 3EK_1\left(\theta_B - \frac{0.096}{20}\right) + 125 - (\tfrac{1}{2})(-125)$$

$$M_{BA} = 3EK_1\theta_B - 0.0144EK_1 + 187.5$$

$$M_{BC} = 2EK_1\left[2\theta_B + \theta_C - 3\frac{0.029}{20}\right] - 100$$

$$M_{BC} = 4EK_1\theta_B + 2EK_1\theta_C - 0.0087EK_1 - 100$$

For support C:

$$M_{CB} = 2EK_1\left[\theta_B + 2\theta_C - (3)\left(\frac{0.029}{20}\right)\right] + 100$$

$$M_{CB} = 2EK_1\theta_B + 4EK_1\theta_C - 0.0087EK_1 + 100$$

$$M_{CD} = 3EK_2\left(\theta_C - \frac{-0.1373}{30}\right) - 177.7 - (\tfrac{1}{2})(+88.8)$$

$$M_{CD} = 3EK_2\theta_C + 0.01373EK_2 - 222.1$$

$$M_{CD} = 2EK_1\theta_C + 0.00915EK_1 - 222.1$$

$$\Sigma M_B = 0 = M_{BA} + M_{BC}$$

$$3EK_1\theta_B - 0.0144EK_1 + 187.5 + 4EK_1\theta_B + 2EK_1\theta_C$$
$$- 0.0087EK_1 - 100 = 0$$

$$7EK_1\theta_B + 2EK_1\theta_C - 0.0231EK_1 + 87.5 = 0 \qquad (1)$$

$$\Sigma M_C = 0 = M_{CB} + M_{CD}$$

$$2EK_1\theta_B + 4EK_1\theta_C - 0.0087EK_1 + 100 + 2EK_1\theta_C$$
$$+ 0.00915EK_1 - 222.1 = 0$$

$$2EK_1\theta_B + 6EK_1\theta_C + 0.00045EK_1 - 122.1 = 0 \qquad (2)$$

By solving Equations (1) and (2) simultaneously for $EK_1\theta_B$ and $EK_1\theta_C$,

$$EK_1\theta_B = 0.00367EK_1 - 20.2$$

$$EK_1\theta_C = -0.00129EK_1 + 27.1$$

$$EK_1 = \frac{(29 \times 10^6)(8147.6)}{(12 \times 20)(12 \times 1000)} = 82{,}040^{'k}$$

$$EK_1\theta_B = (0.00367)(82{,}040) - 20.2 = +280.9$$

$$EK_1\theta_C = -(0.00129)(81{,}970) + 27.1 = -78.7$$

Final moments:

$$M_{BA} = (3)(280.9) - (0.0144)(82{,}040) + 187.5 = -151^{'k}$$

$$M_{BC} = (4)(280.9) + (2)(-78.7) - (0.0087)(82{,}040) - 100 = +152^{'k}$$

$$M_{CB} = (2)(280.9) + (4)(-78.7) - (0.0087)(82{,}040) + 100 = -367^{'k}$$

$$M_{CD} = (2)(-78.7) + (0.00915)(82{,}040) - 222.1 = +371^{'k} \quad \blacksquare$$

16.4 ANALYSIS OF FRAMES—NO SIDESWAY

The slope-deflection equations may be applied to statically indeterminate frames in the same manner as they were to continuous beams if there is no possibility for the frames to lean or deflect asymmetrically. A frame theoretically will not lean to one side or sway if it is symmetrical about the centerline as to dimensions, loads, and moments of inertia, or if it is prevented from swaying by other parts of the structure. On the following page, the frame of Figure 16.8 cannot sway because member AB is restrained against horizontal movement. Example 16.5 illustrates the analysis of a simple frame with no sidesway. Where sidesway occurs, the joints of the frame move, and that affects the θ values and the moments, as discussed in Section 16.5.

Lehigh Shop Center. (Courtesy Bethlehem Steel Corporation.)

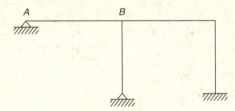

Figure 16.8

EXAMPLE 16.5 _____

Find all moments for the frame shown in Figure 16.9, which has no sidesway because of symmetry.

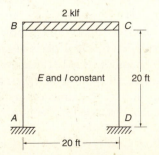

Figure 16.9

Solution.   By computing fixed-end moments,

$$\text{FEM}_{BC} = -\frac{(2)(20)^2}{12} = -66.7^{'k}$$

$$\text{FEM}_{CB} = +\frac{(2)(20)^2}{12} = +66.7^{'k}$$

By writing the equations, and noting $\theta_A = \theta_D = \psi = 0$ and $2EK$ is constant for all members, which permits it to be neglected,

$$M_{AB} = \theta_B$$
$$M_{BA} = 2\theta_B$$
$$M_{BC} = 2\theta_B + \theta_C - 66.7$$
$$M_{CB} = \theta_B + 2\theta_C + 66.7$$
$$M_{CD} = 2\theta_C$$
$$M_{DC} = \theta_C$$
$$\Sigma M_B = 0 = M_{BA} + M_{BC}$$
$$2\theta_B + 2\theta_B + \theta_C - 66.7 = 0$$
$$4\theta_B + \theta_C = 66.7 \tag{1}$$
$$\Sigma M_C = 0 = M_{CB} + M_{CD}$$
$$\theta_B + 2\theta_C + 66.7 + 2\theta_C = 0$$
$$\theta_B + 4\theta_C = -66.7 \tag{2}$$

By solving Equations (1) and (2) simultaneously,

$$\theta_B = +22.2$$
$$\theta_C = -22.2$$

Final end moments

$$M_{AB} = +22.2^{'k} \qquad M_{CB} = +44.4^{'k}$$
$$M_{BA} = +44.4^{'k} \qquad M_{CD} = -44.4^{'k}$$
$$M_{BC} = -44.4^{'k} \qquad M_{DC} = -22.2^{'k} \quad \blacksquare$$

16.5 ANALYSIS OF FRAMES WITH SIDESWAY

The loads, moments of inertia, and dimensions of the frame of Figure 16.10 on page 442 are not symmetrical about the centerline, and the frame will obviously sway to one side. Joints B and C deflect to the right, which causes chord rotations in members AB and CD, there being theoretically no rotation of BC if axial shortening (or lengthening) of AB and CD is neglected. Neglecting axial deformation of BC, each of the joints deflects the same horizontal distance Δ.

The chord rotations of members AB and CD, due to sidesway, can be seen to equal Δ/ℓ_{AB} and Δ/ℓ_{CD}, respectively. (Note that the shorter a member for the same Δ, the larger its chord rotation and thus the larger the effect on its moment.) For this

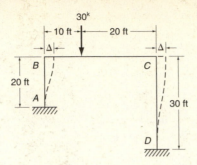

Figure 16.10

frame ψ_{AB} is 1.5 times as large as ψ_{CD} because ℓ_{AB} is only two-thirds of ℓ_{CD}. It is convenient to work with only one unknown chord rotation, and in setting up the slope-deflection equations, relative values are used. The value ψ is used for the two equations for member AB, whereas $\frac{2}{3}\psi$ is used for the equation for member CD.

Example 16.6 presents the analysis of the frame of Figure 16.10. By noting $\theta_A = \theta_D = 0$, it will be seen that the six end-moment equations for the entire structure contain a total of three unknowns: θ_B, θ_C, and ψ. There are present, however, three conditions that permit their determination. They are

1. The sum of the moments at B is zero ($\Sigma M_B = 0 = M_{BA} + M_{BC}$).
2. The sum of the moments at C is zero ($\Sigma M_C = 0 = M_{CB} + M_{CD}$).
3. The sum of the horizontal forces on the entire structure must be zero.

The only horizontal forces are the horizontal reactions at A and D, and they will be equal in magnitude and opposite in direction. Horizontal reactions may be computed for each of the columns by dividing the column moments by the column heights, or, that is, by taking moments at the top of each column. The sum of the two reactions must be zero.

$$H_A = \frac{M_{AB} + M_{BA}}{\ell_{AB}} \qquad H_D = \frac{M_{CD} + M_{DC}}{\ell_{DC}}$$

$$\Sigma H = 0 = \frac{M_{AB} + M_{BA}}{\ell'_{AB}} + \frac{M_{CD} + M_{DC}}{\ell_{CD}}$$

EXAMPLE 16.6 ──

Determine all of the moments for the frame shown in Figure 16.11, for which E and I are constant.

Solution.   By using relative ψ values.

$$\psi_{AB} = \frac{\Delta}{20} \qquad \psi_{CD} = \frac{\Delta}{30} \qquad \frac{\psi_{AB}}{\psi_{CD}} = \frac{3}{2}$$

By using relative $2EK$ values,

for AB $2EK = \dfrac{2E}{20}$ say a relative value of 3 with respect to $2EK$ values of BC and CD

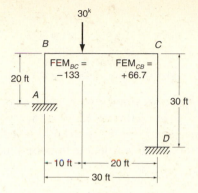

Figure 16.11

$$\text{for } BC \qquad 2EK = \frac{2E}{30} \qquad \text{use relative value} = 2$$

$$\text{for } CD \qquad 2EK = \frac{2E}{30} \qquad \text{use relative value} = 2$$

By writing the equations, and noting $\theta_A = \theta_D = \psi_{BC} = 0$,

$$M_{AB} = 3(\theta_B - 3\psi) = 3\theta_B - 9\psi$$
$$M_{BA} = 3(2\theta_B - 3\psi) = 6\theta_B - 9\psi$$
$$M_{BC} = 2(2\theta_B + \theta_C) - 133 = 4\theta_B + 2\theta_C - 133$$
$$M_{CB} = 2(\theta_B + 2\theta_C) + 66.7 = 2\theta_B + 4\theta_C + 66.7$$
$$M_{CD} = 2[2\theta_C - (3)(\tfrac{2}{3}\psi)] = 4\theta_C - 4\psi$$
$$M_{DC} = 2[\theta_C - (3)(\tfrac{2}{3}\psi)] = 2\theta_C - 4\psi$$
$$\Sigma M_B = 0 = M_{BA} + M_{BC}$$
$$6\theta_B - 9\psi + 4\theta_B + 2\theta_C - 133 = 0$$
$$10\theta_B + 2\theta_C - 9\psi = 133 \qquad (1)$$
$$\Sigma M_C = 0 = M_{CB} + M_{CD}$$
$$2\theta_B + 4\theta_C + 66.7 + 4\theta_C - 4\psi = 0$$
$$2\theta_B + 8\theta_C - 4\psi = -66.7 \qquad (2)$$
$$\Sigma H = 0 = H_A + H_D$$

$$\frac{M_{AB} + M_{BA}}{\ell_{AB}} + \frac{M_{CD} + M_{DC}}{\ell_{CD}} = 0$$

$$\frac{3\theta_B - 9\psi + 6\theta_B - 9\psi}{20} + \frac{4\theta_C - 4\psi + 2\theta_C - 4\psi}{30} = 0$$

$$27\theta_B + 12\theta_C - 70\psi = 0 \qquad (3)$$

By solving Equations (1), (2), and (3) simultaneously,

$$\theta_B = +21.2$$
$$\theta_C = -10.5$$
$$\psi = +6.4$$

Final moments:

$$M_{AB} = +6.0^{'k} \qquad M_{CB} = +67.4^{'k}$$

$$M_{BA} = +69.4^{'k} \qquad M_{CD} = -67.4^{'k}$$

$$M_{BC} = -69.4^{'k} \qquad M_{DC} = -46.6^{'k} \quad \blacksquare$$

Slope deflection can be applied to frames with more than one condition of sidesway, such as the two-story frame of Figure 16.12. Hand analysis of frames of this type usually is handled more conveniently by the moment-distribution method, but a knowledge of the slope-deflection solution is valuable in understanding the moment-distribution solution.

The horizontal loads cause the structure to lean to the right; joints B and E deflect horizontally a distance Δ_1; and joints C and D deflect horizontally $\Delta_1 + \Delta_2$, as shown in Figure 16.12. The chord rotations for the columns ψ_1 and ψ_2 will therefore equal Δ_1/ℓ_{AB} for the lower level and Δ_2/ℓ_{BC} for the upper level.

The slope-deflection equations may be written for the moment at each end of the six members, and the usual joint-condition equations are available as follows:

$$\Sigma M_B = 0 = M_{BA} + M_{BC} + M_{BE} \tag{16.1}$$

$$\Sigma M_C = 0 = M_{CB} + M_{CD} \tag{16.2}$$

$$\Sigma M_D = 0 = M_{DC} + M_{DE} \tag{16.3}$$

$$\Sigma M_E = 0 = M_{ED} + M_{EB} + M_{EF} \tag{16.4}$$

These equations involve six unknowns (θ_B, θ_C, θ_D, θ_E, ψ_1, and ψ_2), noting that θ_A and θ_F are equal to zero. Two more equations are necessary for determining the unknowns, and they are found by considering the horizontal forces or shears on the frame. It is obvious that the sum of the horizontal resisting forces on any level must be equal and opposite to the external horizontal shear on that level. For the bottom level the horizontal shear equals $P_1 + P_2$, and the reactions at the base of each column equal the end moments divided by the column heights. Therefore

$$P_1 + P_2 - H_A - H_F = 0$$

$$H_A = \frac{M_{AB} + M_{BA}}{\ell_{AB}}$$

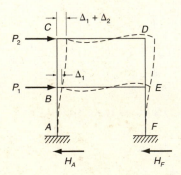

Figure 16.12

$$H_F = \frac{M_{EF} + M_{FE}}{\ell_{EF}}$$

$$\frac{M_{AB} + M_{BA}}{\ell_{AB}} + \frac{M_{EF} + M_{FE}}{\ell_{EF}} + P_1 + P_2 = 0 \qquad (16.5)$$

A similar condition equation may be written for the top level, which has an external shear of P_2. The moments in the columns produce shears equal and opposite to P_2, which permits the writing of the following equation for the level:

$$\frac{M_{BC} + M_{CB}}{\ell_{BC}} + \frac{M_{DE} + M_{ED}}{\ell_{DE}} + P_2 = 0 \qquad (16.6)$$

Alcoa Building, San Francisco. Dominant feature of the topped-out steel framework of the "tiara" of San Francisco's Golden Gateway project, the Alcoa Building, is the diamond-patterned seismic-resisting shear walls. (Courtesy Bethlehem Steel Corporation.)

Six condition equations are available for determining the six unknowns in the end-moment equations, and the problem may be solved as illustrated in Example 16.7. It is desirable to assume all θ and ψ values to be positive when setting up the equations. In this way the signs of the answers will take care of themselves.

No matter how many floors the building has, one shear-condition equation is available for each floor. The slope-deflection procedure is not very practical for multistory buildings. For a six-story building four bays wide there will be 6 unknown ψ values and 30 unknown θ values, or a total of 36 simultaneous equations to solve. (Their solution is not quite as bad as it may seem, because each equation would contain only a few unknowns—not 36.)

EXAMPLE 16.7

Find the moments for the frame of Figure 16.13 by the method of slope deflection.

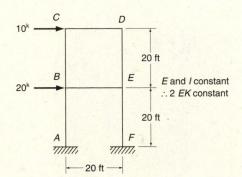

Figure 16.13

Solution. By noting that chords BE and CD do not rotate.

$$\psi_{AB} = \psi_{EF} = \psi_1 \qquad \psi_{BC} = \psi_{DE} = \psi_2$$

By writing the equations, and noting that $\theta_A = \theta_F = 0$,

$$M_{AB} = \theta_B - 3\psi_1 \qquad\qquad M_{DC} = \theta_C + 2\theta_D$$

$$M_{BA} = 2\theta_B - 3\psi_1 \qquad\qquad M_{DE} = 2\theta_D + \theta_E - 3\psi_2$$

$$M_{BC} = 2\theta_B + \theta_C - 3\psi_2 \qquad M_{ED} = \theta_D + 2\theta_E - 3\psi_2$$

$$M_{BE} = 2\theta_B + \theta_E \qquad\qquad M_{EB} = \theta_B + 2\theta_E$$

$$M_{CB} = \theta_B + 2\theta_C - 3\psi_2 \qquad M_{EF} = 2\theta_E - 3\psi_1$$

$$M_{CD} = 2\theta_C + \theta_D \qquad\qquad M_{FE} = \theta_E - 3\psi_1$$

$$\Sigma M_B = 0 = M_{BA} + M_{BC} + M_{BE}$$

$$2\theta_B - 3\psi_1 + 2\theta_B + \theta_C - 3\psi_2 + 2\theta_B + \theta_E = 0$$

$$6\theta_B + \theta_C + \theta_E - 3\psi_1 - 3\psi_2 = 0 \qquad\qquad (1)$$

$$\Sigma M_C = 0 = M_{CB} + M_{CD}$$

$$\theta_B + 2\theta_C - 3\psi_2 + 2\theta_2 + \theta_D = 0$$

$$\theta_B + 4\theta_C + \theta_D - 3\psi_2 = 0 \qquad\qquad (2)$$

$$\Sigma M_D = 0 = M_{DC} + M_{DE}$$
$$\theta_C + 2\theta_D + 2\theta_D + \theta_E - 3\psi_2 = 0$$
$$\theta_C + 4\theta_D + \theta_E - 3\psi_2 = 0 \qquad (3)$$
$$\Sigma M_E = 0 = M_{ED} + M_{EB} + M_{EF}$$
$$\theta_D + 2\theta_E - 3\psi_2 + \theta_B + 2\theta_E + 2\theta_E - 3\psi_1 = 0$$
$$\theta_B + \theta_D + 6\theta_E - 3\psi_1 - 3\psi_2 = 0 \qquad (4)$$

$\Sigma H = -10$, top level:

$$\frac{M_{BC} + M_{CB}}{20} + \frac{M_{DE} + M_{ED}}{20} = -10$$

$$\frac{2\theta_B + \theta_C - 3\psi_2 + \theta_B + 2\theta_C - 3\psi_2}{20}$$

$$+ \frac{2\theta_D + \theta_E - 3\psi_2 + \theta_D + 2\theta_E - 3\psi_2}{20} = -10$$

$$3\theta_B + 3\theta_C + 3\theta_D + 3\theta_E - 12\psi_2 = -200 \qquad (5)$$

$\Sigma H = 30$, bottom level:

$$\frac{M_{AB} + M_{BA}}{20} + \frac{M_{EF} + M_{FE}}{20} = -30$$

$$\frac{\theta_B - 3\psi_1 + 2\theta_B - 3\psi_1}{20} + \frac{2\theta_E - 3\psi_1 + \theta_E - 3\psi_1}{20} = -30$$

$$3\theta_B + 3\theta_E - 12\psi_1 = -600 \qquad (6)$$

By solving equations simultaneously,

$$\theta_B = \theta_E = +52.75 \qquad \psi_1 = +76.37$$
$$\theta_C = \theta_D = +21.82 \qquad \psi_2 = +53.95$$

Final moments:

$$M_{AB} = M_{FE} = -176.4^{'k} \qquad M_{BE} = M_{EB} = +158.2^{'k}$$
$$M_{BA} = M_{EF} = -123.6^{'k} \qquad M_{CB} = M_{DE} = -65.5^{'k}$$
$$M_{BC} = M_{ED} = -34.6^{'k} \qquad M_{CD} = M_{DC} = +65.5^{'k} \quad \blacksquare$$

16.6 ANALYSIS OF FRAMES WITH SLOPING LEGS

Space is not taken in this chapter to present the analysis of frames with sloping legs such as the ones shown in Figures 19.12 and 19.14. A slope-deflection analysis of such frames with a handheld calculator is a rather tedious job, and the moment-distribution procedure of Chapters 18 and 19 is somewhat more practical. Once these kinds of frames have been analyzed by moment distribution the analyst will be able to return to this chapter and use slope deflection for their analysis.

Transit shed, Port of Long Beach, California. (Courtesy of the American Institute of Steel Construction.)

PROBLEMS

For Problems 16.1 through 16.13 compute the end moments of the beams with the slope-deflection equations. Draw shear and moment diagrams. E is constant for all problems.

16.1

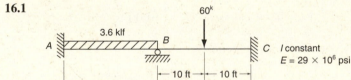

16.2   (Ans. $M_B = 152.2$ ft-k, $M_D = 233.9$ ft-k)

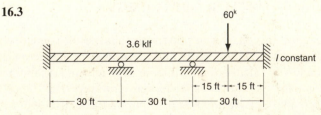

16.3

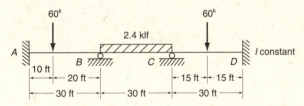

16.4   (*Ans.* $M_B = 560$ ft-k, $M_C = 655.4$ ft-k)

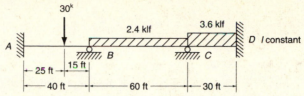

16.5

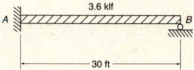

16.6   (*Ans.* $V_A = 67.5^k \uparrow$, $M_{AB} = 405$ ft-k)

16.7

16.8   (*Ans.* $M_A = 205$ kN · m, $M_B = 265$ kN · m)

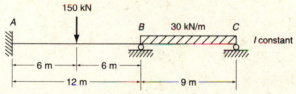

16.9

16.10   (*Ans.* $M_A = 31.1$ ft-k, $M_B = 57.7$ ft-k)

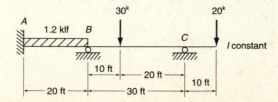

16.11   See Figure 18.5 for fixed-end moments.

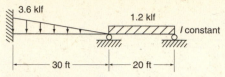

16.12   The beam of Problem 16.2 if both ends are simply supported (*Ans. M_B = 205.1 ft-k, M_C = 252.4 ft-k*)

16.13   The beam of Problem 16.4 if the right end is simply supported

For Problems 16.14 through 16.20 determine the end moments for the members of all of the structures by using the slope deflection equations.

16.14   (*Ans. M_A = 137 ft-k, M_B = 274.1 ft-k*)

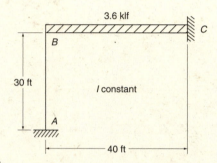

16.15

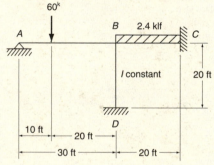

16.16   (*Ans. $M_A = M_C$ = 70 kN · m, $M_B = M_D$ = 380 kN · m*)

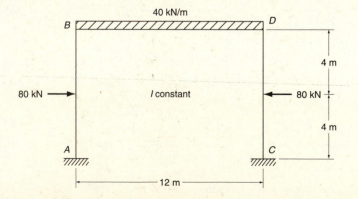

16.17

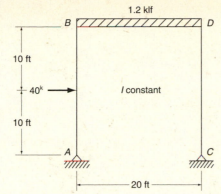

16.18 The frame of Problem 16.17 if the column bases are fixed. (*Ans.* $M_A = 210.5$ ft-k, $M_B = 0.5$ ft-k)

16.19

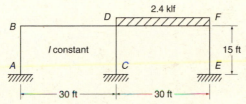

16.20 (*Ans.* $M_A = 138.5$ ft-k, $M_D = 192$ ft-k)

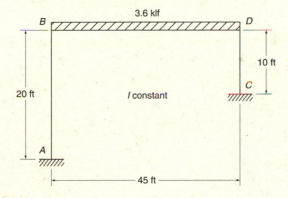

PART THREE

STATICALLY INDETERMINATE STRUCTURES— STRUCTURES— MODERN METHODS

Chapter 17

Approximate Analysis of Statically Indeterminate Structures

17.1 INTRODUCTION

The approximate methods presented in this chapter for analyzing statically indeterminate structures could very well be designated as *classical methods*. The same designation could be made for the moment distribution method presented in Chapters 18 and 19. The authors feel, however, that these methods are frequently very useful to today's analyst. As a result, this material has been included with the so-called *modern methods* of analysis.

Statically indeterminate structures may be analyzed "exactly" or "approximately." Several "exact" methods, which are based on elastic distortions, are discussed in Chapters 13 through 16. Approximate methods involving the use of simplifying assumptions are presented in this chapter. These methods have many practical applications such as the following:

1. When costs are being estimated for alternate design configurations and design types, approximate analyses are often very helpful. Approximate analyses and approximate designs of the various alternatives can be made quickly and used for initial cost estimates.

2. To design the members of a statically indeterminate structure, it is necessary to make an estimate of their sizes before an analysis can be made by an "exact" method. This is necessary because the analysis of a statically indeterminate structure is based on the elastic properties of the members. An approximate analysis of the structure will yield forces from which reasonably good initial estimates can be made as to member sizes.

3. Today computers are available with which "exact" analyses and designs of highly indeterminate structures can be quickly and economically made. To make use of computer programs, it is economically advisable to make some preliminary estimates as to member sizes. If an approximate analysis of the structure has been made, it will be possible to make

very reasonable estimates of member sizes. The result will be appreciable savings of both computer time and money.

4. Approximate analyses are quite useful for rough checking of "exact" computer solutions (a very important matter).

5. An "exact" analysis may be too expensive, particularly when preliminary estimates and designs are being made. (It is assumed that for such a situation an acceptable and applicable approximate method is available to come up with a solution.)

6. An additional advantage of approximate methods is that they provide the analyst with a "feel" for the actual behavior of structures under various loading conditions. This important ability probably will not be developed from computer solutions.

To make an "exact" analysis of a complicated statically indeterminate structure, is it necessary for a qualified analyst to "model" the structure, that is, to make certain assumptions. For instance, the joints are assumed to be simple or semirigid. Certain characteristics of material behavior are assumed, load conditions are assumed, and so on. The result of all these assumptions is that all analyses are approximate (or rather, we apply an "exact" analysis method to a structure that does not really exist). Furthermore, all analysis methods are approximate in the sense that every structure is constructed within certain tolerances—no structure is perfect, and its behavior cannot be determined precisely.

It is hoped that the approximate methods described in this chapter will provide the student with a general overall knowledge or feeling for a wide range of statically indeterminate structures. Not all types of statically indeterminate structures are considered in this chapter; however, based on the ideas herein the student should be able to make reasonable assumptions when other types of statically indeterminate structures are encountered.

Many different methods are available for making approximate analyses. A few of the more common ones are presented here, with consideration being given to trusses, continuous beams, and building frames.

To be able to analyze a structure by statics, there must be no more unknowns than there are equations of statics available. If a truss or frame has 10 more unknowns than equations, it is statically indeterminate to the 10th degree. To analyze it by an approximate method, one assumption for each degree of indeterminacy, or a total of 10 assumptions, must be made. It will be seen that each assumption presents another equation to use in the calculations.

17.2 TRUSSES WITH TWO DIAGONALS IN EACH PANEL

Diagonals Having Little Stiffness

The truss of Figure 17.1 has two diagonals in each panel. If one of these diagonals were to be removed from each of the six panels, the truss would become statically determinate. The structure is statically indeterminate to the sixth degree.

If the diagonals are relatively long and slender, such as those made of a pair of small steel angles, they will be able to carry reasonably large tensile forces but

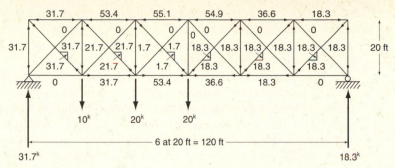

Figure 17.1

negligible compression forces. For this situation it is logical to assume that the shear in each panel is carried entirely by the diagonal that would be in tension for that type of shear. The other diagonal is assumed to have no force. Making this assumption in each panel makes a total of six assumptions for six redundants, and the equations of statics may be used to complete the analysis. The forces in Figure 17.1 were obtained on this basis.

Diagonals Having Considerable Stiffness

In some trusses the diagonals are constructed of sufficient stiffness to resist compressive loads. For panels having two diagonals the shear may be considered to be taken by both of them. The division of shear causes one diagonal to be in tension and the other to be in compression. The usual approximation made is that each diagonal takes 50% of the shear. It is possible to assume some other division, however, such as one-third of shear to compression diagonal and two-thirds to tension diagonal.

The forces calculated for the truss in Figure 17.2 are based on a 50% division of the shear in each panel.

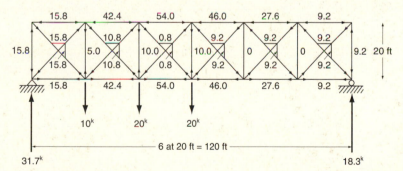

Figure 17.2

17.3 CONTINUOUS BEAMS

Before beginning an "exact" analysis of a building frame, it is necessary to estimate the sizes of the members. Preliminary beam sizes can be obtained by considering their approximate moments. It is frequently practical to remove a portion of a building and analyze that part of the structure. For instance, one or more beam spans may be taken out as a free body and assumptions made as to the moments in those spans. To facilitate such an analysis, moment diagrams are shown in Figure 17.3 for several different uniformly loaded beams.

It is obvious from this figure that the types of supports assumed can have a tremendous effect on the magnitude of the calculated moments. For instance, the uniformly loaded simple beam shown in Figure 17.3(a) will have a maximum moment equal to $w\ell^2/8$, while the uniformly loaded fixed-ended beam of part (b) will have a maximum moment equal to $w\ell^2/12$. For a continuous uniformly loaded beam the designer may very well decide to estimate a maximum moment somewhere between the preceding values, as, say, $w\ell^2/10$, and use that value for approximating the member size.

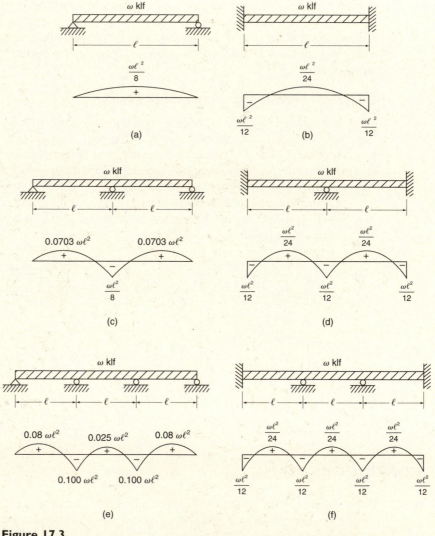

Figure 17.3

A very common method used for the approximate analysis of continuous reinforced-concrete structures involves the use of the American Concrete Institute moment and shear coefficients.[1] These coefficients, which are reproduced in Table 17.1, provide estimated maximum shears and moments for buildings of normal proportions. The values calculated in this manner usually will be somewhat larger than those that would be obtained with an exact analysis. As a result, appreciable economy can normally be obtained by taking the time or effort to make such an analysis. In this regard it should be realized that these coefficients are considered to have their best application for continuous frames having more than three or four continuous spans.

In developing the coefficients the negative moment values were reduced to take into account the usual support widths and also some moment redistribution before collapse. In addition, the positive moment values have been increased somewhat to account for the moment redistribution. It will also be noted that the coefficients account for the fact that in monolithic construction the supports are not simple and moments are present at end supports, such as where those supports are beams or columns.

In applying the coefficients, w is the design load per unit of length, while ℓ_n is the clear span for calculating positive moments and the average of the adjacent clear spans for calculating negative moments. These values were developed for members

TABLE 17.1

ACI Coefficients	
Positive moment	
End spans	
If discontinuous end is unrestrained	$\frac{1}{11} w \ell_n^2$
If discontinuous end is integral with the support	$\frac{1}{14} w \ell_n^2$
Interior spans	$\frac{1}{16} w \ell_n^2$
Negative moment at exterior face of first interior support	
Two spans	$\frac{1}{9} w \ell_n^2$
More than two spans	$\frac{1}{10} w \ell_n^2$
Negative moment at other faces of interior supports	$\frac{1}{11} w \ell_n^2$
Negative moment at face of all supports for (a) slabs with spans not exceeding 10 ft and (b) beams and girders where ratio of sum of column stiffnesses to beam stiffness exceeds eight at each end of the span	$\frac{1}{12} w \ell_n^2$
Negative moment at interior faces of exterior supports for members built integrally with their supports	
Where the support is a spandrel beam or girder	$\frac{1}{24} w \ell_n^2$
Where the support is a column	$\frac{1}{16} w \ell_n^2$
Shear in end members at face of first interior support	$\dfrac{1.15 w \ell_n}{2}$
Shear at face of all other supports	$\dfrac{w \ell_n}{2}$

[1] *Building Code Requirements for Reinforced Concrete*, ACI 318-95 (Detroit: American Concrete Institute) Section 8.3.3, pp. 79–80.

with approximately equal spans (the larger of two adjacent spans not exceeding the smaller by more than 20%) and for cases where the ratio of the uniform service live load to the uniform service dead load is not greater than 3. In addition, the values are not applicable to prestressed-concrete members. Should these limitations not be met, a more precise method of analysis must be used.

For the design of a continuous beam or slab, the moment coefficients provide in effect two sets of moment diagrams for each span of the structure. One diagram is the result of placing the live loads so that they will cause maximum positive moment out in the span, while the other is the result of placing the live loads so as to cause maximum negative moments at the supports. Actually, however, it is not possible to produce maximum negative moments at both ends of a span simultaneously. It takes one placement of the live loads to produce maximum negative moment at one end of the span and another placement to produce maximum negative moment at the other end. The assumption of both maximums occurring at the same time is on the safe side, however, because the resulting diagram will have greater critical values than are produced by either one of the two separate loading conditions.

The ACI coefficients give maximum points for a moment envelope for each span of a continuous frame. Typical envelopes are shown in Figure 17.4 for a continuous slab, which is assumed to be constructed integrally with its exterior supports, which are spandrel girders.

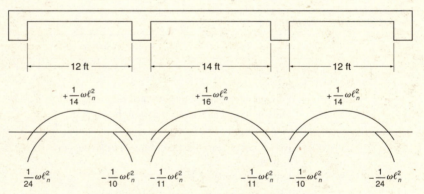

Figure 17.4 Moment envelopes for continuous slab constructed integrally with exterior supports that are spandrel girders.

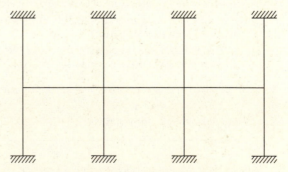

Figure 17.5 A portion of a building frame to be analyzed by the equivalent frame method.

On some occasions the analyst will take out a portion of a structure that includes not only the beams but also the columns for the floor above and the floor below, as shown in Figure 17.5. This procedure, usually called the *equivalent frame method,* is applicable only for gravity loads. The sizes of the members are estimated and an

Multistory steel-frame building, showing structural skeleton; Hotel Sheraton, Philadelphia, Pennsylvania. (Courtesy Bethlehem Steel Corporation.)

analysis is made using an appropriate "exact" method such as the moment-distribution procedure described in Chapters 18 and 19.

17.4 ANALYSIS OF BUILDING FRAMES FOR VERTICAL LOADS

One approximate method for analyzing building frames for vertical loads involves the estimation of the location of the points of zero moment in the girders. These points, which occur where the moment changes from one sign to the other, are commonly

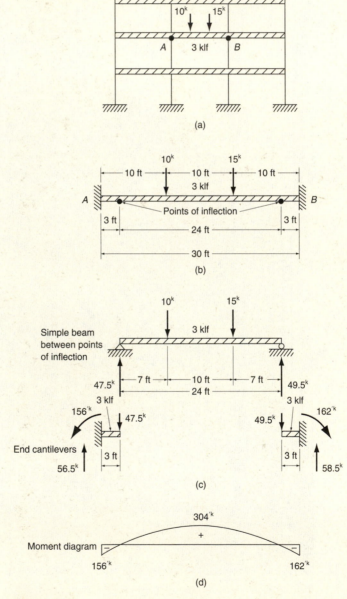

Figure 17.6

called *points of inflection* (PIs) or points of contraflexure. One common practice is to assume the PIs are located at the one-tenth points from each girder end. In addition, the axial forces in the girders are assumed to be zero.[2]

These assumptions have the effect of creating a simple beam between the points of inflection, and the positive moments in the beam can be determined by statics. Negative moments occur in the girders between their ends and the points of inflection. They may be computed by considering the portion of the beam out to the point of inflection to be a cantilever.

The shear at the end of the girders contributes to the axial forces in the columns. Similarly, the negative moments at the ends of the girders are transferred to the columns. For interior columns the girder moments on each side oppose each other and may cancel. Exterior columns have moments on only one side caused by the girders framing into them, and these need to be considered in design.

In Figure 17.6, beam *AB* of the building frame shown is analyzed by assuming points of inflection at one-tenth points and fixed supports at beam ends.

To make reasonable estimates for *PI* locations the reader may find the sketching of the approximate deflected shape of the structure in question very helpful. For illustration, a continuous beam is drawn to scale in Figure 17.7(a) and a sketch of its estimated deflected shape for the loads shown is given in part (b). From the sketch an approximate location of the PIs is estimated. Finally, in part (c) of the figure the part of the beam between the PIs in the center span is taken out as an assumed simple span.

It might be useful for the reader to see where PIs occur for a few types of statically indeterminate beams. These may be helpful in estimating where PIs will

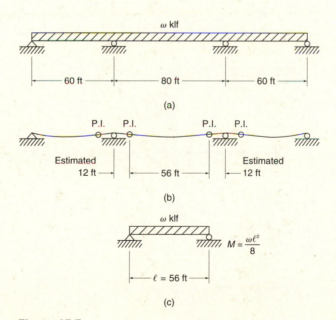

Figure 17.7

[2] C. H. Norris, J. B. Wilbur, and S. Utku, *Elementary Structural Analysis,* 3rd ed. (New York: McGraw-Hill, 1976), 200–201.

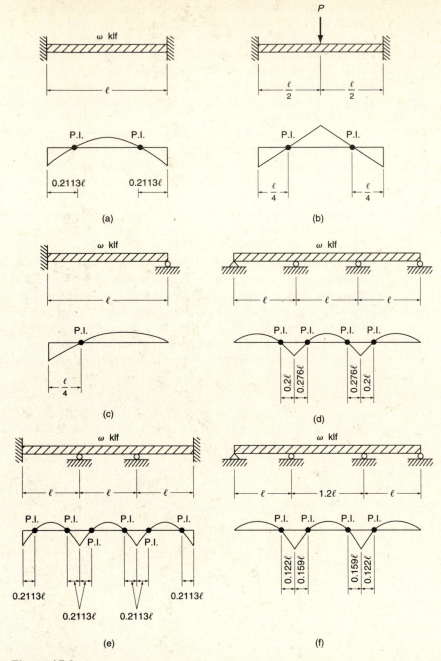

Figure 17.8

occur in other structures. Moment diagrams are shown for several beams in Figure 17.8. The PIs obviously occur where the moment diagrams change from + to −, or vice versa.

17.5 ANALYSIS OF PORTAL FRAMES

Portal frames of the type shown in Figures 17.9 and 17.10 may be fixed at their column bases, may be simply supported, or may be partially fixed.

The columns of the frame of Figure 17.9(a) are assumed to have their bases fixed. As a result there will be 3 unknown reactions at each support, giving a total of 6 unknowns. The structure is statically indeterminate to the third degree, and to analyze it by an approximate method three assumptions must be made.

When a column is rigidly attached to its foundation there can be no rotation at the base. Even though the frame is subjected to wind loads causing the columns to bend laterally, a tangent to the column at the base will remain vertical. If the beam at the tops of the columns is very stiff and rigidly fastened to the columns, the tangents to the columns at the junctions will remain vertical. A column rigidly fixed top and bottom will assume the shape of an S curve when subjected to lateral loads, as shown in part (b) of Figure 17.9.

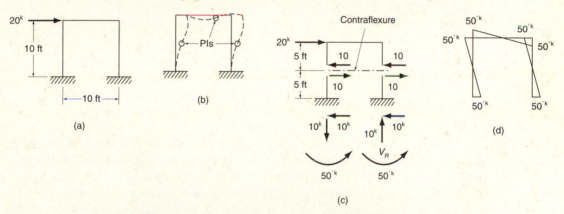

Figure 17.9   Approximate analysis of portal frame with fixed column bases.

At a point midway from the column base to the beam the moment will be zero, because it changes from a moment causing tension on one side of the column to a moment causing tension on the other side. If points of inflection are assumed in each column, two of the necessary three assumptions will have been made, that is, two $\Sigma M = 0$ equations are made available.

The third assumption usually made is that the horizontal shear divides equally between the two columns at the plane of contraflexure. An "exact" analysis proves this to be a very reasonable assumption as long as the columns are approximately the same size. If they are not similar in size, an assumption may be made that the shear splits between them in a little different proportion, with the stiffer column carrying the larger amount of shear. A rather good assumption for such cases is to distribute the shear to the columns in proportion to their I/ℓ^3 values.

To analyze the frame in Figure 17.9, the 20^k lateral load is divided into 10^k shears at the column PIs as shown in part (c) of the figure. The moment at the top and bottom of each column can be computed to equal $10 \times 5 = 50$ ft-k. Then moments are taken about the left column PI to determine the right-hand vertical reaction as follows:

$$(20)(5) - 10V_R = 0$$
$$V_R = 10^k \uparrow$$

Finally, the moment diagrams are drawn with the results shown in Figure 17.9(d).

Should the column bases be assumed to be simply supported, that is, pinned or hinged, the points of contraflexure will occur at those hinges. Assuming the columns are the same size, the horizontal shear is split equally between the columns and the other values are determined as shown in Figure 17.10.

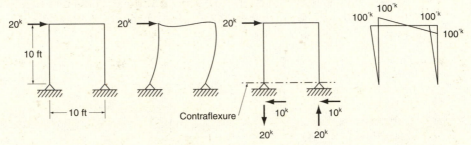

Figure 17.10   Approximate analysis of portal frame with pinned column bases.

The exact location of the PIs varies depending on the relative stiffness of the girders with respect to that of the columns. Such a situation is described in the next section of this chapter. Another factor affecting the location of the PIs is the actual degree of fixity at the column bases. Obviously, these supports are not going to be perfectly pinned or perfectly fixed. Sketching the estimated deflected shape of the frame will be quite helpful in making realistic assumptions as to PI locations.

17.6 LATERAL BRACING FOR BRIDGES

Bridge trusses may be braced laterally by bracing systems in the planes of the top and bottom chords, as well as by vertical or inclined planes of bracing. The planes of bracing tie the main trusses together and cause the entire structure to act as a rigid framework. They prevent excessive vibrations and resist the lateral loads of wind, earthquake, nosing of locomotives, and the centrifugal effect of traffic on curved bridges. Figure 17.11 shows a Warren bridge truss with the bracing systems that might be used in the plane of the top and bottom chords.

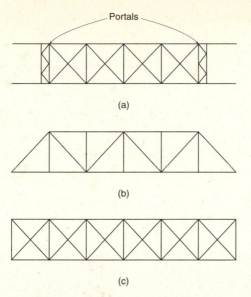

(a)

(b)

(c)

Figure 17.11   (a) Top lateral truss. (b) Plan. (c) Bottom lateral truss.

The loads are applied as concentrated loads at the joints of the lateral trusses, which permits analysis by the approximate methods discussed in Section 17.2. The top lateral bracing usually is subject to light loads, and the diagonals probably will be slender and able to resist only tensile loads. The design lateral loads probably are larger on the bottom of the truss, and the diagonals of the bottom-chord lateral system may be large enough to carry some compressive forces. The chord members of the lateral trusses are the chord members of the main trusses, but the AASHTO specifications do not require a strengthening of these members unless their forces as a part of the lateral system are greater than 25% of their normal forces as parts of the main trusses.

A system of bracing frequently used in the plane of the end posts is similar to the one shown in Figures 17.11 and 17.12. This type of bracing is commonly referred to as "portal bracing." The portal has the purpose of furnishing the end reaction to the top lateral system and transferring it down to the supports. The arrangement of the members of portals is quite similar to that of mill buildings, and the portals may be analyzed exactly as are the mill buildings in the next section.

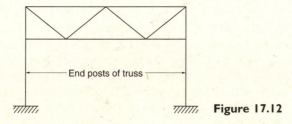

End posts of truss

Figure 17.12

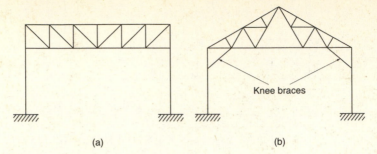

Figure 17.13

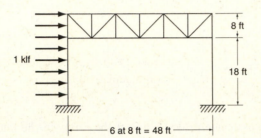

Figure 17.14

Portals for girder bridges may be of the types shown in Figure 17.10, in which the horizontal beam is rigidly connected to the columns (actually the girders). Similar structures are an essential part of steel building frames, and they also may be analyzed on the basis of the same assumptions used for the portals of Figures 17.13 and 17.14.

17.7 ANALYSIS OF MILL BUILDINGS

The building trusses analyzed in previous chapters have been assumed to rest on top of masonry walls or on top of columns on the sides of buildings. A different, but common, type of industrial construction is the mill bent or mill building, the trusses of which are rigidly fastened to the columns so that they act together. Figure 17.13 shows two types of framing commonly used for mill buildings.

For gravity loads these trusses are analyzed as though they were simply supported on walls instead of being rigidly fastened to the columns, but for lateral loads the columns and trusses should be analyzed as a unit. In office buildings of corresponding heights, the inside walls and partitions offer considerable resistance to wind, in many cases supplying sufficient resistance. Mill bents, however, must be analyzed and designed for wind loads, because there are no interior walls to help resist the wind forces. These buildings may have travelling cranes whose operation will cause additional lateral loads that need to be considered in design. The mill building of Figure 17.14 is analyzed in this section for a wind load of 1 kip per foot of height. Should there be crane loads, they would be handled in exactly the same manner.

The trusses will keep the columns straight in the area over which they are attached. Thus, mill buildings can be analyzed using the same assumptions used for the analysis of portals. Practically, it is impossible to construct perfectly fixed or perfectly pinned supports for the columns. As a result, the rotation occurring at the column bases will be somewhere between the theoretical values for the two cases. Then the PIs will fall somewhere between the column bases and halfway to the bottom of the trusses or kneebraces. It is rather common to assume that the PIs fall at approximately the one-third points.

If the column bases are fixed, there will be three unknown reactions at each support, giving a total of six unknowns. The structure is statically indeterminate to the third degree, and to analyze it by an approximate method, three assumptions must be made.

When a column is rigidly attached to the foundation, there can be no column rotation at the base. Even though the building is subjected to wind loads causing the columns to bend laterally, a tangent to the column at the base will remain vertical. If the truss at the tops of the columns is very stiff and rigidly fastened to them, a tangent to the column at the junction will remain vertical. A column rigidly fixed top and bottom will assume the shape of an S curve when it is subjected to lateral loads, as shown in Figure 17.15.

At a point midway from the column base to the bottom of the truss or knee brace the moment is zero because it changes from a moment causing tension on one side of the column to a moment causing tension on the other side. If points of inflection are assumed in each of the columns, two of the necessary three assumptions have been made (two $\Sigma M = 0$ equations are made available).

The discussion of the location of points of inflection has been based on the assumption that the column bases are completely fixed. If the columns are anchored in a deep concrete foundation or a concrete foundation wall, the assumption is good. Frequently, though, the columns are supported by small concrete footings that offer little resistance to rotation, and the column bases act as hinges, in which case the points of inflection are at the bases. The usual situation probably lies in between the two extremes, with the column bases only partially fixed. The points of inflection are commonly assumed to lie about one-fourth to one-third of the distance from the base up to the bottom of the truss, or to the knee brace if one is used.

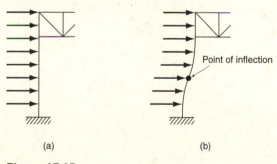

(a)   (b)

Figure 17.15

The third assumption made is that the horizontal shear divides equally between the two columns at the plane of contraflexure. An exact analysis proves this to be a very reasonable assumption if the columns are approximately the same size. If they are not similar in size, an assumption may be made that the shear splits between them in a little different proportion, the stiffer column carrying the larger amount of shear. The methods of analysis discussed in later chapters show that distributing the shear in proportion to the I/l^3 values of the columns is a very good assumption.

The mill building of Figure 17.14 is analyzed in Figure 17.16 on the basis of the following assumptions: The plane of contraflexure is assumed to be at the one-third point, or 6 ft above the column bases, and the total shear above the plane of contraflexure of 20 kips divides equally between the columns.

Moments are taken of the forces above the plane of contraflexure about the point of inflection in the left column to find the vertical force in the right-hand column. By $\Sigma V = 0$ the vertical force in the left-hand column is obtained. With the wind blowing from left to right, the right-hand (or leeward) column is in compression and the left-hand (or windward) column is in tension.

Moments at each column base are determined by taking moments about the base of the forces applied to the column at and below the point of inflection.

Finally, forces in the truss are computed by statics. In computing the forces of members of this truss which are connected to a column, all of the forces acting on the column must be taken into account because the columns are subject to both axial force and bending moment. Section 1.1 is passed through the truss, and moments are taken about joint U_0 to obtain the force in L_0L_1. By using the same section, moments are taken about joint L_1 to find the force in U_0U_1. Similarly, Section 2.2 is passed through the truss and moments are taken about U_6 to find the force in L_5L_6 and about L_5 to obtain the force in U_5U_6. The remaining forces can be obtained by the method of joints.

$$\Sigma M_{\text{left PI}} = 0 \text{ of forces above plane of contraflexure}$$

$$(1)(20)(10) - 48V_R = 0$$

$$V_R = 4.17^k \uparrow = \text{axial force in right column}$$

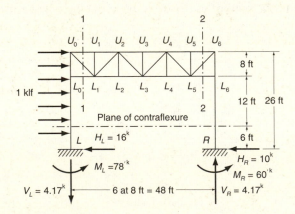

Figure 17.16

By $\Sigma V = 0$,

$$V_L = 4.17^k \downarrow = \text{axial force in left column}$$

Assuming total horizontal shear above plane of contraflexure of $(1)(20)$ divides equally between the two columns (10^k each),

$$H_L = 10 + (1)(6) = 16^k \leftarrow$$
$$H_R = 10^k \leftarrow$$

Moment reactions for each column are found by taking moments at the base of forces on the column up to points of inflection,

$$M_L = (10)(6) + (1)(6)(3) = 78'^k \curvearrowleft$$
$$M_R = (10)(6) = 60'^k \curvearrowleft$$

Force in $L_0 L_1$

$$\Sigma M_{U_0} = 0$$
$$(10)(20) - (20)(1)(10) - 8L_0 L_1 = 0$$
$$L_0 L_1 = 0$$

Force in $U_0 U_1$

$$\Sigma M_{L_1} = 0$$
$$(10)(12) - (20)(1)(2) - (4.17)(8) + 8U_0 U_1 = 0$$
$$U_0 U_1 - 5.83^k$$

17.8 ANALYSIS OF BUILDING FRAMES FOR LATERAL LOADS

Building frames are subjected to lateral loads as well as to vertical loads. The necessity for careful attention to these forces increases as buildings become taller. Not only must a building have sufficient lateral resistance to prevent failure, it also must have sufficient resistance to deflecting to prevent injury to its various parts.

Another item of importance is the provision of sufficient lateral rigidity to give the occupants a feeling of safety. They might not have this feeling in tall buildings that have a great deal of lateral movement in times of high winds. There have been actual instances of occupants of the upper floors of tall buildings complaining of seasickness on very windy days.

Lateral loads can be taken care of by means of X or other types of bracing, by shear walls, or by moment-resisting wind connections. Only the last of these three methods is considered in this chapter.

Large, rigid-frame buildings are highly indeterminate and until the 1960s their analysis by the so-called "exact" methods usually was impractical because of the

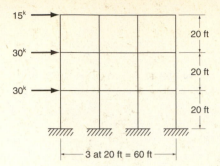

Figure 17.17

Figure 17.18   One level of the frame of
Figure 17.17.

multitude of equations involved. The total degree of indeterminacy of a rigid-frame
building (internal and external) can be determined by considering it to consist of
separate portals. One level of the rigid frame of Figure 17.17 is broken down into a
set of portals in Figure 17.18. Each of the portals is statically indeterminate to the
third degree, and the total degree of indeterminancy of a rigid-frame building equals
three times the number of individual portals in the frame.

Another method of obtaining the degree of indeterminacy is to assume that each
of the girders is cut by an imaginary section. If these values—shear, axial force, and
moment—are known in each girder, the free bodies produced can be analyzed by
statics. The total degree of indeterminacy equals three times the number of girders.

Today such structures usually are analyzed "exactly" with modern computers,
though approximate methods are used occasionally for preliminary analysis and
sizing of members. Approximate methods also provide checks on computer solutions
and give the analyst a better "feel" for and understanding of a structure's behavior
than he or she can obtain by examining the seemingly endless printout of a computer.

The building frame of Figure 17.17 is analyzed by two approximate methods in
the pages to follow, and the results are compared with those obtained by one of the
"exact" methods considered in a later chapter. The dimensions and loading of the
frame are selected to illustrate the methods involved while keeping the computations
as simple as possible. There are nine girders in the frame, giving a total degree of
indeterminacy of 27, and at least 27 assumptions will be needed to permit an approximate solution.

The reader should be aware that with the availability of digital computers today
it is feasible to make "exact" analyses in appreciably less time than that required to
make approximate analyses without the use of computers. The more-accurate values
obtained permit the use of smaller members. The results of computer usage are money

saving in analysis time and in the use of smaller members. It is possible to analyze in minutes structures (such as tall buildings) that are statically indeterminate with hundreds or even thousands of redundants, via the displacement method. The results for these highly indeterminate structures are far more accurate and more economically obtained than from approximate analyses.

The two methods considered here are the portal and cantilever methods. These methods have been used in so many successful building designs that they were almost the unofficial standard procedure for the design profession before the advent of modern computers. No consideration is given in either of these methods to the elastic properties of the members. These omissions can be very serious in asymmetrical frames and in very tall buildings. To illustrate the seriousness of the matter, the changes in member sizes are considered in a very tall building. In such a building probably there will not be a great deal of change in beam sizes from the top floor to the bottom floor. For the same loadings and spans the changed sizes would be due to the large wind moments in the lower floors. The change, however, in column sizes from top to bottom would be tremendous. The result is that the relative sizes of columns and beams on the top floors are entirely different from the relative sizes on the lower floors. When this fact is not considered, it causes large errors in the analysis.

In both the portal and cantilever methods, the entire wind loads are assumed to be resisted by the building frames, with no stiffening assistance from the floors, walls, and partitions. Changes in length of girders and columns are assumed to be negligible. However, they are not negligible in tall slender buildings whose height is five or more times the least horizontal dimension.

If the height of a building is roughly five or more times its least lateral dimension, it is generally felt that a more precise method of analysis should be used than the portal or cantilever methods. There are several excellent approximate methods that make use of the elastic properties of the structures and that give values closely approaching the results of the "exact" methods. These include the Factor method,[3] the Witmer method of K percentages,[4] and the Spurr method.[5] Should an "exact" hand-calculation method be desired, the moment-distribution procedure of Chapters 18 and 19 is convenient.

The Portal Method

The most common approximate method of analyzing building frames for lateral loads is the portal method. Because of its simplicity, it probably has been used more than any other approximate method for determining wind forces in building frames. This method, which was presented by Albert Smith in the *Journal of the Western Society*

[3] C. H. Norris, J. B. Wilbur, and S. Utku, *Elementary Structural Analysis*, 3rd ed. (New York: McGraw-Hill, 1976), 207–212.

[4] "Wind Bracing in Steel Buildings," *Transactions of the American Society of Civil Engineers* 105 (1940): 1725–1727.

[5] *Ibid*, pp. 1723–1725.

of Engineers in April 1915, is said to be satisfactory for most buildings up to 25 stories in height.[6]

At least three assumptions must be made for each individual portal or for each girder. In the portal method, the frame is theoretically divided into independent portals (Figure 17.18) and the following three assumptions are made:

1. The columns bend in such a manner that there is a point of inflection at middepth Figure 17.15(b).
2. The girders bend in such a manner that there is a point of inflection at their centerlines.
3. The horizontal shears on each level are arbitrarily distributed between the columns. One commonly used distribution (and the one illustrated here) is to assume the shear divides among the columns in the ratio of one part to exterior columns and two parts to interior columns. The reason for this ratio can be seen in Figure 17.18. Each of the interior columns is serving two bents, whereas the exterior columns are serving only one. Another common distribution is to assume that the shear V taken by each column is in proportion to the floor area it supports. The shear distribution by the two procedures would be the same for a building with equal bays, but for one with unequal bays the results would differ with the floor area method, probably giving more realistic results.

For this frame there are 27 redundants; to obtain their values, one assumption as to the location of the point of inflection has been made for each of the 21 columns and girders. Three assumptions are made on each level as to the shear split in each individual portal, or the number of shear assumptions equals one less than the number of columns on each level. For the frame, 9 shear assumptions are made, giving a total of 30 assumptions and only 27 redundants. More assumptions are made than necessary, but they are consistent with the solution (that is, if only 27 of the assumptions were used and the remaining values were obtained by statics, the results would be identical).

Frame Analysis

The frame is analyzed in Figure 17.19 on the basis of these assumptions. The arrows shown on the figure give the direction of the girder shears and the column axial forces. The reader can visualize the stress condition of the frame if he or she assumes the wind is tending to push it over from the left to right, stretching the left exterior columns and compressing the right exterior columns. Briefly, the calculations were made as follows.

I. Column Shears
The shears in each column on the various levels were first obtained. The total shear on the top level is 15 kips. Because there are two exterior and two interior columns,

[6] *Ibid*, p. 1723.

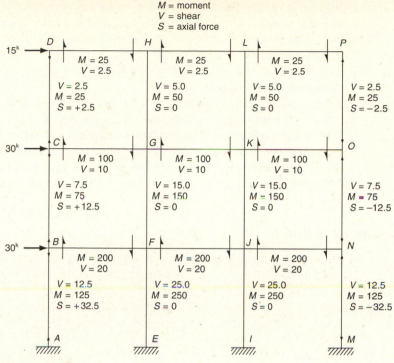

Figure 17.19

the following expression may be written:

$$x + 2x + 2x + x = 15^k$$
$$x = 2.5^k$$
$$2x = 5.0^k$$

The shear in column CD is 2.5 kips; in GH it is 5.0 kips, and so on. Similarly, the shears were determined for the columns on the first and second levels, where the total shears are 75 and 45 kips, respectively.

2. Column Moments
The columns are assumed to have points of inflection at their middepths; therefore, their moments, top and bottom, equal the column shears times half the column heights.

3. Girder Moments and Shears
At any joint in the frame the sum of the moments in the girders equals the sum of the moments in the columns. The column moments have been previously determined. By beginning at the upper left-hand corner of the frame and working across from left to

right, adding or subtracting the moments as the case may be, the girder moments were found in this order: *DH*, *HL*, *LP*, *CG*, *GK*, and so on. It follows that with points of inflection at girder centerlines, the girder shears equal the girder moments divided by half-girder lengths.

4. Column Axial Forces

The axial forces in the columns may be directly obtained from the girder shears. Starting at the upper left-hand corner, the column axial force in *CD* is numerically equal to the shear in girder *DH*. The axial force in column *GH* is equal to the difference between the two girder shears *DH* and *HL*, which equals zero in this case. (If the width of each of the portals is the same, the shears in the girder on one level will be equal, and the interior columns will have no axial force, since only lateral loads are considered.)

The Cantilever Method

Another simple method of analyzing building frames for lateral forces is the cantilever method presented by A. C. Wilson in *Engineering Record,* September 5, 1908. This method is said to be a little more desirable for high narrow buildings than the portal method and may be used satisfactorily for buildings with heights not in excess of 25 to 35 stories.[7] It is not as popular as the portal method.

Mr. Wilson's method makes use of the assumptions that the portal method uses as to locations of points of inflection in columns and girders, but the third assumption differs somewhat. Rather than assume the shear on a particular level to divide between the columns in some ratio, the axial stress in each column is considered to be proportional to its distance from the center of gravity of all the columns on that level. If the columns on each level are assumed to have equal cross-sectional areas (as is done for the cantilever problems of this chapter) then their forces will vary in proportion to their distances from the center of gravity. The wind loads are tending to overturn the building, and the columns on the leeward side will be compressed, whereas those on the windward side will be put in tension. The greater the distance a column is from the center of gravity of its group of columns, the greater will be its axial stress.

The new assumption is equivalent to making a number of axial-force assumptions equal to one less than the number of columns on each level. Again, the structure has 27 redundants, and 30 assumptions are made (21 column and girder point-of-inflection assumptions and 9 column axial-force assumptions), but the extra assumptions are consistent with the solution.

Frame Analysis

The frame previously analyzed by the portal method is analyzed by the cantilever method in Figure 17.20. Briefly, the calculations are made as follows.

[7] *Ibid,* p. 1723.

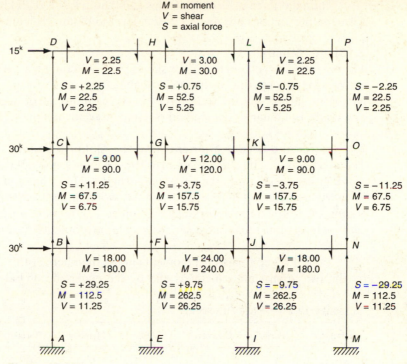

M = moment
V = shear
S = axial force

Figure 17.20

1. Column Axial Forces

Considering first the top level, moments are taken about the point of contraflexure in column CD of the forces above the plane of contraflexure through the columns on that level. According to the third assumption, the axial force in GH will be only one-third of that in CD, and these forces in GH and CD will be tensile, whereas those in KL and OP will be compressive. The following expression is written, with respect to Figure 17.21, to determine the values of the column axial forces on the top level.

$$(15)(10) + (1S)(20) - (1S)(40) - (3S)(60) = 0$$
$$S = 0.75^k$$
$$3S = 2.25^k$$

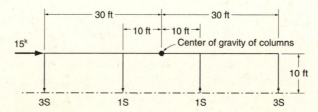

Figure 17.21

The axial force in *CD* is 2.25 kips and that in *GH* is 0.75 kip, and so on. Similar moment calculations are made for each level to obtain the column axial forces.

2. Girder Shears

The next step is to obtain the girder shears from the column axial forces. These shears are obtained by starting at the top left-hand corner and working across the top level, adding or subtracting the axial forces in the columns according to their signs. This procedure is similar to the method of joints used for finding truss forces.

3. Column and Girder Moments and the Column Shears

The final steps can be quickly summarized. The girder moments, as before, are equal to the girder shears times the girder half-lengths. The column moments are obtained by starting at the top left-hand corner and working across each level in succession, adding or subtracting the previously obtained column and girder moments as indicated. The column shears are equal to the column moments divided by half the column heights.

Table 17.2 compares the moments in the members of this frame as determined by the two approximate methods and by the enclosed computer program SABLE.

TABLE 17.2   MEMBER MOMENTS

Member	Portal	Cantilever	SABLE	Member	Portal	Cantilever	SABLE
AB	125	112.5	258.3	IJ	250	262.5	222.6
BA	125	112.5	148.9	JI	250	262.5	161.6
BC	75	67.5	57.4	JF	200	240	162.6
BF	200	180	206.3	JK	150	157.5	140.6
CB	75	67.5	100.7	JN	200	180	139.6
CD	25	22.5	7.1	KJ	150	157.5	146.4
CG	100	90	93.6	KG	100	120	114.5
DC	25	22.5	20.8	KL	50	52.5	54.2
DH	25	22.5	20.8	KO	100	90	86.1
EF	250	262.5	250.3	LK	50	52.5	71.9
FE	250	262.5	185.6	LH	25	30	47.4
FB	200	180	161.6	LP	25	22.5	24.5
FG	150	157.5	142.4	MN	125	112.5	179.8
FJ	200	240	166.3	NM	125	112.5	92.9
GF	150	157.5	152.4	NJ	200	180	165.5
GC	100	90	82.2	NO	75	67.5	72.7
GH	50	52.5	43.9	ON	75	67.5	87.4
GK	100	120	114.1	OK	100	90	103.1
HG	50	52.5	63.2	OP	25	22.5	15.6
HD	25	22.5	18.0	PO	25	22.5	37.4
HL	25	30	45.2	PL	25	22.5	37.4

Wachovia Plaza, Charlotte, North Carolina. (Courtesy Bethlehem Steel Corporation.)

Note that for several members the approximate results vary considerably from the results obtained by the "exact" method. Experience with the "exact" methods for handling indeterminate building frames will show that the points of inflection will not occur exactly at the midpoints. Using more realistic locations for the assumed inflection points will greatly improve results. The Bowman method[8] involves the location of the points of inflection in the columns and girders according to a specified set of rules depending on the number of stories in the building. In addition, the shear is divided between the columns on each level according to a set of rules that are based on the moments of inertia of the columns as well as the bay widths. Application of the Bowman method gives much better results than the portal and cantilever procedures.

[8] H. Sutherland and H. L. Bowman, *Structural Theory* (New York: Wiley, 1950), 295–301.

17.9 MOMENT DISTRIBUTION

The moment distribution method described in the next two chapters for the analysis of statically indeterminate beams and frames involves successive cycles of computation, each cycle drawing closer to the "exact" answers. When the computations are carried out until the changes in the numbers are very small, it is considered to be an "exact" method of analysis. Should the number of cycles be limited, the method becomes a splendid approximate method. As such, it is discussed in Section 18.7 of the next chapter.

17.10 ANALYSIS OF VIERENDEEL "TRUSSES"

In previous chapters a truss has been defined as a structure assembled with a group of ties and struts connected at their joints with frictionless pins so the members are subjected only to axial tension or axial compression. A special type of truss (it's not really a truss by the preceding definition) is the Vierendeel "truss," examples of which are shown in Figure 17.22. There you can see that they are actually rigid frames or, as some people say, they are girders with big holes in them. The Vierendeel "truss," which was developed in 1896 by the Belgian engineer and builder Arthur Vierendeel, is used rather frequently in Europe, but only occasionally in the United States.

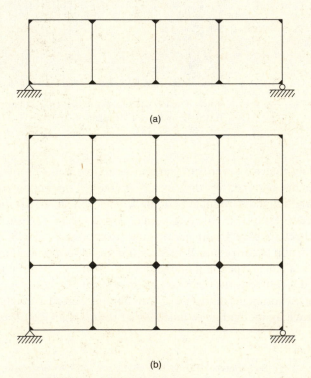

(a)

(b)

Figure 17.22   Vierendeel "trusses."

Vierendeel truss, Clinic Inn Pedestrian Bridge, Cleveland, Ohio. (Courtesy of the American Institute of Steel Construction, Inc.)

Vierendeel "trusses" usually are constructed with reinforced concrete, but also may be fabricated with structural steel. The external loads are supported by means of the bending resistances of the short heavy members. The continuous moment-resisting joints cause the "trusses" to be highly statically indeterminate. Though Vierendeels are rather inefficient, they are nicely used in certain situations in which their large clear openings are desirable. Also, they are particularly convenient in buildings where they can be constructed with depths equal to the story heights.

These highly statically indeterminate structures can be approximately analyzed by the portal and cantilever methods described in the preceding section. On the following page Figure 17.23 shows the results obtained by applying the portal method to a Vierendeel truss. (For this symmetrical case the cantilever method will yield the same results.) To follow the calculations the reader may find it convenient to turn the structure up on its end because the shear being considered for the Vierendeel is vertical, whereas it was horizontal for the building frames previously considered.

An "exact" analysis of this truss by the computer program SABLE made available on disks with this textbook yields the values shown in Figure 17.24 on page 482. For this computer program the members were assumed to have constant areas and moments of inertia. As the points of inflection are not exactly at the midpoints of the members, the moments at the member ends vary somewhat, and both moments are given in the figure. Though the results obtained with the portal method seem quite reasonable for this particular truss, they could have been improved by adjusting the assumed locations of the points of inflection a little.

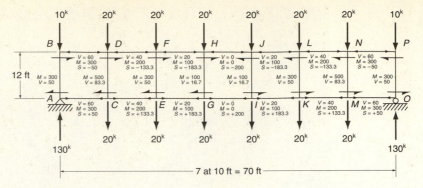

Figure 17.23

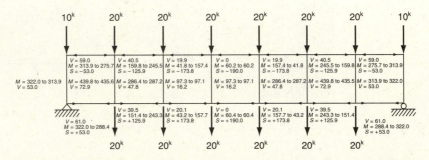

Note: Analysis made assuming $A = 27.7$ in.^{2}, $I = 2700$ in.^{4}, $E = 29 \times 10^3$ ksi for all members.

Figure 17.24

For many Vierendeels—particularly those of several stories—the lower horizontal members may be much larger and stiffer than the other horizontal members. To obtain better results with the portal or cantilever methods, the nonuniformity of sizes should be taken into account by assuming more of the shear is carried by the stiffer members.

PROBLEMS

17.1   Compute the forces in the members of the truss shown in the accompanying illustration if the diagonals are unable to carry compression.

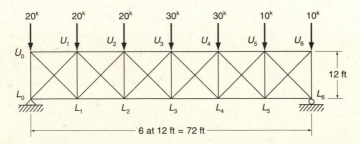

17.2   Repeat Problem 17.1 if the diagonals that theoretically are in compression can resist half of the shear in each panel. (*Ans.* $L_0L_1 = +28.33^k$, $L_1U_2 = -25.92^k$, $U_3U_4 = -103.35^k$)

17.3   Repeat Problem 17.1 if the diagonals that theoretically are in compression can resist one-third of the shear in each panel.

17.4   Compute the forces in the members of the cantilever truss shown if the diagonals are unable to carry any compression. (*Ans.* $U_0U_1 = +25^k$, $U_1L_1 = -60^k$, $U_1L_2 = +32.02^k$)

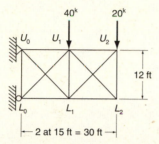

17.5   Repeat Problem 17.4 if the diagonals that would normally be in compression can resist only one-third of the shear in each panel.

17.6   Compute the forces in the members of the truss shown if the interior diagonals that would normally be in compression can resist no compression. (*Ans.* $AF = +32.02^k$, $FG = -41.23^k$, $CG = -20^k$)

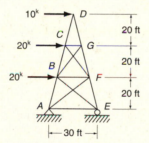

17.7   It is desired to prepare the shear and moment diagrams for the continuous beam shown using the ACI coefficients of Table 17.1. Assume the beam is constructed integrally with girders at all of its supports. The total load is to be 5 k/ft. Draw the moment diagrams as envelopes.

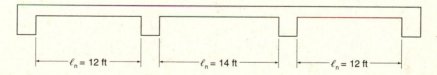

17.8   Analyze the portal shown if PIs are assumed to be located at column mid-depths. (*Ans.* Column moments top and bottom = 150 ft-k, right-hand vertical reaction = 10^k ↑)

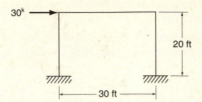

17.9   Repeat Problem 17.8 if the PIs are assumed to be located 8 ft above the column bases.

17.10   Repeat Problem 17.8 if the column bases are assumed to be pinned and the PIs located there. (*Ans.* Moment at top of columns = 300 ft-k, right-hand vertical reaction = 20^k ↑)

17.11   Determine the forces for all of the members of the mill-building truss shown in the accompanying illustration if points of inflection are assumed to be 12 ft above the column bases.

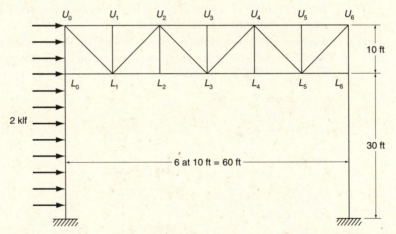

17.12   Repeat Problem 17.11 if points of inflection are assumed to be 15 ft above column bases. (*Ans.* $U_0U_1 = -14.58^k$, $L_2L_3 = -20.84^k$, $U_2L_3 + 14.74^k$)

17.13   Assuming points of inflection to be 10 ft from the column bases, determine the forces in all members of the structure shown in the accompanying illustration.

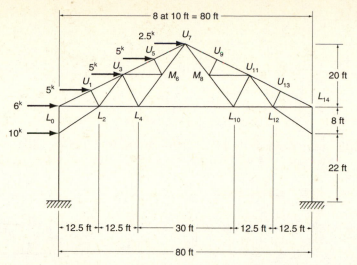

17.14   Rework Problem 17.13 with points of inflection assumed to be 6 ft above the column bases. (*Ans.* $L_0 L_2 = -11.08^k$, $L_4 L_{10} = -13.75^k$, $U_7 U_9 = +10.13^k$)

For Problems 17.15 through 17.22 compute moments, shears, and axial forces for all of the members of the frames shown (a) by using the portal method and (b) by using the cantilever method.

17.15

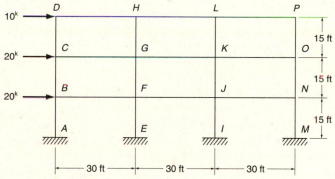

17.16   (*Ans.* Portal method for *CD*: $V = 10$ kN, $M = 20$ kN $\cdot$ m, $S = +6.67$ kN; Cantilever method: $V = 6$ kN, $M = 12$ kN $\cdot$ m, $S = +4$ kN)

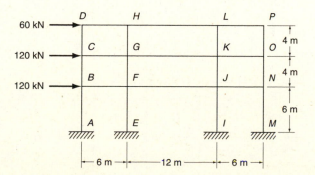

17.17

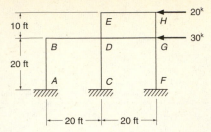

17.18 (*Ans.* Portal method for FG: $V = 50 \text{ kN}$, $M = 125 \text{ kN} \cdot \text{m}$, $S = -6.95 \text{ kN}$; Cantilever method: $V = 41.52 \text{ kN}$, $M = 103.8 \text{ kN} \cdot \text{m}$, $S = -3.66 \text{ kN}$)

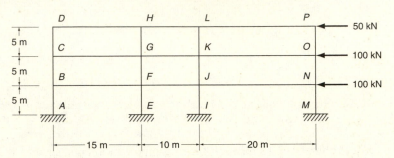

17.19

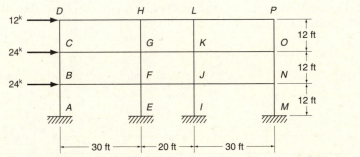

17.20 Rework Problem 17.19 if the column bases are assumed to be pinned. (*Ans.* for *BC*; Portal method: $V = 6^k$, $M = 36$ ft-k, $S = +4.0^k$. Cantilever method: $V = 6.35^k$, $M = 38.3$ ft-k, $S = +4.24^k$)

17.21

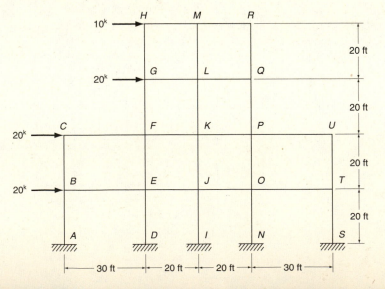

17.22   (*Ans.* Portal method for *DH*: V = 3.33 kN, M = 13.33 kN · m; Cantilever method: V = 1.81 kN, M = 7.24 kN · m)

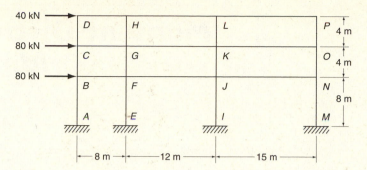

For Problems 17.23 and 17.24 compute moments, shears, and axial forces for all of the members of the Vierendeel trusses using the portal method.

17.23

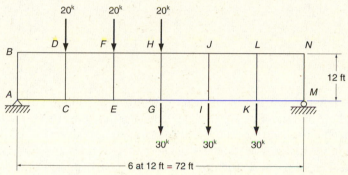

17.24   (*Ans.* for *BC*: V = 12.5 k, M = 75 ft-k, for *GK*: V = 15 k, M = 90 ft-k, S = 0)

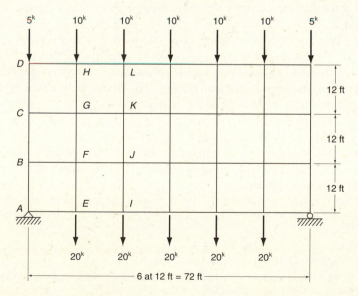

Chapter 18

Moment Distribution for Beams

18.1 GENERAL

The late Hardy Cross published the moment-distribution method in the Proceedings of the American Society of Civil Engineers in May 1930[1,2] after having taught the subject to his students at the University of Illinois since 1924. His paper began a new era in the analysis of statically indeterminate frames and gave added impetus to their use. The moment-distribution method of analyzing continuous beams and frames involves little more labor than the approximate methods but yields accuracy equivalent to that obtained from the infinitely more laborious "exact" methods previously studied.

The analysis of statically indeterminate structures in the preceding chapters frequently involved the solution of inconvenient simultaneous equations. Simultaneous equations are not necessary in solutions by moment distribution, except in a few rare situations for complicated frames. The Cross method involves successive cycles of computation, each cycle drawing closer to the "exact" answers. The calculations may be stopped after two or three cycles, giving a very good approximate analysis, or they may be carried on to whatever degree of accuracy is desired. When these advantages are considered in the light of the fact that the accuracy obtained by the lengthy "classical" methods is often questionable, the true worth of this quick and practical method is understood.

From the 1930s until the 1960s moment distribution was the dominant method used for the analysis of continuous beams and frames. Since the 1960s, however, there

[1] H. Cross, "Continuity as a Factor in Reinforced Concrete Design," *Proceedings of the American Society of Civil Engineers* 25(1929), 669–708.

[2] H. Cross, "Analysis of Continuous Frames by Distributing Fixed-End Moments," *Proceedings of the American Society of Civil Engineers* (May 1930): 919–928.

has been an ever increasing use of computers for the analysis of all types of structures. Computers are extremely efficient for solving the simultaneous equations that are generated by other methods of analysis. Generally, the software used is developed from the matrix-analysis procedures described in Chapters 20 and 21 of this book.

Moment distribution remains the most important hand-calculation method for the analysis of continuous beams and frames. The structural engineer can quickly make approximate analyses for preliminary designs and can also check computer results (very important). In addition, moment distribution may be solely used for the analysis of small structures.

18.2 INTRODUCTION

The beauty of moment distribution lies in its simplicity of theory and application. Readers will be able to grasp quickly the principles involved and will clearly understand what they are doing and why they are doing it.

The following discussion pertains to structures having members of constant cross section throughout their respective lengths (that is, prismatic members). It is assumed that there is no joint translation where two or more members frame together, but that there can be some joint rotation (that is, the members may rotate as a group but may not move with respect to each other). Finally, axial deformation of members is neglected.

Considering the frame of Figure 18.1(a), joints A to D are seen to be fixed. Joint E, however, is not fixed, and the loads on the structure will cause it to rotate slightly, as represented by the angle θ_E in Figure 18.1(b).

If an imaginary clamp is placed at E, fixing it so that it cannot be displaced, the structure will under load take the shape of Figure 18.1(c). For this situation, with all ends fixed, the fixed-end moments can be calculated with little difficulty by the usual expressions ($w\ell^2/12$ for uniform loads and Pab^2/ℓ^2 or Pa^2b/ℓ^2 for concentrated loads) as shown in Figure 18.5.

If the clamp at E is removed, the joint will rotate slightly, twisting the ends of the members meeting there and causing a redistribution of the moments in the member ends. The changes in the moments or twists at the E ends of members AE, BE, CE, and DE cause some effect at their other ends. When a moment is applied to one end of a member, the other end being fixed, there is some effect or *carryover* to the fixed end.

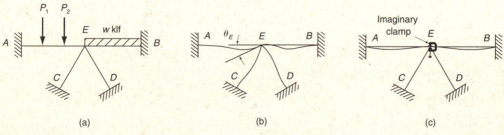

(a)   (b)   (c)

Figure 18.1

Largest curved "horizontal skyscraper" in the United States, Boston, Massachusetts. (Courtesy Bethlehem Steel Corporation.)

After the fixed-end moments are computed, the problem to be handled may be briefly stated as consisting of the calculation of (1) the moments caused in the E ends of the members by the rotation of joint E, (2) the magnitude of the moments carried over to the other ends of the members, and (3) the addition or subtraction of these latter moments to the original fixed-end moments.

These steps can be simply written as being the fixed-end moments plus the moments due to the rotation of joint E.

$$M = M_{\text{fixed}} + M_{\theta_E}$$

18.3 BASIC RELATIONS

There are two questions that must be answered in order to apply the moment-distribution method to actual structures. They are

1. What is the moment developed or carried over to a fixed end of a member when the other end is subjected to a certain moment?
2. When a joint is unclamped and rotates, what is the distribution of the unbalanced moment to the members meeting at the joint, or how much resisting moment is supplied by each member?

Carryover Moment

To determine the carryover moment, the unloaded beam of constant cross section in Figure 18.2(a) is considered. If a moment M_1 is applied to the left end of the beam, a moment M_2 will be developed at the right end. The left end is at a joint that has been

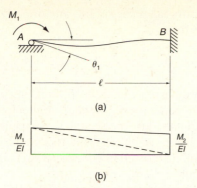

Figure 18.2

released and the moment M_1 causes it to rotate an amount θ_1. There will, however, be no deflection or translation of the left end with respect to the right end.

The second moment-area theorem may be used to determine the magnitude of M_2. The deflection of the tangent to the elastic curve of the beam at the left end with respect to the tangent at the right end (which remains horizontal) is equal to the moment of the area of the M/EI diagram taken about the left end and is equal to zero. By drawing the M/EI diagram in Figure 18.2(b) and dividing it into two triangles to facilitate the area computations, the following expression may be written and solved for M_2:

$$\delta_A = \frac{(\frac{1}{2} \times M_1 \times \ell)(\frac{1}{3}\ell) + (\frac{1}{2} \times M_2 \times \ell)(\frac{2}{3}\ell)}{EI} = 0$$

$$\frac{M_1\ell^2}{6EI} + \frac{M_2\ell^2}{3EI} = 0$$

$$M_2 = -\tfrac{1}{2}M_1$$

A moment applied at one end of a prismatic beam, the other end being fixed, will cause a moment half as large and of opposite sign at the fixed end. The carryover factor is $-\frac{1}{2}$. The minus sign refers to strength-of-materials sign convention: A distributed moment on one end causing tension in bottom fibers must be carried over so that it will cause tension in the top fibers of the other end. A study of Figures 18.2 and 18.3 shows that carrying over with a $+\frac{1}{2}$ value with the moment-

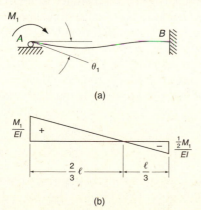

Figure 18.3

distribution sign convention automatically takes care of the situation, and it is unnecessary to change signs with each carryover.

Distribution Factors

Usually a group of members framed together at a joint have different stiffnesses. When a joint is unclamped and begins to rotate under the unbalanced moment, the resistance to rotation varies from member to member. The problem is to determine how much of the unbalanced moment will be taken up by each of the members. It seems reasonable to assume the unbalance will be resisted in direct relation to the respective resistance to end rotation of each member.

The beam and M/EI diagram of Figure 18.2 are redrawn in Figure 18.3, with the proper relationship between M_1 and M_2, and an expression is written for the amount of rotation caused by moment M_1.

Using the first moment-area theorem, the angle θ_1 may be represented by the area of the M/EI diagram between A and B, the tangent at B remaining horizontal.

$$\theta_1 = \frac{(\frac{1}{2})(M_1)(\frac{2}{3}\ell) - (\frac{1}{2})(\frac{1}{2}M_1)(\frac{1}{3}\ell)}{EI}$$

$$= \frac{M_1\ell}{4EI}$$

Assuming that all the members consist of the same material, having the same E values, the only variables in the foregoing equation affecting the amount of end rotation are the ℓ and I values. The amount of rotation occurring at the end of a member obviously varies directly as the ℓ/I value for the member. The larger the rotation of the member, the less moment it will carry. The moment resisted varies inversely as the amount of rotation or directly as the I/ℓ value. This latter value is referred to as the *stiffness factor K*.

$$K = \frac{I}{\ell}$$

To determine the unbalanced moment taken by each of the members at a joint, the stiffness factors at the joint are totaled, and each member is assumed to carry a proportion of the unbalanced moment equal to its K value divided by the sum of all the K values at the joint. These proportions of the total unbalanced moment carried by each of the members are the *distribution factors*.

$$DF_1 = \frac{K_1}{\Sigma K} \qquad DF_2 = \frac{K_2}{\Sigma K}$$

18.4 DEFINITIONS

The following terms are constantly used in discussing moment distribution.

Fixed-End Moments

When all of the joints of a structure are clamped to prevent any joint rotation, the external loads produce certain moments at the ends of the members to which they are applied. These moments are referred to as fixed-end moments.

Unbalanced Moments

Initially the joints in a structure are considered to be clamped. When a joint is released, it rotates if the sum of the fixed-end moments at the joint is not zero. The difference between zero and the actual sum of the end moments is the unbalanced moment.

Distributed Moments

After the clamp at a joint is released, the unbalanced moment causes the joint to rotate. The rotation twists the ends of the members at the joint and changes their moments. In other words, rotation of the joint is resisted by the members and resisting moments are built up in the members as they are twisted. Rotation continues until equilibrium is reached—when the resisting moments equal the unbalanced moment—at which time the sum of the moments at the joint is equal to zero. The moments developed in the members resisting rotation are the distributed moments.

Carryover Moments

The distributed moments in the ends of the members cause moments in the other ends, which are assumed fixed, and these are the carryover moments.

18.5 SIGN CONVENTION

The moments at the end of a member are assumed to be negative when they tend to rotate the member end clockwise about the joint (the resisting moment of the joint would be counterclockwise). The continuous beam of Figure 18.4, with all joints assumed to be clamped, has clockwise (or −) moments on the left end of each span and counterclockwise (or +) moments on the right end of each span. (The usual sign convention used in strength of materials shows fixed-ended beams to have negative moments on both ends for downward loads, because tension is caused in the top fibers of the beams at those points.) It should be noted that this sign convention, to be used for moment distribution, is the same one used in Chapter 16 for slope deflection.

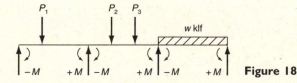

Figure 18.4

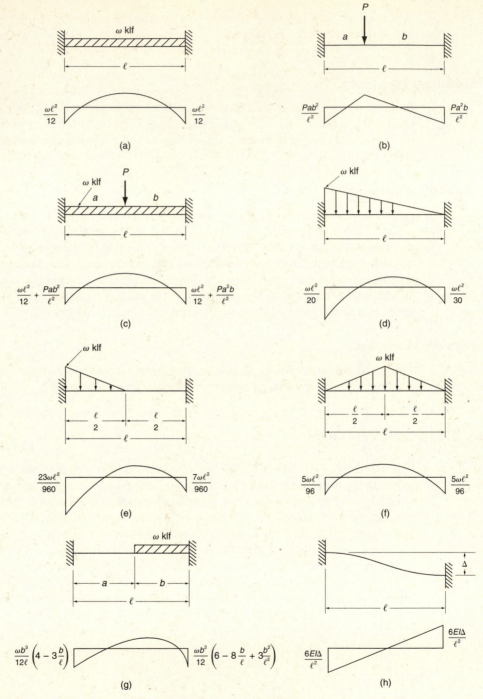

Figure 18.5 Fixed-end moments for various loadings.

18.6 FIXED-END MOMENTS FOR VARIOUS LOADS

Figure 18.5 presents expressions for computing the fixed-end moments for several types of loading conditions. The first three parts of the figure cover the large percentage of practical cases, whereas the other parts are for a little more unusual loading conditions. Particular attention should be given to part (c) of the figure, in which the superposition idea is presented. If fixed-end moments are needed in the same span for different loading conditions, they are each computed separately and added together as shown in the figure.

For cases not shown in Figure 18.5, fixed-end moments can be obtained from various tables, or they may be calculated with the moment area method as illustrated with Example 10.6.

18.7 APPLICATION OF MOMENT DISTRIBUTION

The very few tools needed for applying moment distribution are now available, and the method of applying them is described, reference being made to Figure 18.6.

Figure 18.6(a) shows a continuous beam and the several loads applied to it. In (b) the interior joints B and C are assumed to be clamped, and the fixed-end moments

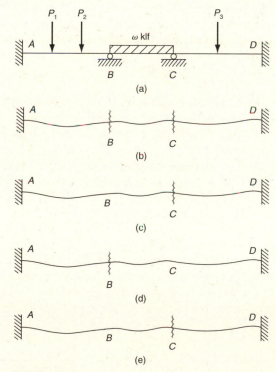

Figure 18.6

are computed. At joint B the unbalanced moment is computed and the clamp is removed, as seen in (c). The joint rotates, thus distributing the unbalanced moment to the B ends of BA and BC in proportion to their distribution factors. The values of these distributed moments are carried over at the one-half rate to the other ends of the members. When equilibrium is reached, joint B is clamped in its new rotated position and joint C is released, as shown in (d). Joint C rotates under its unbalanced moment until it reaches equilibrium, the rotation causing distributed moments in the C ends of members CB and CD and their resulting carryover moments. Joint C is now clamped and joint B is released, in Figure 18.6(e).

The same procedure is repeated again and again for joints B and C, the amount of unbalanced moment quickly diminishing, until the release of a joint causes only negligible rotation. This process, in brief, is moment distribution.

Examples 18.1 to 18.3 illustrate the procedure used for analyzing relatively simple continuous beams. The stiffness factors and distribution factors are computed as follows for Example 18.1.

$$DF_{BA} = \frac{K_{BA}}{\Sigma K} = \frac{\frac{1}{20}}{\frac{1}{20} + \frac{1}{15}} = 0.43$$

$$DF_{BC} = \frac{K_{BC}}{\Sigma K} = \frac{\frac{1}{15}}{\frac{1}{20} + \frac{1}{15}} = 0.57$$

A simple tabular form is used for Examples 18.1 and 18.2 to introduce the reader to moment distribution. For subsequent examples a slightly varying but quicker solution much preferred by the authors is used. The tabular procedure may be summarized as follows:

1. The fixed-end moments are computed and recorded on one line (line FEM in Examples 18.1 and 18.2).
2. The unbalanced moments at each joint are balanced in the next line (Dist 1).
3. The carryovers are made from each of the joints on the next line (CO 1).
4. The new unbalanced moments at each joint are balanced (Dist 2), and so on. (As the beam of Example 18.1 has only one joint to be balanced, only one balancing cycle is necessary.)

When the distribution has reached the accuracy desired, a double line is drawn under each column of figures. The final moment in the end of a member equals the sum of the moments at its position in the table. Unless a joint is fixed, the sum of the final end moments in the ends of the members meeting at the joint must total zero.

EXAMPLE 18.1 ⎯⎯⎯⎯⎯⎯⎯⎯⎯⎯⎯⎯⎯⎯⎯⎯⎯⎯⎯⎯⎯⎯⎯⎯⎯⎯⎯⎯⎯⎯⎯

Determine the end moments of the structure shown in Figure 18.7 by moment distribution.

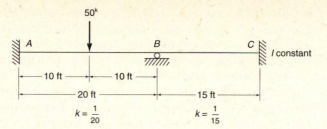

Figure 18.7

Solution.

	0.43	0.57		
−125	+125			FEM
	− 53.8	−71.2		Dist 1
− 26.9			−35.6	CO 1
−151.9	+ 71.2	−71.2	−35.6	Final moments ∎

EXAMPLE 18.2

Determine the end moments of the structure shown in Figure 18.8.

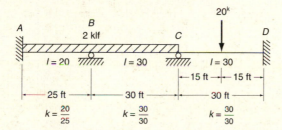

Figure 18.8

Solution.

	0.44	0.56		0.5	0.5		
−104.2	+104.2	−150.0	+150.0	− 75.0	+75.0	FEM	
	+ 20.2	+ 25.6	− 37.5	− 37.5		Dist 1	Cycle 1
+ 10.1		− 18.8	+ 12.8		−18.8	CO 1	
	+ 8.3	+ 10.5	− 6.4	− 6.4		Dist 2	Cycle 2
+ 4.2		− 3.2	+ 5.3		− 3.2	CO 2	
	+ 1.4	+ 1.8	− 2.7	− 2.7		Dist 3	Cycle 3
+ 0.7		− 1.3	+ 0.9		− 1.3	CO 3	
	+ 0.6	+ 0.7	− 0.4	− 0.4		Dist 4	Cycle 4
+ 0.3		− 0.2	+ 0.4		− 0.2	CO 4	
	+ 0.1	+ 0.1	− 0.2	− 0.2		Dist 5	Cycle 5
− 88.9	+134.8	−134.8	+122.2	−122.2	+51.5	Final moments ∎	

In Chapter 17 several methods for approximately analyzing statically indeterminate structures were introduced. Moment distribution is one of the "exact" methods of analysis if it's carried out until the moments to be distributed and carried over become quite small. It can, however, be used as a superb approximate method for statically indeterminate structures if only a few cycles of distribution are made.

The beam of Example 18.2 is considered for this discussion. In this problem each cycle of distribution is said to end when the unbalanced moments are balanced. In Table 18.1 the ratios of the total moments up through each cycle at joints A and C to the final moments at those joints after all the cycles are completed are shown. These ratios should give you an idea of how good partial moment distribution can be as an approximate method.

TABLE 18.1   ACCURACY OF MOMENT DISTRIBUTION AFTER VARIOUS CYCLES OF BALANCING FOR THE BEAM OF EXAMPLE 18.2

Values given after cycle no.	Moments at support	Ratio of approximate moment to "exact" moment support A	Moment at support C	Ratio of approximate moment to "exact" moment support C
1	104.2	1.172	112.5	0.921
2	94.1	1.058	118.9	0.973
3	89.9	1.011	121.5	0.994
4	89.2	1.003	122.0	0.998
5	88.9	1.000	122.2	1.000

The reader should clearly understand that for many cases where the structure and/or the loads are very unsymmetrical there will be a few cycles of major adjustment of the moments throughout the structure. Until these adjustments are substantially made, moment distribution will not serve as a very good approximate method. The student will easily be able to recognize, however, when the process can be stopped with good approximate results. This will occur when the unbalanced moments and carryover values become rather small as compared to the initial values.

Beginning with Example 18.3, a slightly different procedure is used for distributing the moments. Only one joint at a time is balanced and the required carryovers are made from that joint. Generally speaking it is desirable (but not necessary) to balance the joint that has the largest unbalance, make the carryovers, balance the joint with the largest unbalance, and so on, because such a process will result in the quickest convergence. This procedure is quicker than the tabular method used for Examples 18.1 and 18.2, and it follows along exactly with the description of the behavior of a continuous beam (with imaginary clamps) as pictured by the authors in Figure 18.6.

EXAMPLE 18.3

Compute the end moments in the beam shown in Figure 18.9.

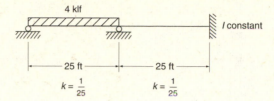

Figure 18.9

Solution.

		0.5	0.5	
-208.3	$+208.3$			
$+208.3$	$+104.2$			
$-\ 78.1$	-156.2	-156.2	-78.1	
$+\ 78.1$	$+\ 39.0$			
$-\ \ 9.8$	$-\ 19.5$	$-\ 19.5$	$-\ 9.8$	
$+\ \ 9.8$	$+\ \ 4.9$			
$-\ \ 1.2$	$-\ \ 2.5$	$-\ \ 2.5$	$-\ 1.2$	
$+\ \ 1.2$	$+\ \ 0.6$			
$-\ \ 0.2$	$-\ \ 0.3$	$-\ \ 0.3$	$-\ 0.2$	
$+\ \ 0.2$	$+178.5$	-178.5	-89.3	
0				

18.8 MODIFICATION OF STIFFNESS FOR SIMPLE ENDS

The carryover factor was developed for carrying over to fixed ends, but it is applicable to simply supported ends, which must have final moments of zero. The simple end of Example 18.3 was considered to be clamped; the carryover was made to the end; and the joint was freed and balanced back to zero. This procedure repeated over and over is absolutely correct, but it involves a little unnecessary work, which may be eliminated by studying the stiffness of members with simply supported ends.

On the following page Figure 18.10(a) and (b) compares the relative stiffness of a member subjected to a moment M_1 when the far end is fixed and when it is simply supported. In part (a) the conjugate beam for the beam with the far end fixed is loaded with the M/EI diagram, and the reactions are determined. The slope at the left end is represented by θ_1 and equals the shear when the conjugate beam is loaded with the M/EI diagram. Its value is $M_1\ell/4EI$.

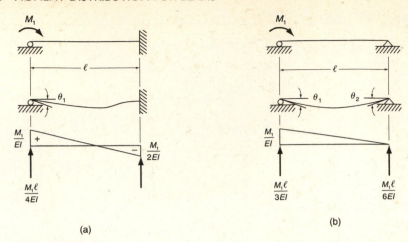

Figure 18.10

The conjugate beam for the simple end-supported beam is loaded with the M/EI diagram and its reactions are determined in Figure 18.10(b). The moment M_1 is found to cause a slope of $\theta_1 = M_1\ell/3EI$; therefore, the slope caused by the moment M_1 when the far end is fixed is only three-fourths as large $(M_1\ell/4EI \div M_1\ell/3EI = \frac{3}{4})$ when the far end is simply supported. The beam simply supported at the far end is only three-fourths as stiff as the one that is fixed. If the stiffness factors for end spans that are simply supported are modified by three-fourths, the simple end is initially balanced to zero, no carryovers are made to that end afterward, and the same results will be obtained. The stiffness modification is used for the beam of Example 18.3 in Example 18.4.

EXAMPLE 18.4 _____

Determine the end moments of the structure shown in Figure 18.11 by using the simple end-stiffness modification for the left end.

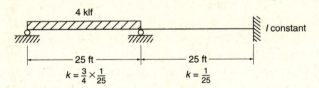

Figure 18.11

Solution.

	0.43	0.57	
−208.3	+208.3		
+208.3	+104.2		
═════	−134.4	−178.1	−89.1
0	═════	═════	═════
	+178.1	−178.1	−89.1

∎

18.9 SHEAR AND MOMENT DIAGRAMS

The drawing of shear and moment diagrams is an excellent way to check the final moments computed by moment distribution and to obtain an overall picture of the stress condition in the structure.

Before preparing the diagrams, it is necessary to consider a few points relating the shear and the moment diagram sign convention to the one used for moment distribution. The usual conventions for drawing the diagrams will be used (tension in bottom fibers of beam is positive moment and upward shear to left is positive shear).

The relationship between the signs of the moments for the two conventions is shown with the beams of Figure 18.12. Part (a) of the figure illustrates a fixed-end beam for which the result of moment distribution is a negative moment. The clockwise moment bends the beam as shown, causing tension in the top fibers or a negative moment for the shear and moment diagram convention. In Figure 18.12(b) the result of moment distribution is a positive moment, but again the top beam fibers are in tension, indicating a negative moment for the moment diagram.

An interior simple support is represented by Figure 18.12(c) and (d). In part (c) moment distribution gives a negative moment to the right and a positive moment to the left, which causes tension in the top fibers. Part (d) shows the effect of moments of opposite sign at the same support considered in part (c).

From the preceding discussion it can be seen that the sign convention used herein for moment distribution for continuous beams agrees with the one used for drawing moment diagrams on the right-hand sides of supports, but disagrees on the left-hand sides.

To draw the diagrams for a vertical member, the right side is often considered to be the bottom side. Moments are distributed for continuous beams in Examples 18.5 and 18.6, and shear and moment diagrams are drawn.

The reactions shown in the solution of these problems were obtained by computing the reactions as though each span were simply supported and by adding them to the reactions due to the moments at the beam supports.

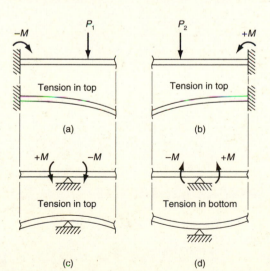

Figure 18.12

EXAMPLE 18.5

Distribute moments and draw shear and moment diagrams for the structure shown in Figure 18.13.

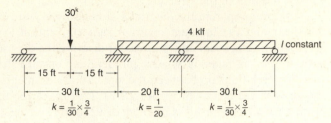

Figure 18.13

Solution.

	0.33	0.67		0.67	0.33	
−112.5	+112.5	−133.3	+133.3	−300.0	+300.0	
+112.5	+ 56.2			−150.0	−300.0	
0		+105.6	+211.1	+105.6	0	
	− 47.0	− 94.0	− 47.0			
		+ 15.7	+ 31.3	+ 15.7		
	− 5.2	− 10.5	− 5.2			
		+ 1.7	+ 3.5	+ 1.7		
	− 0.6	− 1.1	− 0.6			
		+ 0.2	+ 0.4	+ 0.2		
	− 0.1	− 0.1	+326.8	−326.8		
	+115.8	−115.8				

Reactions						
↑15.00	15.00 ↑	↑ 40.00	40.00 ↑	↑ 60.00	60.00 ↑	
↓ 3.86	3.86 ↑	↓ 10.55	10.55 ↑	↑ 10.89	10.89 ↓	
11.14	18.86 ↑	↑ 29.45	50.55 ↑	↑ 70.89	49.11 ↑	

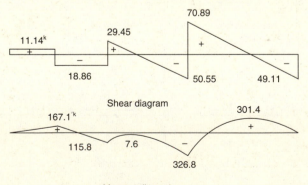

Shear diagram

Moment diagram

EXAMPLE 18.6 —————————————————————————————————

Distribute moments and draw shear and moment diagrams for the continuous beam shown in Figure 18.14.

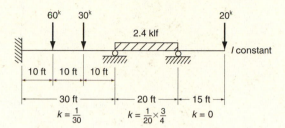

Figure 18.14

Solution.

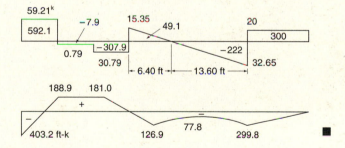

18.10 COMPUTER SOLUTIONS

The computer program SABLE enables the reader to very quickly analyze all of the problems contained in this chapter. One such example follows.

EXAMPLE 18.7 —————————————————————————————————

Determine the final moments for the beam of Figure 18.15 on page 504. This is the beam analyzed by moment distribution in Example 18.6.

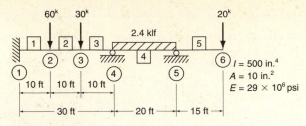

Figure 18.15

Solution.

Structural Data

1. Nodal Location and Restraint Data

Node	Coordinates			Restraints		
	X	Y	Z	X	Y	Rot
1	0.000E+00	0.000E+00	0.000E+00	Y	Y	Y
2	1.200E+02	0.000E+00	0.000E+00	N	N	N
3	2.400E+02	0.000E+00	0.000E+00	N	N	N
4	3.600E+02	0.000E+00	0.000E+00	N	Y	N
5	6.000E+02	0.000E+00	0.000E+00	N	Y	N
6	7.800E+02	0.000E+00	0.000E+00	N	N	N

2. Beam Location and Property Data

Beam	i	j	Type	Area	Beam properties	
					IZZ	E
1	1	2	F-F	1.000E+01	5.000E+02	2.900E+04
2	2	3	F-F	1.000E+01	5.000E+02	2.900E+04
3	3	4	F-F	1.000E+01	5.000E+02	2.900E+04
4	4	5	F-F	1.000E+01	5.000E+02	2.900E+04
5	5	6	F-F	1.000E+01	5.000E+02	2.900E+04

3. Applied Joint Loads

Node	Case	Force-X	Force-Y	Moment-Z
1	1	0.000E+00	0.000E+00	0.000E+00
2	1	0.000E+00	−6.000E+01	0.000E+00
3	1	0.000E+00	−3.000E+01	0.000E+00
4	1	0.000E+00	0.000E+00	0.000E+00
5	1	0.000E+00	0.000E+00	0.000E+00
6	1	0.000E+00	−2.000E+01	0.000E+00

4. Applied Beam Loads

Beam	Case	P	a	W
1	1	0.000E+00	0.000E+00	0.000E+00
2	1	0.000E+00	0.000E+00	0.000E+00
3	1	0.000E+00	0.000E+00	0.000E+00
4	1	0.000E+00	0.000E+00	−2.000E−01
5	1	0.000E+00	0.000E+00	0.000E+00

Results

Calculated beam end forces (*Note: units are* in.-k *for moments*)

Beam	Case	End	Axial	Shear-Y	Moment-Z
1	1	i	$0.000E{+}00$	$5.920E{+}01$	$4.838E{+}03$
		j	$0.000E{+}00$	$-5.920E{+}01$	$2.267E{+}03$
2	1	i	$0.000E{+}00$	$-7.974E{-}01$	$-2.267E{+}03$
		j	$0.000E{+}00$	$7.974E{-}01$	$2.171E{+}03$
3	1	i	$0.000E{+}00$	$-3.080E{+}01$	$-2.171E{+}03$
		j	$0.000E{+}00$	$3.080E{+}01$	$-1.525E{+}03$
4	1	i	$0.000E{+}00$	$1.535E{+}01$	$1.525E{+}03$
		j	$0.000E{+}00$	$3.265E{+}01$	$-3.600E{+}03$
5	1	i	$0.000E{+}00$	$2.000E{+}01$	$3.600E{+}03$
		j	$0.000E{+}00$	$-2.000E{+}01$	$7.276E{-}04$ ■

PROBLEMS

For Problems 18.1 through 18.22 analyze the structures by the moment-distribution method and draw shear and moment diagrams.

18.1

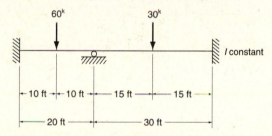

18.2   (*Ans. M_A = 380 ft-k, M_B = 200 ft-k*)

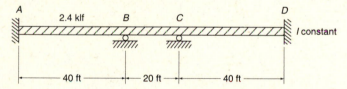

18.3

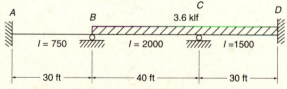

18.4   (*Ans. M_B = 380 ft-k, M_C = 458 ft-k*)

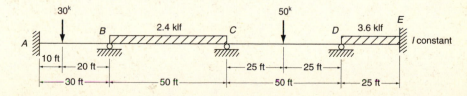

18.5

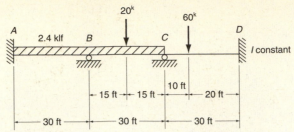

18.6   Rework Problem 18.3 if a 40-kip load is added at the centerline of each of the right two spans. (*Ans.* $M_A = 135.5$ ft-k, $M_C = 685.5$ ft-k)

18.7

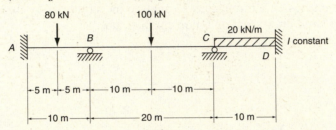

18.8   (*Ans.* $M_B = 212.3$ kN · m, $M_D = 129.6$ kN · m)

18.9

18.10   (*Ans.* $M_A = 53.6$ kN · m, $M_B = 267.8$ kN · m)

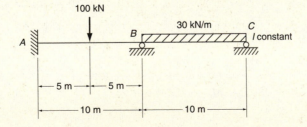

18.11

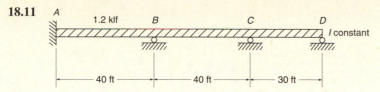

18.12   (*Ans.* $M_A = 150.3$ ft-k, $M_C = 95.6$ ft-k)

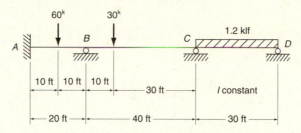

18.13

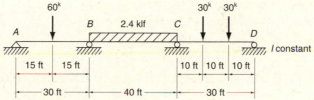

18.14   Repeat Problem 18.2 if the end supports A and D are made simple supports. (*Ans.* $M_B = M_C = 308.5$ ft-k)

18.15

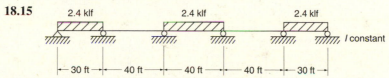

18.16   (*Ans.* $M_B = 499.8$ ft-k, $M_D = 267.2$ ft-k)

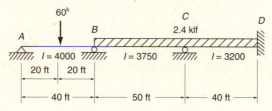

18.17

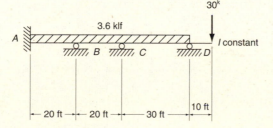

18.18 *(Ans. $M_C = 126.8$ ft-k, $M_D = 130.4$ ft-k)*

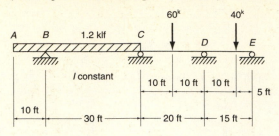

18.19

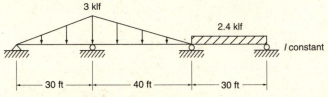

18.20 *(Ans. $M_A = 151.1$ ft-k, $M_B = 258$ ft-k)*

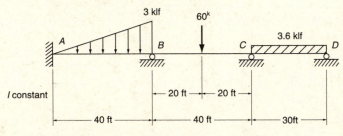

18.21

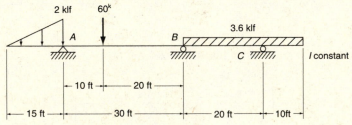

18.22 *(Ans. $M_B = 277$ ft-k, $M_D = 362.4$ ft-k)*

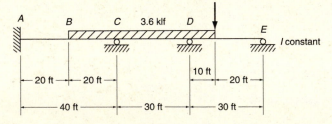

For Problems 18.23 through 18.26 rework the problems using SABLE.

18.23   Problem 18.1

18.24   Problem 18.3 (*Ans.* $M_B = 193.7$ ft-k, $M_D = 169.1$ ft-k)

18.25   Problem 18.17

18.26   Problem 18.22 (*Ans.* $M_A = 176.4$ ft-k, $M_C = 362.4$ ft-k)

Chapter 19

Moment Distribution
for Frames

19.1 FRAMES WITH SIDESWAY PREVENTED

Moment distribution for frames is handled as it is for beams if sidesway or movement laterally is prevented. Analysis of frames without sidesway is illustrated by Examples 19.1 and 19.2. Where sidesway is possible, however, it must be taken into account because the movements or deflections of the joints cause rotations and affect the moments in all the members.

As the structures being analyzed become more complex, it is necessary to use some method of recording the calculations so they will not interfere or run into each other. In this chapter a system is used for frames whereby the moments are recorded below beams on their left ends and above them on their right ends. For columns the same system is used, the right sides being considered the bottom sides.

EXAMPLE 19.1

Determine the end moments for the frame shown in Figure 19.1.

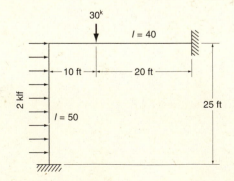

Figure 19.1

Solution.

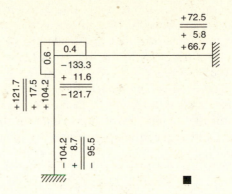

EXAMPLE 19.2 _____

Compute the end moments of the structure shown in Figure 19.2.

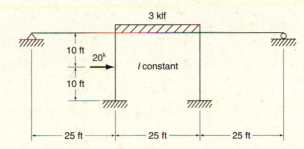

Figure 19.2

Solution.

19.2 FRAMES WITH SIDESWAY

Structural frames, similar to the one shown in Figure 19.3, are usually so constructed that they may possibly sway to one side or the other under load. The frame in this figure is symmetrical, but it will tend to sway because the load P is not centered. Analysis of the frame by the usual procedure, illustrated in the figure, gives inconsistent results.

The fixed-end moments are balanced and the horizontal reaction components are calculated at the supports. Each column is taken out as a free body. Moments are taken at the top of a particular column of the forces and moments applied to that column and the horizontal reaction component is determined. With reference to the

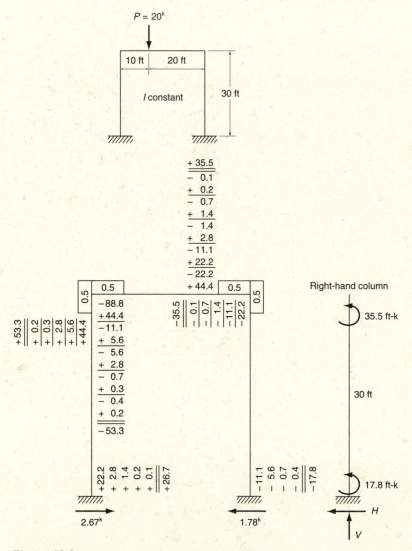

Figure 19.3

sketch of the right-hand column in Figure 19.3, the following expression can be written noting that the vertical reaction component passes through the point where moments are being taken if the column is itself vertical.

$$\underline{\Sigma M_{\text{top of column}} = 0}$$
$$-35.5 - 17.8 + 30H = 0$$
$$H = +1.78^k \leftarrow$$

From the results obtained it can be seen that the sum of the horizontal forces on the structure is not equal to zero. The sum of the forces to the right is 0.89 kip more than the sum of the forces to the left. If the structure were subjected to an unbalanced force system such as this one, it would not be in equilibrium.

The usual analysis does not yield consistent results because the structure actually sways or deflects to one side, and the resulting deflections affect the moments. One possible solution is to compute the deflections caused by applying a force of 0.89 kip acting to the right at the top of the bent. The moments caused could be obtained for the computed deflections and added to the originally distributed fixed-end moments, but the method is rather difficult to apply.

A much more convenient method is to assume the existence of an imaginary support that prevents the structure from swaying, as shown in Figure 19.4. The fixed-end moments are distributed, and the force the imaginary support must supply to hold the frame in place is computed. For the frame of Figure 19.3 the fictitious support must supply 0.89 kip pushing to the left.

The support is imaginary and if removed will allow the frame to sway to the right. As the structure sways to the right the joints are assumed to be locked against rotation. The ends of the columns rotate in a clockwise direction and produce clockwise or negative moments at the joints (Figure 19.5). Assumed values of these sidesway moments can be placed in the columns. The necessary relations between these moments are discussed in Section 19.3.

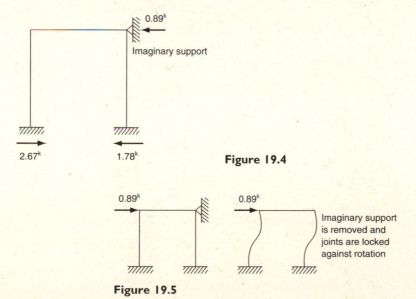

Figure 19.4

Figure 19.5

19.3 SIDESWAY MOMENTS

Should a frame have columns all of the same length and the same moments of inertia, the assumed sidesway moments will be the same for each column. However, should the columns have differing lengths and/or moments of inertia, this will not be the case. It will be proved in the following paragraphs that the assumed sidesway moments should vary from column to column in proportion to their I/ℓ^2 values.

If the frame of Figure 19.6 is pushed laterally an amount Δ by the load P, it will take the deflected shape of Figure 19.7.

Theoretically, both columns will become perfect S curves, if the beam is considered to be rigid and unbending. At their middepths the deflection for both columns will equal $\Delta/2$. Middepths of the columns may be considered points of contraflexure; the bottom halves may be handled as though they were cantilevered beams. The expression for deflection of a cantilevered beam with a concentrated load at its end is $P\ell^3/3EI$. Since the deflections are the same for both columns, the following expressions may be written:

$$\frac{\Delta}{2} = \frac{(P_1)(\ell_1/2)^3}{3EI_1} = \frac{P_1\ell_1^3}{24EI_1}$$

$$\frac{\Delta}{2} = \frac{(P_2)(\ell_2/2)^3}{3EI_2} = \frac{P_2\ell_2^3}{24EI_2}$$

By solving these deflection expressions for P_1 and P_2, the forces pushing on the cantilevers, we have

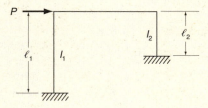

Figure 19.6

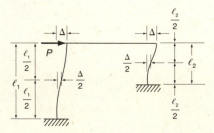

Figure 19.7

$$P_1 = \frac{12EI_1 \, \Delta}{\ell_1^3}$$

$$P_2 = \frac{12EI_2 \, \Delta}{\ell_2^3}$$

The moments caused by the two forces at the ends of their respective cantilevers are equal to the force times the cantilever length. These moments are written and the values of P_1 and P_2 are substituted in them.

$$M_1 = P_1 \frac{\ell_1}{2}$$

$$= \left(\frac{12EI_1 \, \Delta}{\ell_1^3} \right) \frac{\ell_1}{2} = \frac{6EI_1 \, \Delta}{\ell_1^2}$$

$$M_2 = P_2 \frac{\ell_2}{2}$$

$$= \left(\frac{12EI_2 \, \Delta}{\ell_2^3} \right) \frac{\ell_2}{2} = \frac{6EI_2 \, \Delta}{\ell_2^2}$$

From these expressions a proportion may be written between the moments as follows because Δ is the same in each column.

$$\frac{M_1}{M_2} = \frac{6EI_1 \, \Delta / \ell_1^2}{6EI_2 \, \Delta / \ell_2^2}$$

$$= \frac{I_1 / \ell_1^2}{I_2 / \ell_2^2}$$

This relationship must be used for assuming sidesway moments for the columns of a frame. Any convenient moments may be assumed, but they must be in proportion to each other, as their I/ℓ^2 values. Should their I and ℓ values be equal, the assumed moments will be equal.

On the following page the analysis of the frame of Figure 19.3 is completed in Figure 19.8 with the sidesway method. First, I/ℓ^2 values are computed for each of the columns (equal in this case) in part (a) of the figure. Then, in part (b) sidesway moments in proportion to the I/ℓ^2 values are assumed and distributed throughout the frame, after which the horizontal reactions at the column bases are calculated. The assumed sidesway moments of $-10^{\cdot k}$ each are found to produce horizontal reactions to the left totaling 0.92 kip. Only 0.89 kip was needed, and if 0.89/0.92 times the values of the distributed moments are added to the originally distributed fixed-end moments, the final moments will be obtained. The results are shown in Figure 19.8(c). The horizontal reactions at the column bases also are calculated and shown.

Examples 19.3 to 19.5 present the solutions for additional sidesway problems. It will be noted in Example 19.4 that the column bases are pinned. Such a situation does not alter the method of solution. For the balancing of fixed-end moments or for balancing assumed sidesway moments, the bases are balanced to their correct zero values as was done in continuous beams. The three-quarter factor was used in computing the column stiffnesses for this example.

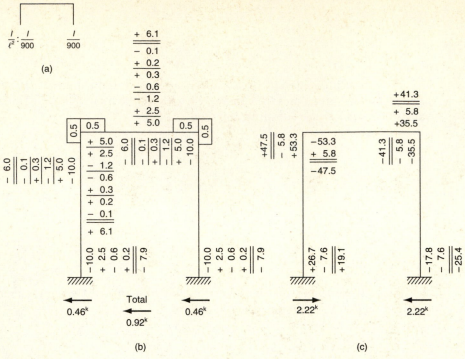

Figure 19.8

EXAMPLE 19.3 _____

Find all moments in the structure shown in Figure 19.9. Use the sidesway method.

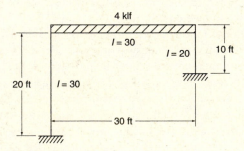

Figure 19.9

Solution. Distribute fixed-end moments and compute horizontal reactions at column bases as shown on the following page.

The imaginary support must supply a force of 20.93 kips to the right. Removal of the support will allow the frame to sway to the left; therefore, counterclockwise or positive moments are assumed in the columns in proportion to their I/ℓ^2 values.

The assumed moments develop total reactions to the right of 10.58 kips; however, 20.93 kips was needed, and 20.93/10.58 times these moments is added to the results obtained initially by distributing the fixed-end moments.

Distribute fixed-end
moments.

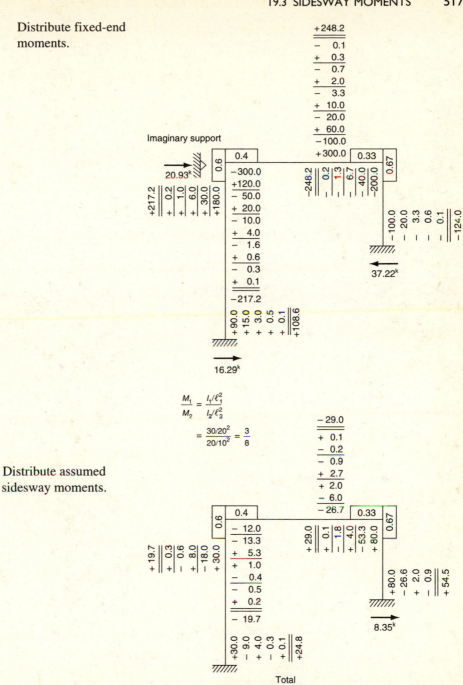

$$\frac{M_1}{M_2} = \frac{l_1/\ell_1^2}{l_2/\ell_2^2}$$

$$= \frac{30/20^2}{20/10^2} = \frac{3}{8}$$

Distribute assumed
sidesway moments.

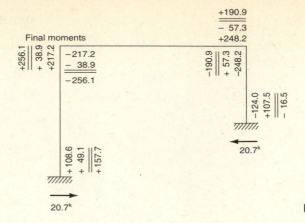

Final moments

+256.1 | +38.9 | +217.2

−217.2
− 38.9
−256.1

+190.9
− 57.3
+248.2

−190.9 | + 57.3 | −248.2

−124.0 | +107.5 | − 16.5

20.7ᵏ

+108.6 | + 49.1 | +157.7

20.7ᵏ

EXAMPLE 19.4 ――――――――――――――――――――――――――――――――――――

Determine the final end moments for the frame shown in Figure 19.10.

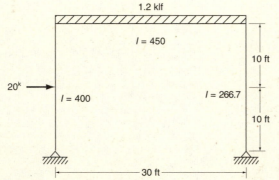

1.2 klf

l = 450

10 ft

20ᵏ

l = 400

l = 266.7

10 ft

30 ft

Figure 19.10

Distribute fixed-end moments.

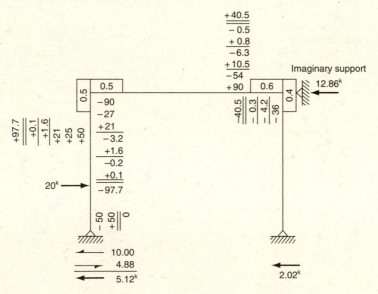

+40.5
− 0.5
+ 0.8
−6.3
+10.5
−54
+90

Imaginary support   12.86ᵏ

0.5 | 0.5 | 0.6 | 0.4

−90
−27
+21
−3.2
+1.6
−0.2
+0.1
−97.7

−40.5 | − 0.3 | − 4.2 | −36

+97.7 | +0.1 | +1.6 | +21 | +25 | +50

20ᵏ

− 50 | +50 | 0

10.00
4.88
5.12ᵏ

2.02ᵏ

518

Solution.   Distribute fixed-end moments and compute horizontal reactions at column bases.

The structure sways to the right and assumed sidesway moments are negative and in the following proportion:

$$M_1 : M_2$$

$$\frac{400}{(20)^2} : \frac{266.7}{(20)^2}$$

$$1.00 : 0.667.$$

Distribute assumed sidesway moments.

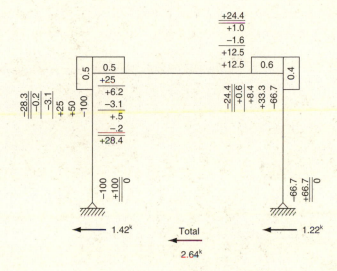

Final moments equal distributed fixed-end moments plus 12.86/2.64 times the distributed assumed sidesway moments.

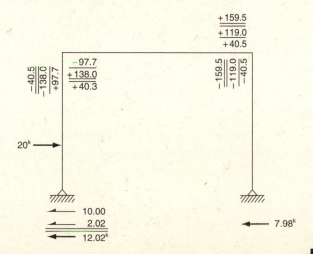

EXAMPLE 19.5

Compute the final end moments for the frame shown in Figure 19.11.

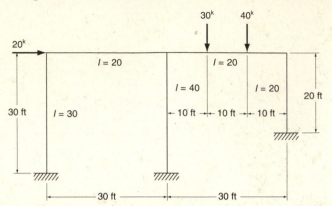

Figure 19.11

Solution. Distribute fixed-end moments and compute horizontal reactions at column bases.

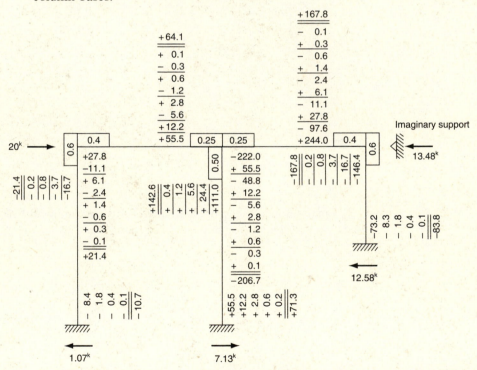

The structure sways to the right; therefore negative moments are assumed in the columns in proportion to the I/ℓ^2 values.

$$M_1 : M_2 : M_3$$

$$\frac{30}{30^2} : \frac{40}{30^2} : \frac{20}{20^2}$$

$$30 : 40 : 45$$

Distribute assumed sidesway moments.

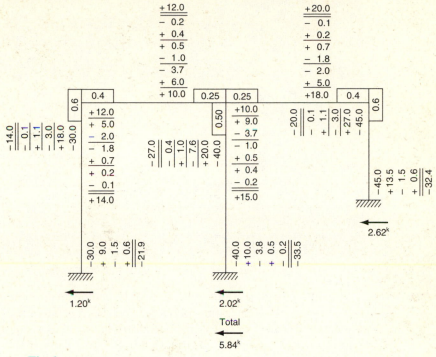

Final moments:

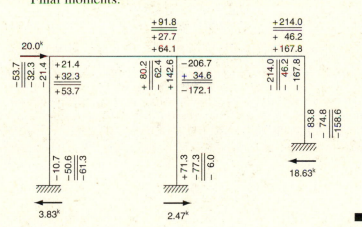

19.4 FRAMES WITH SLOPING LEGS

The frames considered up to this point have been made up of vertical and horizontal members. It was proved earlier in this chapter that when sidesway occurs in such frames, it causes fixed-end moments in the columns proportional to their $6EI\Delta/\ell^2$ values. (As $6E\Delta$ was a constant for these frames, moments were assumed proportional to their I/ℓ^2 values.) Furthermore, lateral swaying did not produce fixed-end moments in the beams.

I-91 bridge, Lyndon, Vermont. (Courtesy of the Vermont Agency of Transportation.)

The sloping-leg frame of Figure 19.12 can be analyzed in much the same manner as the vertical-leg frames previously considered. The fixed-end moments due to the external loads are calculated and distributed; the horizontal reactions are computed; and the horizontal force needed at the imaginary support is determined.

Once again sidesway causes moments in the frame members proportional to their $\dfrac{6EI\Delta}{\ell^2}$ values. For sloping leg frames the Δ values often will be unequal for the various members, as will be the I and ℓ values. Therefore, sidesway moments are assumed in proportion to the $\dfrac{I\Delta}{\ell^2}$ values. These assumed moments are distributed, the horizontal reactions are computed, and the necessary moments needed for balancing are calculated and superimposed on the distributed fixed-end moments.

For this discussion the frame of Figure 19.12(a) is considered and is assumed to sway to the right as shown in part (b) of the figure. It can be seen that lateral movement

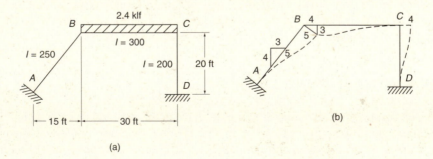

Figure 19.12

of the frame will cause Δs, and thus moments in the beam as well as in the columns. To determine the $I\Delta/\ell^2$ value for each member it is necessary to determine the relative Δ value in each member.

As the frame sways to the right, joint B moves in an arc about joint A and joint C moves in an arc about joint D. As these movements or arcs are very short, they are considered to consist of straight lines perpendicular to the respective members. Rather than attempting to develop complex trigonometric formulas for the relative Δs, the authors have merely drawn deformation triangles on the figure.

Column AB has a slope of four vertically, three horizontally, or five inclined. If the relative movement of joint B perpendicular to AB is assumed to be five, then its vertical movement will be three and its horizontal movement will be four.

Joint C moves in a horizontal direction to the right perpendicular to member CD. If the change in length of member BC is neglected, joint C must move horizontally the same distance as joint B, or a distance of four. The relative Δ values are now available as follows: $\Delta_{AB} = 5$, clockwise; $\Delta_{BC} = 3$, counterclockwise; and $\Delta_{CD} = 4$, clockwise. The clockwise rotation produces a counterclockwise or negative resisting moment. These values are given at the end of this paragraph together with the $I\Delta/\ell^2$ values.

Relative Δ values	Relative sidesway moments $I\Delta/\ell^2$
$\Delta_{AB} = -5$	$M_{AB} = \dfrac{(250)(-5)}{(25)^2} = -2$
$\Delta_{BC} = +3$	
$\Delta_{CD} = -4$	$M_{BC} = \dfrac{(300)(+3)}{(30)^2} = +1$
	$M_{CD} = \dfrac{(200)(-4)}{(20)^2} = -2$

Example 19.6 illustrates the analysis of the frame of Figure 19.12. The same procedure used for determining the horizontal reaction components at the bases of the sloping columns is used for the vertical columns. (That is, each column is considered to be a free body, and moments are taken at its top to determine the horizontal reaction at its base.) For vertical columns, the vertical reactions pass through the points where moments are taken and thus may be neglected. *This is not the case for sloping columns, and the vertical reactions will appear in the moment equations.*

The left column AB of the frame of Example 19.6 is considered as a free body after the fixed-end moments are distributed and shown in Figure 19.13. Taking moments at B to determine H_A, the following equation results:

$$+ 59.9 + 120 + 15V_A - 20H_A = 0$$

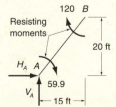

Figure 19.13

Note that it is necessary to compute V_A before the equation can be solved for H_A. The authors find it convenient to remove the beam as a free body and compute the vertical reaction applied at each end by the columns. Once the value at B is determined, the sum of the vertical forces on column AB can be equated to zero and V_A can be determined. The same procedure is followed after the assumed sidesway moments are distributed.

EXAMPLE 19.6 _____

Determine the final end moments for the frame shown in Figure 19.12.

Solution.   Distribute fixed-end moments and determine support reactions.

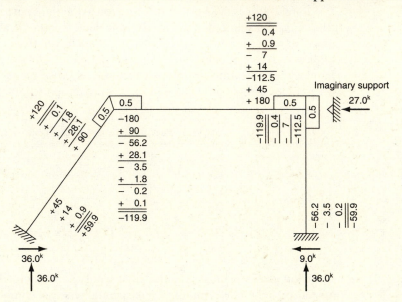

Distribute assumed sidesway moments.

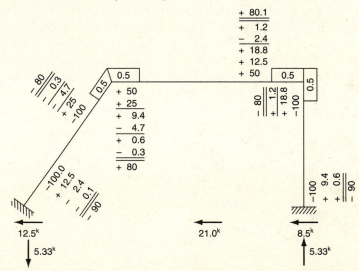

Final moments:

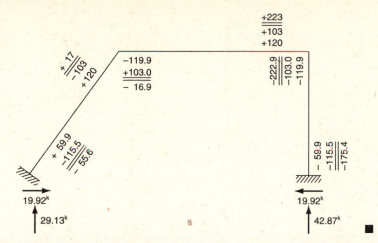

In Figure 19.14 another illustration of the determination of relative Δ values for a sloping leg frame is presented. With the numbers shown initially on the deformation triangles, joint B moves three units to the right and joint C moves four units to the right, but the horizontal movement of joints B and C must be equal. For this reason the initial values at C are marked through and multiplied by three quarters so the horizontal values will be equal. The resulting values are shown below.

Relative Δ values

$\Delta_{AB} = -3.605$

$\Delta_{BC} = +4.25$

$\Delta_{CD} = -3.75$

From the analyses considered in this chapter it is quite obvious that frame members are subject to axial forces (and thus axial deformations) as well as moments and shears. *The reader should clearly understand that the effects of axial deformations (which usually are negligible) are neglected in the moment distribution procedures described herein.*

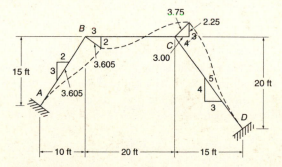

Figure 19.14

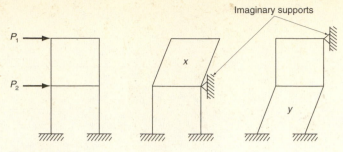

Figure 19.15

19.5 MULTISTORY FRAMES

These are two possible ways in which the frame of Figure 19.15 may sway. The loads P_1 and P_2 obviously will cause both floors of the structure to sway to the right, but it is not known how much of the swaying is going to occur in the top floor (x condition) or how much will occur in the bottom floor (y condition). There are two sidesway conditions that need to be considered.

To analyze the frame by the usual sidesway procedure would involve (1) an assumption of moments in the top floor for the x condition and the distribution of the moments throughout the frame, and (2) an assumption of moments in the lower floor for the y condition and the distribution of those moments throughout the frame. One equation could be written for the top floor by equating x times the horizontal forces caused by the x moments plus y times the horizontal forces caused by the y moments to the actual total shear on the floor, P_1. A similar equation could be written for the lower floor by equating the horizontal forces caused by the assumed moments to the shear on that floor, $P_1 + P_2$. Simultaneous solution of the two equations would yield the values of x and y. The final moments in the frame equal x times the x distributed moments plus y times the y distributed moments.

The sidesway method is not difficult to apply for a two-story frame, but for multistory frames it becomes unwieldy because each additional floor introduces another sidesway condition and another simultaneous equation.

Professor C. T. Morris of Ohio State University introduced a much simpler method for handling multistory frames, which involves a series of successive corrections.[1] His method also is based on the total horizontal shear along each level of a building. In considering the frame of Figure 19.16(a), shown on page 528, which is being deflected laterally by the loads P_1 and P_2, each column is assumed to take the approximate S shape shown in (b).

At the middepth of the columns there is assumed to be a point of contraflexure. The column may be considered to consist of a pair of cantilevered beams, one above the point and one below the point, as shown in (c). The moment in each cantilever will equal the shear times $h/2 = Vh/2$, and the total moment top and bottom is equal to

[1] C. T. Morris, "Morris on Analysis of Continuous Frames," *Transactions of the American Society of Civil Engineers* 96 (1932): 66–69.

Office building, 99 Park Avenue, New York City. (Courtesy of the American Institute of Steel Construction, Inc.)

$Vh/2 + Vh/2 = Vh$. As such, the total moment in a column is equal to the shear carried by the column times the column height.

Similarly, on any one level the total moments top and bottom of all the columns will equal the total shear on that level multiplied by the column height. The method consists in initially assuming this total for the column moments on a floor and distributing it between the columns in proportion to their I/ℓ^2 values.

The moment taken by each column is distributed half to top end and half to

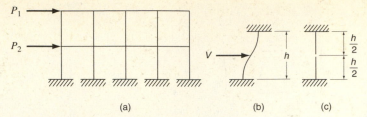

Figure 19.16

bottom end. The joints are balanced, including the fixed-end moments, making no carryovers until all the joints are balanced. At this time the sum of the column moments doesn't equal the correct final value—thus, they are corrected to their initial and final total, and the joints are balanced again. Successive corrections work exceptionally well for multistory frames, as illustrated by Examples 19.7 and 19.8. The procedure used is as follows:

1. Compute fixed-end moments.
2. Compute total moments in columns (equal to shear on the level times column height) for each level and distribute between the columns in proportion to their I/ℓ^2 values, then divide each by half—one half to top of column and one half to bottom.
3. Balance all joints throughout the structure, making no carryovers.
4. Make carryovers for the entire frame.
5. The total of the column moments on each level has been changed and will not equal the shear times the column height. Determine the difference and add or subtract the amount back to the columns in proportion to their I/ℓ^2 values.
6. Steps 3 to 5 are repeated over and over until the amount of the corrections to be made is negligible.

EXAMPLE 19.7

Determine final moments for the structure in Figure 19.17. Use the successive correction method developed by Professor Morris.

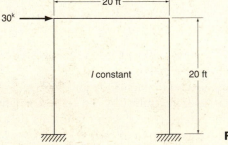

Figure 19.17

Solution.

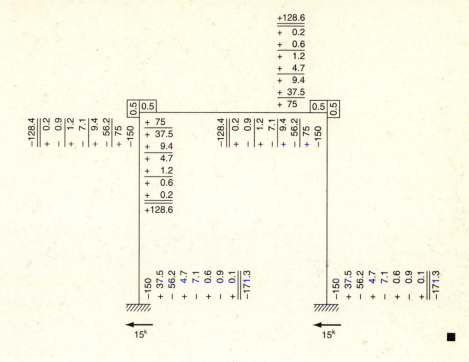

EXAMPLE 19.8

Determine the final moments for the frame of Example 16.7, reproduced in Figure 19.18 by successive corrections.

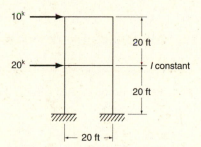

Figure 19.18

Solution.

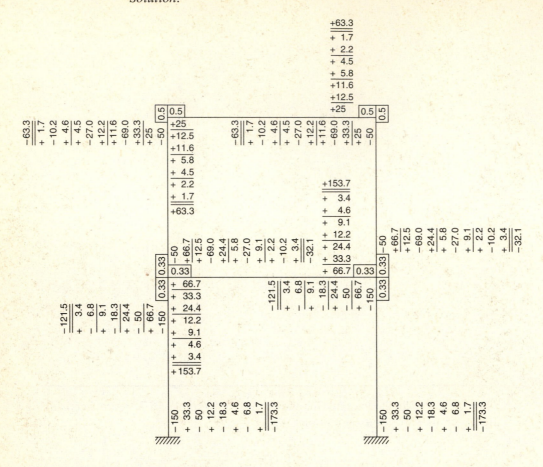

19.6 COMPUTER EXAMPLE

In Example 19.9 the computer program SABLE is used to rework Example 19.6.

EXAMPLE 19.9 ───

Determine the end moments in the sloping leg frame of Figure 19.19. This structure was previously analyzed in Example 19.6.

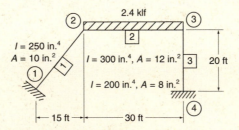

Figure 19.19

Solution.

Structural Data

1. Nodal location and restraint data

Node	Coordinates X	Coordinates Y	Coordinates Z	Restraints X	Restraints Y	Restraints Rot
1	0.000E+00	0.000E+00	0.000E+00	Y	Y	Y
2	1.800E+02	2.400E+02	0.000E+00	N	N	N
3	5.400E+02	2.400E+02	0.000E+00	N	N	N
4	5.400E+02	0.000E+00	0.000E+00	Y	Y	Y

2. Beam location and property data

Beam	i	j	Type	Beam properties Area	Beam properties Izz	Beam properties E
1	1	2	F-F	1.000E+01	2.500E+02	2.900E+04
2	2	3	F-F	1.200E+01	3.000E+02	2.900E+04
3	4	3	F-F	8.000E+00	2.000E+02	2.900E+04

3. Applied beam loads

Beam	Case	P	a	W
1	1	0.000E+00	0.000E+00	0.000E+00
2	1	0.000E+00	0.000E+00	−2.000E−01
3	1	0.000E+00	0.000E+00	0.000E+00

Results

Calculated beam end forces
(*Note: units are* in.−k *for moments*)

Beam	Case	End	Axial	Shear-Y	Moment-Z
1	1	i	3.524E+01	1.586E+00	6.771E+02
		j	−3.524E+01	−1.586E+00	−2.014E+02
2	1	i	1.988E+01	2.914E+01	2.014E+02
		j	−1.988E+01	4.286E+01	−2.670E+03
3	1	i	4.286E+01	1.988E+01	2.100E+03
		j	−4.286E+01	−1.988E+01	2.670E+03 ∎

PROBLEMS

Balance moments and calculate horizontal reactions at bases of columns for Problems 19.1 through 19.6.

19.1

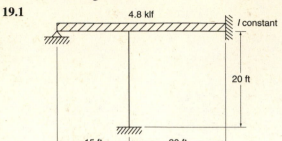

19.2   (*Ans.* $M_B = 14.44$ ft-k, $M_D = 15.12$ ft-k, $H_B = 2.2^k \rightarrow$)

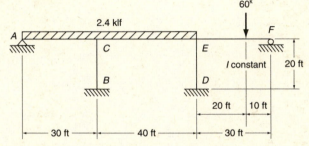

19.3

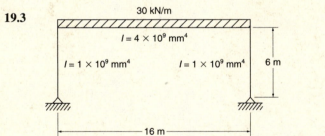

19.4   (*Ans.* $M_A = -172.4$ ft-k, $M_D = 105.1$ ft-k, $H_L = 33.37$ k $\leftarrow H_R = 33.37^k \rightarrow$)

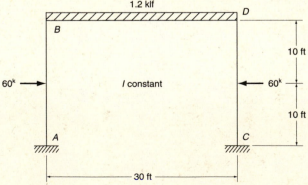

19.5

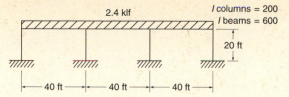

19.6   (*Ans.* $M_B = 389.7$ kN · m, $M_{DB} = 537.6$ kN · m, $M_E = 39.6$ kN · m)

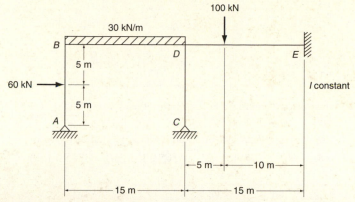

For Problems 19.7 through 19.20, determine final moments with the sidesway method.

19.7

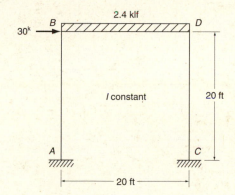

19.8   (*Ans.* $M_A = 101.3$ ft-k, $M_D = 179.9$ ft-k)

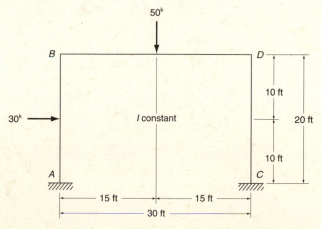

19.9

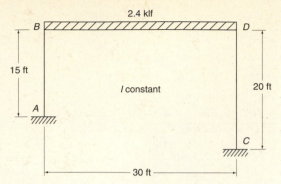

19.10   Rework Problem 19.8 if column bases are pinned. (*Ans.* $M_B = 2.9$ ft-k, $M_D = 297$ ft-k)

19.11   Rework Problem 19.7 if column *CD* is pinned at its base.

19.12   (*Ans.* $M_A = 22.9$ ft-k, $M_B = 305.0$ ft-k, $M_D = 315.6$ ft-k)

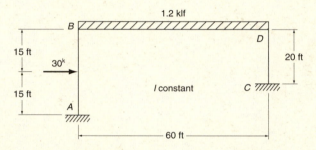

19.13

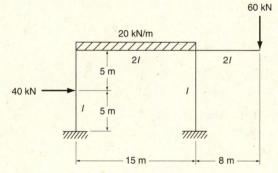

19.14   (*Ans.* $M_A = 221.2$ ft-k, $M_{DF} = 426.8$ ft-k, $H_A = 24.92^k \rightarrow$)

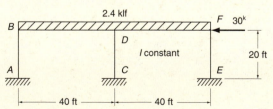

19.15

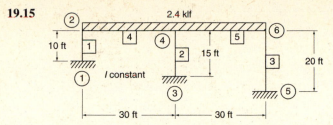

19.16   (*Ans.* M_A = 147.8 ft-k, M_{CD} = 83.5 ft-k, M_D = 133.3 ft-k, M_E = 102.8 ft-k)

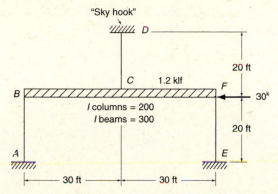

19.17

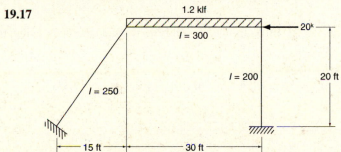

19.18   (*Ans.* M_A = 21.1$^{'k}$, M_B = 14.7$^{'k}$, M_C = 41.0$^{'k}$, M_D = 54.4$^{'k}$)

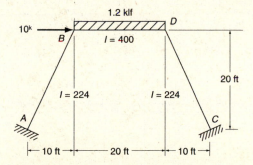

19.19

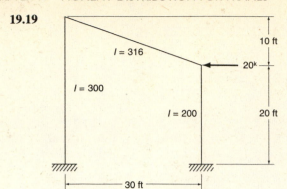

19.20   (*Ans.* $M_A = 18.9^{'k}$, $M_B = 21.1^{'k}$, $M_C = 24.4^{'k}$, $M_D = 23.9^{'k}$)

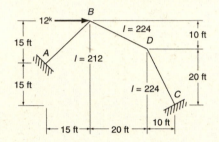

For Problems 19.21 through 19.27, analyze these structures using the successive correction method.

19.21   Rework Problem 19.7

19.22   (*Ans.* $M_A = 1.5^{'k}$, $M_B = 7.6^{'k}$, $M_{DF} = 106.3^{'k}$, $M_F = 67.5^{'k}$)

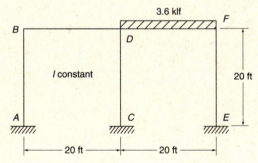

19.23

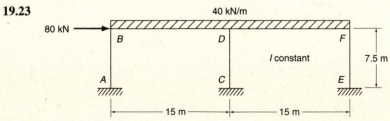

19.24   (*Ans. $M_{BA} = 151^{'k}$, $M_{BE} = 222^{'k}$, $M_{CF} = 96^{'k}$, $M_{ED} = 94^{'k}$*)

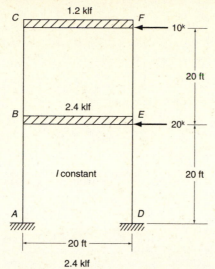

19.25

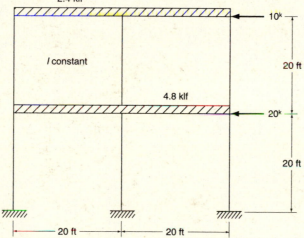

19.26   (*Ans. $M_A = 131^{'k}$, $M_B = 86^{'k}$, $M_{DC} = 94^{'k}$, $M_{JG} = 86^{'k}$*)

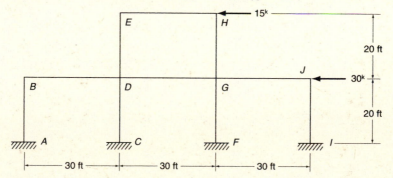

19.27

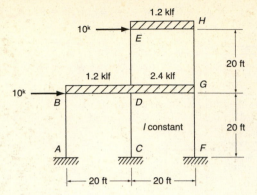

For Problems 19.28 through 19.30 rework the problems given using SABLE.

19.28   Problem 19.8 (*Ans.* $M_B = 120$ ft-k, $M_C = 180$ ft-k)

19.29   Problem 19.20

19.30   Problem 19.26 (*Ans.* $M_E = 76.5$ ft-k, $M_F = -136$ ft-k)

Chapter 20

Introduction to Matrix Methods

20.1 REASONS FOR MATRIX METHODS

During the past three decades there have been tremendous changes in the structural analysis methods used in engineering practice. These changes have occurred primarily because of the great developments made with high-speed digital computers and the increasing use of very complex structures. Matrix methods of structural analysis are a convenient mathematical representation of a structural system that can be easily solved using digital computers. Prior to the advent of the computer these methods were of little use because the equations were too cumbersome for solution using hand calculation methods.

The change in analysis methods has been accelerated during the past decade with the ready availability of very powerful personal computers and workstations. Today a majority of structural analysis is conducted using computer-based methods. Therefore, knowledge of the fundamental principles of matrix structural analysis and an appreciation of the strengths and weaknesses of the analysis procedure are very important for students of structural engineering.

Structural analysis using matrix methods involves no new concepts of structural engineering. We use principles learned in this book and in courses in mechanics of materials. We will simply organize the procedures and the resulting equations in a format conducive to analysis with a digital computer. The computer is capable of extraordinary feats of arithmetic, but it can do only those tasks that can be described with simple, precise, and unambiguous instructions. In the following chapters the authors have attempted to explain the principles of matrix structural analysis and the application of these principles to the analysis of structural systems.

Knowledge of elementary matrix algebra is very important to an understanding of matrix structural analysis—virtually all of the equations involved in the procedures are matrix equations. Although most engineers have studied matrix algebra at some time, they may not have used it often and their memory of the principles may have become a little rusty. For this reason an Appendix about matrix algebra has been included in this textbook and is suggested for review. If a student can understand the information presented there, he or she will have no problem understanding the discussions that follow.

20.2 USE OF MATRIX METHODS

Structural engineers have been attempting to handle analysis problems for a good many years by applying the methods used by mathematicians in linear algebra. Although many structures could be analyzed with the resulting equations, the work was extremely tedious, at least until large-scale computers became available. In fact, the usual matrix equations are not manageable with hand-held calculators unless the most elementary structures are involved.

Today matrix analysis (using computers) has substantially replaced the classical methods of analysis in engineering offices. As a result, engineering educators and

Erecting the world's largest radio telescope, Greenbank, West Virginia. (Courtesy of Lincoln Electric Company.)

writers of structural analysis textbooks are faced with a difficult decision. Should they require a thorough study of the classical methods followed by a study of modern matrix methods; should they require students to study both at the same time in an integrated approach; or should they present only a study of the modern methods? The reader can see from the preceding chapters that the authors feel that an initial study of some of the classical methods followed by a study of the matrix methods will result in an engineer who has a better understanding of structural behavior.

Any method of analysis involving linear algebraic equations can be put into matrix notation, and matrix operations then can be used to obtain their solution. The possibility of the application of matrix methods by the structural engineer is very important because all linearly elastic, statically determinate and indeterminate structures are governed by systems of linear equations.

The simple numerical examples presented in this chapter and the next could be solved more quickly by classical methods using a pocket calculator rather than with a matrix approach. However, as structures become more complex and as more loading patterns are considered, matrix methods using computers become increasingly useful.

20.3 FORCE AND DISPLACEMENT REPRESENTATIONS

The methods presented in earlier chapters for analyzing statically indeterminate structures can be placed in two general classes. These are the force methods and the displacement methods. Both of these methods have been developed to a stage where they can be applied to almost any structure—trusses, beams, frames, plates, shells, and so on. The displacement procedures, however, are much more commonly used today since they can be more easily programmed for solution by computers. These two methods of analysis, which were previously discussed in Section 13.1, are redefined here, as it is felt that the material presented in the last few chapters will enable the reader to better understand the definitions.

Force Method of Analysis   With the force method, also called the *flexibility* or *compatibility method,* redundants are selected and removed from the structure so that a stable and statically determinate structure remains. An equation of deformation compatibility is written at each location where a redundant has been removed. These equations are written in terms of the redundants and the resulting equations are solved for the numerical values of the redundants. After the redundants are determined, statics can be used to compute all other desired internal forces, moments, and so on. The Method of Consistent Distortions (see Chapter 13) is a force method.

Displacement Method of Analysis   With the displacement method of analysis, also called the *stiffness* or *equilibrium method,* the displacements of the joints necessary to describe fully the deformed shape of the structure are used in a set of simultaneous equations. When the equations are solved for these displacements, they are substituted into the force-deformation relations of each member to determine the various internal forces. The Slope Deflection method (see Chapter 16) is a displacement method.

The number of unknowns in the displacement method is generally greater than the number of unknowns in flexibility methods. A greater number of simultaneous equations must therefore be evaluated. Despite this fact, displacement methods are the commonly used method for matrix analysis because representation of the structure is significantly easier. Further, there are no differences in application of the method to statically determinate or statically indeterminate systems—the solution for both types of systems is formulated in the same manner. For this reason we will discuss only the displacement method in this book. Excellent discussions of the flexibility method have been presented by Rubinstein[1] and Przemieniecki.[2] Readers having need for, or interest in, flexibility methods should consult these references.

20.4 SOME NECESSARY DEFINITIONS

Before beginning a discussion of matrix methods an understanding of the following basic terms is necessary.

- *Elements*—These are the pieces that comprise the structural system that is being represented. In the types of structures that have been analyzed in this book, the elements are the beams and columns.
- *Nodes*—The nodes are the locations in the structure at which the elements are connected. In structures composed of beams and columns, the nodes usually are the joints.
- *Coordinate*—A structural coordinate is a possible displacement, a degree of freedom, in a structure at a node.[3] In planar structures there are three coordinates at each node. These possible displacements, or degrees of freedom, are two orthogonal translations and a rotation about an axis perpendicular to the plane defined by the translations. In a space structure, there are six coordinates at each node—three orthogonal translations and three rotations. Please note that a structural coordinate should not be confused with a Cartesian coordinate, which describes the location of a point in space.
- *Force*—The term "force" in matrix analysis is a general term that refers either to a force acting at a translational coordinate or to a moment acting at a rotational coordinate. There is no distinction as to whether the force is a known structural load or an unknown force of reaction.
- *Displacement*—Like "force," "displacement" is a general term that refers either to a translation at a translational coordinate or to a rotation at a

---

[1] Moshe F. Rubinstein, *Matrix Computer Analysis of Structures* (Englewood Cliffs, New Jersey: Prentice-Hall, Incorporated, 1966).

[2] J. S. Przemieniecki, *Theory of Matrix Structural Analysis* (New York: McGraw-Hill, Incorporated, 1968).

[3] In the context of this book, a structural coordinate and a degree of freedom are really the same. In advanced courses, though, generalized coordinates often are used instead of degrees of freedom. For this reason, we will use the term *coordinate* here.

rotational coordinate. There is no distinction as to whether the displacement is an unknown displacement at an unconstrained degree-of-freedom or a known displacement at a constrained degree-of-freedom.

- *Stiffness*—Stiffness is the force required to cause a unit deformation in an elastic material.

20.5 THE FUNDAMENTAL CONCEPT

In this book we have been studying structural systems that behave in a linearly elastic manner. Further, the structures considered have undergone small displacements. You should remember that in a structure undergoing small displacements the undeformed geometry can be used throughout the calculations. Such structural systems behave in accordance with Hooke's law,[4] which we recall has the general form

$$F = kx \tag{20.1}$$

The basic form of the equation used in matrix analysis, which is analogous to the basic expression of Hooke's law, is

$$\{F\} = [K]\{X\} \tag{20.2}$$

The basic difference is that now we are dealing with a system of linear simultaneous equations. The term $\{F\}$ is the force vector and is analogous to the term F in Hooke's law. The force vector contains all of the forces that are acting at the structural coordinates. The vector $\{X\}$ is the displacement vector and is analogous to the displacement in Hooke's law. It contains the displacements at each of the structural coordinates. Lastly, $[K]$ is the stiffness matrix. It is analogous to the spring constant k in Hooke's law. The stiffness matrix is a representation of the ability of a structural system to resist load. In Section 20.8 we will see how the stiffness matrix for a structure will be developed.

Before going on, the concept of stiffness deserves a little thought and discussion. By definition, the constant k for a spring—its stiffness—is the force required to cause a unit displacement. Consider the spring in Figure 20.1, which has a stiffness $K = 50$ lb/in. If we prevent Point A from moving and place a force of 50 lbs at Point B, Point B will displace one inch to the right. Conversely, to cause a unit displacement at Point B, we must apply a force of 50 lbs at that point. At Point A there is a force of 50 lbs acting to the left that causes the system to be in a state of static equilibrium.

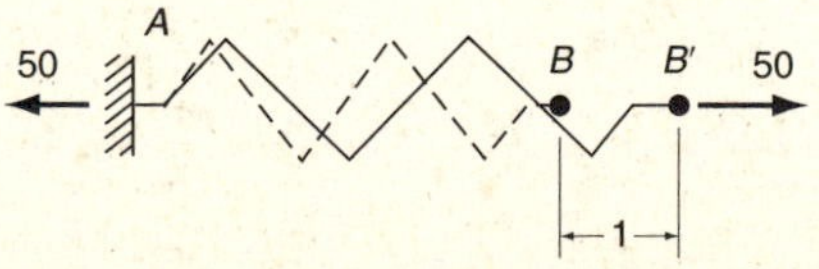

Figure 20.1

---

[4] A common misconception is that Hooke's law is $\sigma = E\epsilon$. Hooke was an English scientist who worked with springs. This relationship was derived from his observation for springs, which in Latin is *Ut Tensio sic Vis*—as the force, so the stretch. Nevertheless, $\sigma = E\epsilon$ often is referred to as Hooke's law.

The coefficients in the stiffness matrix $[K]$ follow the same general idea. Each of the coefficients in the matrix is identified by subscripts that indicate the row and column in which it is located. For example, K_{34} is the stiffness coefficient in the third row and fourth column of the stiffness matrix. The definition of a stiffness coefficient is:

Definition.   *The stiffness coefficient K_{ij} is the force at coordinate i when a unit displacement is imposed at coordinate j and the displacement at all of the other coordinates is constrained to zero.*

This definition can be used to develop the stiffness matrix for a structural system. To see how this is done, let's return to the spring in Figure 20.1. Consider two coordinates directed along the axis of the spring; one of the coordinates will be located at each end of the spring at the two possible displacement degrees of freedom. These coordinates are shown in Figure 20.2(a). There is a row and column in the stiffness matrix for each coordinate that is being considered in the structure. The stiffness matrix for this spring will have two rows and two columns.

To determine the first column of the stiffness matrix, displace Coordinate 1 one unit and prevent displacement at Coordinate 2, as shown in Figure 20.2(b). Then determine the forces at all of the coordinates necessary to maintain this configuration. By inspection we see that a force of 50 lbs must be applied to the right at Coordinate 1 and a force of 50 lbs must be applied to the left at Coordinate 2 to

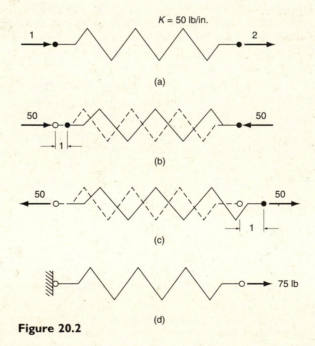

Figure 20.2

maintain this configuration and satisfy the equations of equilibrium. We now have the first column of the stiffness matrix—K_{11}, the force at Coordinate 1 to have a unit displacement there and zero displacement elsewhere is equal to 50 lbs/in. and K_{21}, the force at Coordinate 2 to have a unit displacement at Coordinate 1 and zero displacement elsewhere, is equal to -50 lbs/in.

This process is repeated for Coordinate 2 to generate the second column of the stiffness matrix. If we displace it one unit, as shown in Figure 20.2(c), and hold Coordinate 1 to zero displacement, it will be found that K_{12} is equal to -50 lbs/in. and K_{22} is equal to 50 lbs/in. The stiffness matrix, then, for this spring is

$$[K] = \begin{bmatrix} K_{11} & K_{12} \\ K_{21} & K_{22} \end{bmatrix} = \begin{bmatrix} 50 & -50 \\ -50 & 50 \end{bmatrix} \tag{20.3}$$

Now assume that the spring is used in a structure that is loaded and configured as shown in Figure 20.2(d). There is a force of 75 lbs applied to the right at Coordinate 2 and Coordinate 1 is constrained to zero displacement. The resulting system equation, in the form of Equation 20.2, becomes

$$\begin{bmatrix} 50 & -50 \\ -50 & 50 \end{bmatrix} \begin{Bmatrix} 0 \\ x_2 \end{Bmatrix} = \begin{Bmatrix} F_1 \\ 75 \end{Bmatrix} \tag{20.4}$$

Speaking in terms of structures we would say that Coordinate 1 is at a support (there is no displacement) and that the 75 lb force is an applied load. If the system of simultaneous linear equations implied by Equation 20.4 is solved for the unknowns x_2 and F_1, using whatever means is convenient, we find that F_1 is equal to -75 lbs and that x_2 is equal to 1.5 in. F_1 is the unknown force of reaction and x_2 is the unknown structural displacement. This concludes our first analysis using matrix methods.

20.6 SYSTEMS WITH SEVERAL ELEMENTS

Structures are rarely composed of single elements such as the single spring in Figure 20.2. There are usually many more elements, but the principles for developing the stiffness matrices are the same. Let's investigate a more complicated spring structure. Consider the three-spring structure shown in Figure 20.3. The system has three elements and four nodes. Each spring has the same stiffness, which is 50 lbs/in. The coordinates we will use are shown in the figure. Because a spring can direct force only along its axis, the only applicable coordinates are those parallel to the axis of the springs. For reasons that will become clear later, we have numbered the unconstrained coordinates first and the constrained coordinates last.

$K = 50$ lb/in.   **Figure 20.3**

The structural stiffness matrix is developed one column at a time. To develop the stiffness coefficients for the first column, displace Coordinate 1 one unit to the right and hold the displacement at all other coordinates to zero. To do this a force of 100 lbs must be applied at Coordinate 1. A force of 50 pounds is required to move the right end of the left spring to the right one unit, and a force of 50 pounds is required to move the left end of the middle spring to the right one unit. As such, a total force of 100 pounds is required at Coordinate 1. Because Coordinate 1 was displaced one unit and the other coordinates were not displaced, there is a tensile force of 50 lbs in the left spring and a compressive force of 50 lbs in the middle spring. For equilibrium to be achieved, there must be forces acting to the left at Coordinates 2 and 3. The right spring did not deform because the coordinates to which this spring was attached did not displace. As such, there are no forces in the system as a result of deformation of the right spring. The stiffness coefficients in the first column, then, are: $K_{11} =$ 100 lbs/in., $K_{21} = -50$ lbs/in., $K_{31} = -50$ lbs/in., and $K_{41} = 0$.

The stiffness coefficients in the other columns are determined in the same manner as was the first column. The coordinate associated with each column is displaced 1 unit, the other coordinates are held to zero displacement, and the forces necessary to maintain this configuration are computed. After all of the stiffness coefficients have been found, the stiffness matrix for this three-spring structure is

$$[K] = \begin{bmatrix} 100 & -50 & -50 & 0 \\ -50 & 100 & 0 & -50 \\ -50 & 0 & 50 & 0 \\ 0 & -50 & 0 & 50 \end{bmatrix} \qquad (20.5)$$

Notice that the stiffness matrix is symmetrical about the main diagonal (the diagonal from upper left to lower right), that there are no zero terms on the main diagonal at rows or columns associated with unconstrained coordinates, and that the terms on the main diagonal are all positive. These are characteristics of the structural stiffness matrix that will always be true.

In the structural analysis, the forces acting at unconstrained coordinates are the known quantities and the displacements at the constrained coordinates are the unknown quantities. The known displacements are the boundary conditions and the known forces are the structural loads. Taking this example a little further, a force of 100 lbs is applied to the left at Coordinate 2. The complete system equation becomes

$$\begin{bmatrix} 100 & -50 & -50 & 0 \\ -50 & 100 & 0 & -50 \\ -50 & 0 & 50 & 0 \\ 0 & -50 & 0 & 50 \end{bmatrix} \begin{Bmatrix} x_1 \\ x_2 \\ 0 \\ 0 \end{Bmatrix} = \begin{Bmatrix} 0 \\ -100 \\ F_3 \\ F_4 \end{Bmatrix} \qquad (20.6)$$

The displacements at Coordinates 1 and 2 are not known. The displacements at Coordinates 3 and 4 are known because these are the structural supports. Likewise, the forces at Coordinates 3 and 4 are not known because these are the structural reactions. There is no applied load at Coordinate 1, so F_1 is zero. The force at

Coordinate 2 is negative because it is acting to the left, which is opposite to the assumed direction of the coordinate.

To solve for the unknowns in Equation 20.6, we will need to partition the matrices. The partitioning will be between the constrained and unconstrained coordinates (this is why the unconstrained coordinates were numbered first and the constrained coordinates second). The partitioning lines are shown in Equation 20.7.

$$\begin{bmatrix} 100 & -50 & -50 & 0 \\ -50 & 100 & 0 & -50 \\ -50 & 0 & 50 & 0 \\ 0 & -50 & 0 & 50 \end{bmatrix} \begin{Bmatrix} x_1 \\ x_2 \\ 0 \\ 0 \end{Bmatrix} = \begin{Bmatrix} 0 \\ -100 \\ F_3 \\ F_4 \end{Bmatrix} \tag{20.7}$$

A partition of a matrix is really just another matrix. A partitioned matrix, then, is nothing more than a matrix of matrices. So Equation 20.7 is really

$$\begin{bmatrix} \begin{bmatrix} 100 & -50 \\ -50 & 100 \end{bmatrix} & \begin{bmatrix} -50 & 0 \\ 0 & -50 \end{bmatrix} \\ \begin{bmatrix} -50 & 0 \\ 0 & -50 \end{bmatrix} & \begin{bmatrix} 50 & 0 \\ 0 & 50 \end{bmatrix} \end{bmatrix} \begin{Bmatrix} \begin{Bmatrix} x_1 \\ x_2 \end{Bmatrix} \\ \begin{Bmatrix} 0 \\ 0 \end{Bmatrix} \end{Bmatrix} = \begin{Bmatrix} \begin{Bmatrix} 0 \\ -100 \end{Bmatrix} \\ \begin{Bmatrix} F_3 \\ F_4 \end{Bmatrix} \end{Bmatrix} \tag{20.8}$$

A shorthand way for representing this equation is

$$\begin{bmatrix} [K_{11}] & [K_{12}] \\ [K_{21}] & [K_{22}] \end{bmatrix} \begin{Bmatrix} \{X_1\} \\ \{X_2\} \end{Bmatrix} = \begin{Bmatrix} \{F_1\} \\ \{F_2\} \end{Bmatrix} \tag{20.9}$$

There are really two matrix equations represented by Equation 20.9

$$[K_{11}]\{X_1\} + [K_{12}]\{X_2\} = \{F_1\} \tag{20.10}$$

and

$$[K_{21}]\{X_1\} + [K_{22}]\{X_2\} = \{F_2\} \tag{20.11}$$

The partition $\{X_2\}$ contains all zeros; therefore, the second term on the left side of Equation 20.10 is equal to zero. The partition $\{X_1\}$ can be solved for from Equation 20.10 as

$$\{X_1\} = \begin{Bmatrix} x_1 \\ x_2 \end{Bmatrix} = [K_{11}]^{-1}\{F_1\} = \begin{Bmatrix} -0.667 \\ -1.333 \end{Bmatrix} \tag{20.12}$$

This result can be substituted into 20.11 and the unknown forces, $\{F_2\}$, the forces of reaction, can be found to be

$$\{F_2\} = \begin{Bmatrix} F_3 \\ F_4 \end{Bmatrix} = \begin{Bmatrix} 33.33 \\ 66.67 \end{Bmatrix} \tag{20.13}$$

The analysis for this more complicated structure is now complete. We determined values for all of the forces and displacements that were initially unknown.

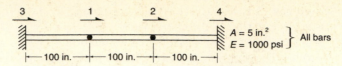

Figure 20.4

20.7 BARS INSTEAD OF SPRINGS

Structures are not really composed of springs. They are composed of beams, bars, and columns. What would the result be if the system that was shown in Figure 20.3 contained bars instead of the springs? Such a system is shown in Figure 20.4.

   To evaluate this system we need to determine the stiffness of each of the bars. This can be done using the equation for the deformation of a prismatic axially loaded bar, which is recalled to be

$$\delta = \left(\frac{L}{AE}\right)P \tag{20.14}$$

Equation 20.14 can be rearranged to resemble Equation 20.1.

$$\left(\frac{AE}{L}\right)\delta = P \tag{20.15}$$

In this form we observe that the term (AE/L) is analogous to the spring constant in Hooke's law. Indeed, (AE/L) is the stiffness for an axial bar. For the bars in Figure 20.4, their stiffness is found to equal

$$k = \frac{5(1000)}{100} = 50 \text{ lbs/in.} \tag{20.16}$$

This is the same value that was used for the spring stiffness in Section 20.6. Because of that, the stiffness matrix, the equations used for analysis, and the results obtained are the same as were computed in that section.

20.8 SOLUTION FOR A TRUSS

Before finishing in this chapter, let's evaluate a structure that looks a little more like the other structures that we have analyzed in this textbook. Consider the truss in Figure 20.5. This is a simple two-bar truss that has forces acting at the unconstrained node. The coordinates to be used in the analysis and the member properties are shown on the figure. This is a pin-connected truss, so there are only two applicable coordinates at each node. The solution begins by first determining the stiffnesses for each of

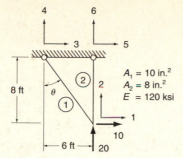

$A_1 = 10$ in.^{2}
$A_2 = 8$ in.^{2}
$E = 120$ ksi

Figure 20.5

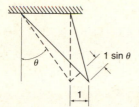

1 sin θ

Figure 20.6

the members in the structure. From Equation 20.15 we find those to be:

$$k_1 = \frac{10(120)}{10(12)} = 10 \text{ k/in.}$$

$$k_2 = \frac{8(120)}{8(12)} = 10 \text{ k/in.} \tag{20.17}$$

The procedures for building the stiffness matrix are a little more involved now than they were for the simpler structures in the previous sections. Regardless, the process for building the stiffness matrix is the same. We begin by displacing the first coordinate one unit and holding the other coordinates to zero displacement. This configuration is shown in Figure 20.6. Notice that Element 1 has undergone a rigid body rotation and has been elongated. The amount by which the bar has been elongated is 1 sin θ. Recall that we are considering small displacements so the undeformed geometry can be used in calculations with the deformed geometry. The force[5] in Bar 1 caused by this elongation, then, is

$$F_1^1 = k_1 \sin \theta = 10(.6) = 6 \text{ kips tension} \tag{20.18}$$

This is the force in the bar for every inch of displacement of Coordinate 1. Components of this force are acting at each of the coordinates to which that bar is attached, as shown in Figure 20.7 on the next page. The contribution to the stiffness of Coordinate 1 from this bar, then, is as follows[6]

---

[5] Here we use the notation F_m^ℓ where m indicates the coordinate that was displaced one unit and ℓ indicates the bar being considered.

[6] The term K_{ij}^l as used here is the contribution to the stiffness coefficient K_{ij} from element l.

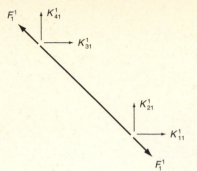

Figure 20.7

$$K_{11}^1 = F_1^1 \sin \theta = 6(0.6) = 3.6 \text{ k/in.}$$
$$K_{21}^1 = -F_1^1 \cos \theta = 6(0.8) = -4.8 \text{ k/in.}$$
$$K_{31}^1 = -F_1^1 \sin \theta = 6(0.6) = -3.6 \text{ k/in.}$$
$$K_{41}^1 = F_1^1 \cos \theta = 6(0.8) = 4.8 \text{ k/in.} \qquad (20.19)$$

Element 2 has undergone only a rigid body rotation; it has not changed length. As such, there is no force in the bar. Therefore, Element 2 does not contribute to the stiffness of Coordinate 1. Column 1 of the stiffness matrix is simply the contribution from Element 1, which is

$$K_{11} = 3.6 \text{ k/in.}$$
$$K_{21} = -4.8 \text{ k/in.}$$
$$K_{31} = -3.6 \text{ k/in.}$$
$$K_{41} = 4.8 \text{ k/in.}$$
$$K_{51} = 0 \text{ k/in.}$$
$$K_{61} = 0 \text{ k/in.} \qquad (20.20)$$

The second column of the stiffness matrix is determined by displacing only Coordinate 2 one unit and determining the forces necessary to maintain this configuration. The geometry of this situation is shown in Figure 20.8. As before, the first bar has undergone a rigid body rotation and has become shorter by an amount 1 cos θ. The force in the bar is:

$$F_2^1 = k_1 \cos \theta = 10(0.8) = 8 \text{ kips compression} \qquad (20.21)$$

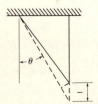

Figure 20.8

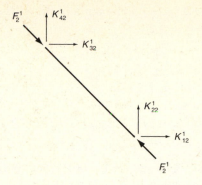

Figure 20.9

Again, this is the force in Bar 1 for every unit of displacement at Coordinate 2. As before, the forces at the coordinates to which the bar is attached are components of this compressive force, shown in Figure 20.9. The contribution to the stiffness of Coordinate 2 caused by Bar 1 is

$$K_{12}^1 = F_2^1 \sin \theta = -8(0.6) = -4.8 \text{ k/in.}$$
$$K_{22}^1 = F_2^1 \cos \theta = 8(0.8) = 6.4 \text{ k/in.}$$
$$K_{32}^1 = F_2^1 \sin \theta = 8(0.6) = 4.8 \text{ k/in.}$$
$$K_{42}^1 = -F_2^1 \cos \theta = 8(0.8) = -6.4 \text{ k/in.} \tag{20.22}$$

When Coordinate 2 was displaced one unit, Bar 2 became shorter by one unit. The force in Bar 2, then, caused by the unit displacement of Coordinate 2 is

$$F_2^2 = 10(1) = 10 \text{ kips compression} \tag{20.23}$$

As with Bar 1, this is the force in Bar 2 for every unit of deformation of Coordinate 2. The contributions to the stiffness at Coordinate 2 from Bar 2 are the components of this force in the coordinate directions as shown on Figure 20.10. The components of force are

$$K_{12}^2 = 0 \text{ k/in.}$$
$$K_{22}^2 = F_2^2 = 10 \text{ k/in.}$$
$$K_{52}^2 = 0 \text{ k/in.}$$
$$K_{62}^2 = -F_2^2 = -10 \text{ k/in.} \tag{20.24}$$

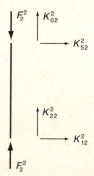

Figure 20.10

The total stiffness at Coordinate 2 is the combination of the stiffness provided by all of the components. This result is

$$K_{12} = -4.8 + 0 = -4.8 \text{ k/in.}$$
$$K_{22} = 6.4 + 10 = 16.4 \text{ k/in.}$$
$$K_{32} = 4.8 + 0 = 4.8 \text{ k/in.}$$
$$K_{42} = -6.4 + 0 = -6.4 \text{ k/in.}$$
$$K_{52} = 0 + 0 = 0 \text{ k/in.}$$
$$K_{62} = 0 - 10 = -10 \text{ k/in.} \tag{20.25}$$

The same procedures as these are followed for the other four coordinates. The resulting system equation is shown in Equation 20.26 along with the applied loads and the displacement boundary conditions.

$$\begin{bmatrix} 3.6 & -4.8 & -3.6 & 4.8 & 0 & 0 \\ -4.8 & 16.4 & 4.8 & -16.4 & 0 & -10 \\ -3.6 & 4.8 & 3.6 & -4.8 & 0 & 0 \\ 4.8 & -6.4 & -4.8 & 6.4 & 0 & 0 \\ 0 & 0 & 0 & 0 & 0 & 0 \\ 0 & -10 & 0 & 0 & 0 & -10 \end{bmatrix} \begin{Bmatrix} x_1 \\ x_2 \\ 0 \\ 0 \\ 0 \\ 0 \end{Bmatrix} = \begin{Bmatrix} 10 \\ 20 \\ F_3 \\ F_4 \\ F_5 \\ F_6 \end{Bmatrix} \tag{20.26}$$

Using Equations 20.10 and 20.11, the unknown forces and displacements are found to be

$$\begin{Bmatrix} F_3 \\ F_4 \\ F_5 \\ F_6 \end{Bmatrix} = \begin{Bmatrix} -9.97 \\ 13.29 \\ 0 \\ -33.3 \end{Bmatrix} \quad \text{and} \quad \begin{Bmatrix} x_1 \\ x_2 \end{Bmatrix} = \begin{Bmatrix} 7.21 \\ 3.33 \end{Bmatrix} \tag{20.27}$$

Using principles learned previously in this textbook, the interested student could easily demonstrate that these are indeed the correct displacements and forces of reaction for this structure. The analysis of trusses containing more members would be handled in the same manner as the truss that we have just analyzed.

20.9 SYSTEM MATRICES USING STRAIN ENERGY

The system stiffness matrix could have been prepared through use of strain energy. Consider the single spring that was shown in Figure 20.2. The strain energy in the spring in terms of the displacement at the coordinates is

$$U = \frac{k}{2} (x_2 - x_1)^2 \tag{20.28}$$

Using Castigliano's first theorem it can be shown that

$$\frac{\partial U}{\partial x_1} = F_1 = \frac{k}{2}(2x_1 - 2x_2)$$

$$\frac{\partial U}{\partial x_2} = F_2 = \frac{k}{2}(2x_2 - 2x_1) \tag{20.29}$$

These two equations can be expressed in matrix form as

$$\begin{Bmatrix} F_1 \\ F_2 \end{Bmatrix} = \begin{bmatrix} k & -k \\ -k & k \end{bmatrix} \begin{Bmatrix} x_1 \\ x_2 \end{Bmatrix} \tag{20.30}$$

If the stiffness of the spring is 50 lbs/in. and the force at F_2 is 75 lbs, this is the same equation as that given in Equation 20.4.

Strain energy and Castigliano's first theorem can also be used for systems composed of multiple elements. Consider the spring system that was shown in Figure 20.3 as an example. The strain energy in this system in terms of displacements at the coordinates is

$$U = \frac{k_1}{2}(x_1 - x_3)^2 + \frac{k_2}{2}(x_2 - x_1)^2 + \frac{k_3}{2}(x_4 - x_2)^2 \tag{20.31}$$

Using Castigliano's first theorem, we can use this expression for strain energy to obtain the force at each of the structural coordinates. The result is the following four equations:

$$\frac{\partial U}{\partial x_1} = F_1 = \frac{k_1}{2}(2x_1 - 2x_3) + \frac{k_2}{2}(2x_1 - 2x_2)$$

$$\frac{\partial U}{\partial x_2} = F_2 = \frac{k_2}{2}(2x_2 - 2x_1) + \frac{k_3}{2}(2x_2 - 2x_4)$$

$$\frac{\partial U}{\partial x_3} = F_3 = \frac{k_1}{2}(2x_3 - 2x_1)$$

$$\frac{\partial U}{\partial x_4} = F_4 = \frac{k_3}{2}(2x_4 - 2x_2) \tag{20.32}$$

If these equations are expressed in matrix form, we obtain the system equation

$$\begin{Bmatrix} F_1 \\ F_2 \\ F_3 \\ F_4 \end{Bmatrix} = \begin{bmatrix} k_1 + k_2 & -k_2 & -k_1 & 0 \\ -k_2 & k_2 + k_3 & 0 & -k_3 \\ -k_1 & 0 & k_1 & 0 \\ 0 & -k_3 & 0 & k_3 \end{bmatrix} \begin{Bmatrix} x_1 \\ x_2 \\ x_3 \\ x_4 \end{Bmatrix} \tag{20.33}$$

Again notice that this is the same result as obtained in Equation 20.6 if we substitute for the stiffness of the spring, the loads, and the boundary conditions.

Before leaving this discussion, there are two observations that should be made. First, as can be seen in Equation 20.31, each component in the system contributes to the total strain energy in the system. The first term on the right side of that equation is the contribution of the left spring, the second term is the contribution of the middle spring, and the last term is the contribution of the right spring.

Secondly, from Equations 20.32, and the representation of those equations in Equation 20.33, it can clearly be seen that each component connected to a structural coordinate contributes to the total stiffness of that coordinate. This observation will be used as we develop a generalized procedure to develop the system stiffness matrix in the next chapter.

20.10 LOOKING AHEAD

In this chapter, the basic concepts of matrix structural analysis have been introduced. When structural systems become very complicated and are composed of many elements, the direct procedures discussed in this chapter become very difficult and cumbersome to apply. In the next chapter we will study a more general procedure for developing the system matrices. We also will look at how loads can be applied to the elements themselves. In that chapter actual beams and bars instead of the springs used in this chapter will be considered. It is very interesting, though, that structural elements can be represented by springs.

PROBLEMS

For Problems 20.1 through 20.5, using matrix methods presented in this chapter, determine the displacements at each joint in the structure and all of the forces of reaction.

20.1

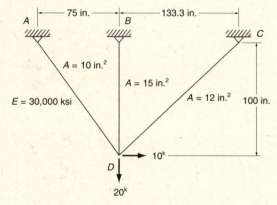

20.2   (Ans. $R_{AH} = -17.68^k$, $R_{DV} = 17.68^k$)

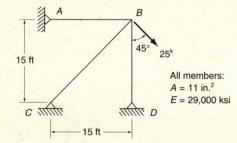

20.3

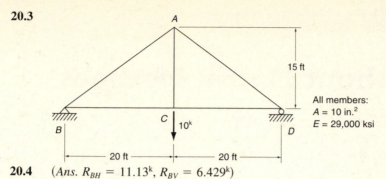

20.4   (*Ans.* $R_{BH} = 11.13^k$, $R_{BV} = 6.429^k$)

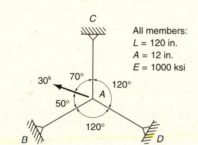

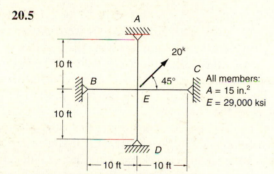

For Problems 20.6 through 20.10 use the computer program SABLE to find the displacements at each joint in the structure, all of the forces of reaction, and the force in each bar. Suggestion: When using SABLE, select the element type to be a PIN-PIN beam. Because such a beam cannot provide rotational stiffness, the rotation about the z axis will need to be constrained at each joint in the structure.

20.6   Repeat Problem 20.1   (*Ans.* $\delta_{DH} = 0.004305$ in., $\delta_{DV} = -0.002863$)

20.7   Repeat Problem 20.2

20.8   Repeat Problem 20.3   (*Ans.* $\delta_{CV} = -0.02793$ in.)

20.9   Repeat Problem 20.4

20.10   Repeat Problem 20.5   (*Ans.* $\delta_{EH} = 0.00195$ in., $\delta_{EV} = 0.00195$ in.)

Chapter 21

More About Matrix Methods

21.1 INTRODUCTION

The fundamental concepts of matrix structural analysis were presented in the last chapter by employing very simple principles and applying them to the development of system matrices. By the time you finished studying that material, however, it probably was painfully clear to you that those methods would be very unwieldy for large structures. Furthermore, the preparation of system matrices in such a manner is not conducive to preparation using the computer.

In this chapter we will develop a general procedure that can be easily turned into algorithms solvable with computers. To do so, we will first develop the stiffness matrices for a general truss element and beam element in what we will call the element coordinate system. By changing the physical properties (area, length, etc.), the truss element can represent any member in a truss structure and the beam can represent any element in a frame. We then go through a transformation from the element coordinate system to the global coordinate system—the coordinate system that is associated with the entire structure. Once the elemental stiffness matrices have been transformed to the global coordinate system, we can then assemble the system stiffness matrix and determine the forces and displacements in the structure.

At this point, this discussion and the operations implied may be confusing and vague. To help your understanding of the material and where it fits into the big picture of structural analysis using the computer, please refer to the analysis flowchart in Figure 21.1. Shown in this flowchart are all of the steps necessary for a computer-based structural analysis program. In this chapter we will be considering the analysis section of the flowchart and the operations performed in each step. The procedures discussed are the very same procedures that have been used in the program SABLE. Although important for a structural analysis program, the other sections are for the specification of the structural data and presentation of the computed results. As such, these sections are not important in the study of the principles of structural analysis. Many commercial software packages that are available for structural analysis also use these principles. These packages include SAP90, STRUDL, NASTRAN, and ANSYS, among others.

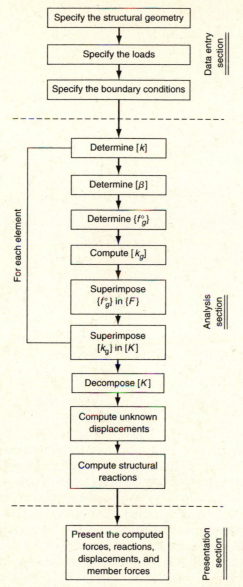

Figure 21.1   Flowchart showing the steps required for a computer-based structural analysis program.

21.2 DEFINITION OF COORDINATE SYSTEMS

In the previous chapter we were working in what is called the global coordinate system. In this context, when we speak of coordinates recall that we are speaking of the structural coordinates. The global structural coordinates are used to define the

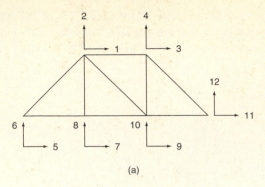

(a)

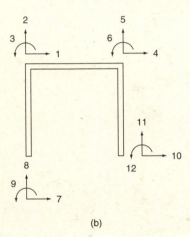

(b)

Figure 21.2   Global structural coordinates.

displacements and forces acting on the entire structure. A coordinate is assigned to each relevant degree-of-freedom at each joint in the structure. Such coordinates are shown in Figure 21.2a for a plane truss and in Figure 21.2b for a plane frame. The joints in a truss only can translate so there are two translational coordinates at each joint. The joints in a plane frame, on the other hand, can translate and rotate. As such, there are three coordinates at each joint—two translations and a rotation. Generally, the global coordinates are aligned with the principal axes of the structure. In the case of frames and trusses, these axes usually are the horizontal and vertical axes of the structure.

The members in the structure each have a coordinate system associated with them. These element coordinate systems do not have to be oriented in the same direction as global coordinates. Typical element coordinates for a truss member and a frame member are shown on page 560 in Figure 21.3(a) and (b), respectively. Notice that the element coordinates are oriented to be parallel and perpendicular to the

U.S. Customs Court and Federal Office Building, New York City. The 41-story office building (rear) of the U.S. Customs Court and Federal Office Building is a typical steel-framed tower, with columns supporting beams. But the eight-story courthouse (foreground), designed for column-free courtrooms and a pedestrian concourse at street level, literally hangs from two parallel steel trusses above the roof level. (Courtesy Bethlehem Steel Corporation.)

longitudinal axis of the member. Also notice that the axial coordinates are directed from what has been named the i end of the member to what has been named the j end of the member. This establishes the positive direction of the element coordinates (which is important when interpreting the results) and the direction in which the angle ϕ is measured. The reference for the angle is the horizontal axis of the structure.

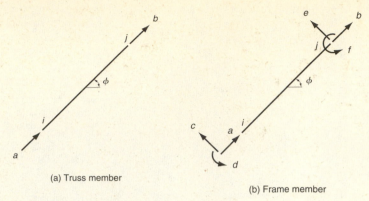

(a) Truss member

(b) Frame member

Figure 21.3   Typical element coordinates.

21.3 THE ELEMENTAL STIFFNESS RELATIONSHIP

In Chapter 20 we presented the basic relationship between forces and displacements for the entire structure. That relationship was given in Equation 20.2. An analogous relationship exists when looking at the forces and displacements at the ends of a component of the structure. That relationship is

$$\{f\} = [k]\{x\} \tag{21.1}$$

In Equation 21.1 $\{f\}$ is the vector of forces, in local coordinates, acting at the ends of the member. The term $\{x\}$ is the vector of displacements, also in local coordinates, of the end of the member. The matrix $[k]$ is the elemental stiffness matrix in local coordinates. This relationship of Equation 21.1 for the member could have been expressed in terms of global coordinates as

$$\{f_g\} = [k_g]\{x_g\} \tag{21.2}$$

In Equation 21.2 each term of the equation has the same significance as it did in Equation 21.1, except that now the relationship is expressed in terms of global coordinates—in reality, the global coordinates at the end of the member.

In the next two sections we will develop the elemental stiffness matrix for a general truss element and a general beam element, respectively. Also, we will develop the transformation matrix that is necessary to transform the quantities from local coordinates to global coordinates.

21.4 TRUSS ELEMENT MATRICES

When we develop the matrices for an element, there are two matrices that need to be developed. These are the stiffness matrix and the transformation matrix. In Chapter 20 we have already developed the stiffness matrix for a truss element. As you recall from that discussion, the element coordinates are f_a and f_b as shown on Figure 21.4

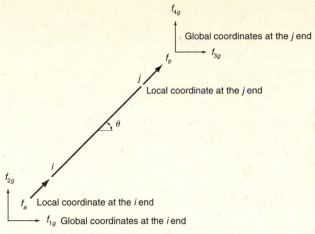

Figure 21.4 Element and global coordinates.

and the stiffness matrix is

$$[k] = \begin{bmatrix} \dfrac{AE}{L} & -\dfrac{AE}{L} \\[2mm] -\dfrac{AE}{L} & \dfrac{AE}{L} \end{bmatrix} \tag{21.3}$$

This is nothing more than Equation 20.3 with the substitution in Equation 20.15 for the stiffness. Please notice that we are now using small letters to denote the element matrices and that we will continue to use large letters to denote the global matrices.

The transformation matrix for the truss element can be developed from principles of equilibrium. At each node on the element, the resultant forces acting at the node must be the same whether they are viewed in element coordinates (f_a and f_b) or in global coordinates (f_{1g}, f_{2g}, f_{3g}, and f_{4g}). As such, the following four equations can be written by considering the geometry and coordinates shown in Figure 21.4

$$f_a \cos \theta = f_{1g}$$
$$f_a \sin \theta = f_{2g}$$
$$f_b \cos \theta = f_{3g}$$
$$f_b \sin \theta = f_{4g} \tag{21.4}$$

These four equations can be represented in matrix form as

$$\begin{Bmatrix} f_{1g} \\ f_{2g} \\ f_{3g} \\ f_{4g} \end{Bmatrix} = \begin{bmatrix} \cos \theta & 0 \\ \sin \theta & 0 \\ 0 & \cos \theta \\ 0 & \sin \theta \end{bmatrix} \begin{Bmatrix} f_a \\ f_b \end{Bmatrix} \tag{21.5}$$

The left side of this equation is the vector of element end forces in global coordinates. The last term on the right-hand side is the vector of element end forces in element coordinates. The first term on the right-hand side is the matrix that maps the element coordinates onto the global coordinates. This matrix is the transpose of the transformation matrix $[\beta]$ for the truss element. Similar equations could have been developed if we considered displacement at the ends of the member instead of force. The transformation matrix maps forces and displacements from one coordinate system to the other; it does not, however, map the stiffness of the element. The procedure to transform element stiffness from the local to the global coordinate system is discussed in Section 21.6.

21.5 BEAM ELEMENT MATRICES

The stiffness matrix for the beam element is composed of two parts. These parts are the axial force effects and the flexural force (shear and moment) effects. Accepting the basic assumption that displacements are small, these two effects will not have an affect on each other.

The stiffness matrix for the beam element has the general form

$$[k] = \begin{bmatrix} [\text{Axial}] & 0 \\ 0 & [\text{Flexural}] \end{bmatrix} \tag{21.6}$$

If the elements coordinates shown in Figure 21.5 are used, the axial effects partition of the stiffness matrix is the same as we have been using for the truss element—that which is contained in Equation 21.3.

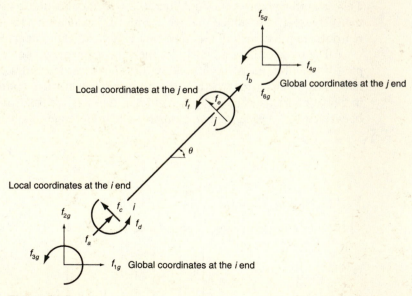

Figure 21.5

The flexural partition of the stiffness matrix is found in the same manner that we found the axial partition of the stiffness matrix. A unit displacement is imposed separately at each flexural coordinate, the displacement at the other flexural coordinates is constrained to zero, and the flexural forces necessary to hold this configuration are computed. These are the forces and displacements that are shown in Figure 21.6. These forces can be computed using any of the procedures discussed earlier in this book, such as virtual work or slope deflection.

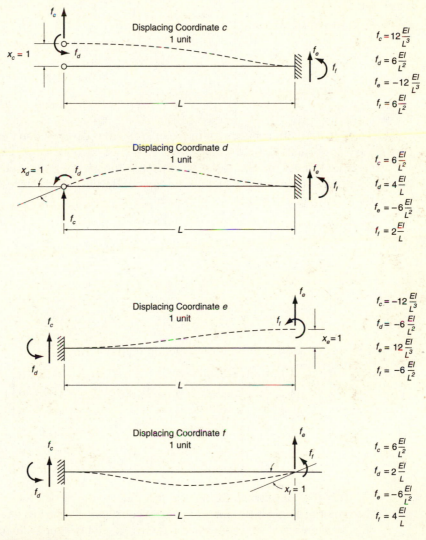

$$f_c = 12\frac{EI}{L^3}$$

$$f_d = 6\frac{EI}{L^2}$$

$$f_e = -12\frac{EI}{L^3}$$

$$f_f = 6\frac{EI}{L^2}$$

$$f_c = 6\frac{EI}{L^2}$$

$$f_d = 4\frac{EI}{L}$$

$$f_e = -6\frac{EI}{L^2}$$

$$f_f = 2\frac{EI}{L}$$

$$f_c = -12\frac{EI}{L^3}$$

$$f_d = -6\frac{EI}{L^2}$$

$$f_e = 12\frac{EI}{L^3}$$

$$f_f = -6\frac{EI}{L^2}$$

$$f_c = 6\frac{EI}{L^2}$$

$$f_d = 2\frac{EI}{L}$$

$$f_e = -6\frac{EI}{L^2}$$

$$f_f = 4\frac{EI}{L}$$

Figure 21.6

The forces necessary to maintain the displaced configuration are the stiffness coefficients in the column represented by the coordinate with the unit displacement. The flexural partition of the element stiffness matrix, then, is

$$[\text{Flexural}] = \begin{bmatrix} \frac{12EI}{L^3} & \frac{6EI}{L^2} & -\frac{12EI}{L^3} & \frac{6EI}{L^2} \\ \frac{6EI}{L^2} & \frac{4EI}{L} & -\frac{6EI}{L^2} & \frac{2EI}{L} \\ -\frac{12EI}{L^3} & -\frac{6EI}{L^2} & \frac{12EI}{L^3} & -\frac{6EI}{L^2} \\ \frac{6EI}{L^2} & \frac{2EI}{L} & -\frac{6EI}{L^2} & \frac{4EI}{L} \end{bmatrix} \tag{21.7}$$

and the final stiffness matrix for the beam element in local coordinates is then

$$[k] = \begin{bmatrix} \frac{AE}{L} & -\frac{AE}{L} & 0 & 0 & 0 & 0 \\ -\frac{AE}{L} & \frac{AE}{L} & 0 & 0 & 0 & 0 \\ 0 & 0 & \frac{12EI}{L^3} & \frac{6EI}{L^2} & -\frac{12EI}{L^3} & \frac{6EI}{L^2} \\ 0 & 0 & \frac{6EI}{L^2} & \frac{4EI}{L} & -\frac{6EI}{L^2} & \frac{2EI}{L} \\ 0 & 0 & -\frac{12EI}{L^3} & -\frac{6EI}{L^2} & \frac{12EI}{L^3} & -\frac{6EI}{L^2} \\ 0 & 0 & \frac{6EI}{L^2} & \frac{2EI}{L} & -\frac{6EI}{L^2} & \frac{4EI}{L} \end{bmatrix} \tag{21.8}$$

The transformation matrix for the beam element is developed in the same manner that the transformation matrix was developed for the truss element. The resultant force at each node must be the same whether the forces are expressed in terms of local coordinates or in terms of global coordinates. Because of this necessary consideration, the following six equations can be written

$$f_a \cos \theta - f_c \sin \theta = f_{1g}$$
$$f_a \sin \theta + f_c \cos \theta = f_{2g}$$
$$f_d = f_{3g}$$
$$f_b \cos \theta - f_e \sin \theta = f_{4g}$$
$$f_b \sin \theta + f_e \cos \theta = f_{5g}$$
$$f_f = f_{6g} \tag{21.9}$$

The six equations can be represented in matrix form as

$$\begin{Bmatrix} f_{1g} \\ f_{2g} \\ f_{3g} \\ f_{4g} \\ f_{5g} \\ f_{6g} \end{Bmatrix} = \begin{bmatrix} \cos \theta & 0 & -\sin \theta & 0 & 0 & 0 \\ \sin \theta & 0 & \cos \theta & 0 & 0 & 0 \\ 0 & 0 & 0 & 1 & 0 & 0 \\ 0 & \cos \theta & 0 & 0 & -\sin \theta & 0 \\ 0 & \sin \theta & 0 & 0 & \cos \theta & 0 \\ 0 & 0 & 0 & 0 & 0 & 1 \end{bmatrix} \begin{Bmatrix} f_a \\ f_b \\ f_c \\ f_d \\ f_e \\ f_f \end{Bmatrix} \tag{21.10}$$

As with the transformation for the truss element, the left side of this equation is the vector of element end forces in global coordinates. The last term on the right-hand side is the vector of element end forces in element coordinates. The first term on the right-hand side is the matrix that maps the element coordinates onto the global coordinates. This matrix is the transpose of the transformation matrix $[\beta]$

for the beam element. Again, this matrix maps the forces or displacements in local coordinates to global coordinates.

21.6 TRANSFORMATION TO GLOBAL COORDINATES

In the previous sections we developed the elemental stiffness and transformation matrices for the truss element and the beam element. Using these matrices we can compute the elemental stiffness matrices in global coordinates. The complete transformation from elemental to global coordinates can be developed from energy principles. The strain energy in the element in terms of local coordinates can be computed from

$$U = \frac{1}{2}\{x\}^T\{f\} \tag{21.11}$$

and the strain energy in the element in terms of global coordinates can be written as

$$U = \frac{1}{2}\{x_g\}^T\{f_g\} \tag{21.12}$$

The strain energy in the element must be the same whether it is written in terms of local coordinates or in terms of global coordinates. Therefore, the following must be true

$$\{x\}^T\{f\} = \{x_g\}^T\{f_g\} \tag{21.13}$$

From the relationship developed in Equation 21.5 or Equation 21.10, observe that

$$\{x\} = [\beta]\{x_g\} \tag{21.14}$$

Using Equation 21.14 and the law of the transpose of a product, it can be shown that

$$\{x_g\}^T = \{[\beta]\{x_g\}\}^T = \{x_g\}^T[\beta]^T \tag{21.15}$$

We can then substitute Equations 21.1, 21.2, and 21.15 into Equation 21.13 to obtain

$$\{x_g\}^T[\beta]^T[k][\beta]\{x_g\} = \{x_g\}^T[k_g]\{x_g\} \tag{21.16}$$

Notice in Equation 21.16 that both sides of the equation are premultiplied by $\{x_g\}^T$ and are post multiplied by $\{x_g\}$. As such, both of these terms can be canceled from the equation. When doing so, Equation 21.16 reduces to

$$[\beta]^T[k][\beta] = [k_g] \tag{21.17}$$

which is the relationship that we wanted to develop. It is the transformation of the elemental stiffness in local coordinates to the elemental stiffness matrix in global coordinates.

21.7 ASSEMBLING THE GLOBAL STIFFNESS MATRIX

After we have developed the elemental stiffness for each component in the structure in terms of global coordinates, these matrices can be assembled to form the global stiffness matrix for the entire structure. Recall from our discussion in Section 20.9

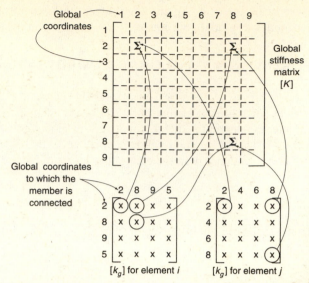

Figure 21.7

that the total stiffness at a coordinate is the sum of the stiffnesses at that coordinate contributed by each element attached to that coordinate. As such, we can superimpose the element stiffness matrices to obtain the total global stiffness matrix for the structure. Unfortunately, all of the elements in the structure are not connected to all of the global coordinates. Generally, any one element is connected to only a few of the global coordinates. Because of this situation, we cannot simply add all of the element stiffness matrices $[k_g]$ to obtain the global stiffness matrix; there is not a one-to-one correspondence between the stiffness coefficients.

To develop the global stiffness matrix from the elemental stiffness matrices, though, we can employ a scheme such as that shown in Figure 21.7. The elements in a structure are connected to a few of the global coordinates. Hypothetical coordinates for two elements are shown in the bottom part of the figure. The coefficients in the element stiffness matrices are added to the coefficients in the global stiffness matrix corresponding to the same coordinate rows and columns. This process is shown for a few of the coefficients in the two element matrices. The summation continues for all of the elements in the structure. After the summation has been performed for all of the elements in the structure, the global stiffness matrix for that structure will result.

21.8 LOADS ACTING ON THE SYSTEM

Loads that act on the structure can emanate from two sources. These are forces applied directly to the joints and forces applied to the members. The loads that act directly on the global coordinates are easy to deal with. These are the same forces that were dealt with in Chapter 20. For our discussion here those forces will be denoted as $\{F_j\}$. If all of the loads on the structure are acting on the joints, the global load vector is simply

$$\{F\} = \{F_j\} \tag{21.18}$$

But all of the loads are not always acting only on the joints. Usually there also are forces acting on the members in the structure. These loads can include distributed loads and concentrated forces that act along the length of the member. To deal with the forces acting on the members, we must determine an equivalent set of global forces. This can be accomplished by again using principles of strain energy. In previous sections we said that the strain energy in the element computed using local coordinates is equal to the strain energy in the element computed using global coordinates. Similarly, the work done on the member by element forces in local coordinates is equal to the work done on the structure by the equivalent forces in global coordinates. This concept can be represented as

$$\tfrac{1}{2}\{x\}^T\{f^0\} = \tfrac{1}{2}\{x_g\}^T\{f_g^0\} \tag{21.19}$$

In Equation 21.19 we are using the notation $\{f^0\}$ to indicate the fixed-end forces from the member loads. These are the same fixed-end forces that we used in our discussions about moment distribution and slope-deflection methods of analysis. Fixed-end forces for common member loads were shown in Figure 18.5. The term $\{f_g^0\}$ is the equivalent force in global coordinates.

Recalling the relationship between local and global element displacements given by Equation 21.14, we can show that

$$\{x_g\}^T[\beta]^T\{f^0\} = \{x_g\}^T\{f_g^0\} \tag{21.20}$$

which reduces to

$$\{\beta\}^T\{f^0\} = \{f_g^0\} \tag{21.21}$$

This is the expression for the equivalent global forces caused by forces acting on the length of the member. By rearranging this equation to the form

$$\{f_g^0\} - [\beta]^T\{f^0\} = 0 \tag{21.22}$$

observe that the fixed-end forces on the element caused by loads acting on the element are in equilibrium with their global effect on the structure. In other words, the fixed-end forces at the ends of the members act in the opposite direction when they are applied to the joints. The final global load vector is then

$$\{F\} = \{F_j\} - [\beta]^T\{f^0\} \tag{21.23}$$

This is the load vector that will be used when solving the matrix equations. Please note that when adding the element forces to the global force vector, the element forces are superimposed in the global force vector in the same way that the element stiffness matrices were superimposed to obtain the global stiffness matrices. This process was discussed in the previous section.

21.9 COMPUTING FINAL BEAM END FORCES

After computing the global forces and displacements using the procedures that were discussed in Chapter 20, we need to compute the forces that are acting at the ends of the element. There are two components to the final beam end forces: forces caused by the response of the structure and forces caused by loads applied directly to the element. The final beam end forces in local coordinates are

$$\{f\} = \{f_s\} + \{f^0\} = [k][\beta]\{x_g\} + \{f^0\} \tag{21.24}$$

These are the final forces that act on the end of the element in the coordinates of the element. The term $\{x_g\}$ is the vector of global displacements at the global coordinates to which the member is connected.

21.10 PUTTING IT ALL TOGETHER

We will now use the principles that we have developed in this chapter for the analysis of a complete structure. The structure to be analyzed is the T-Bar truss shown in Figure 21.8. The two rods act in tension and serve to stiffen the system. The global coordinates that we will use in the analysis are shown in Figure 21.9. Notice that we have numbered the unconstrained degrees of freedom first and the constrained de-

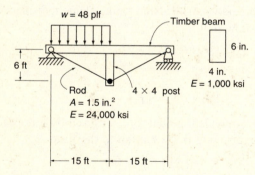

Figure 21.8

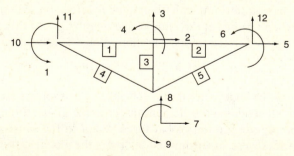

Figure 21.9

grees of freedom last. This caused the unconstrained coordinates to be in the upper partitions of the matrices and the constrained coordinates to be in the lower partitions.

To begin the analysis, we must first compute the elemental stiffness and transformation matrices for each component in the structure. Because of the nature of this structure, some of these matrices are identical. The two horizontal components have the same stiffness and transformation matrices. The two rods have the same stiffness matrix.

Using Equations 21.8 and 21.10, the stiffness and transformation matrices, respectively, for the two horizontal beams are found to be

$$[k]_{1,2} = \begin{bmatrix} 133 & -133 & 0 & 0 & 0 & 0 \\ -133 & 133 & 0 & 0 & 0 & 0 \\ 0 & 0 & 0.148 & 13.3 & -0.148 & 13.3 \\ 0 & 0 & 13.3 & 1600 & -13.3 & 800 \\ 0 & 0 & -0.148 & -13.3 & 0.148 & -13.3 \\ 0 & 0 & 13.3 & 800 & -13.3 & 1600 \end{bmatrix} \quad (21.25)$$

$$\{\beta\}_{1,2}^T = \begin{bmatrix} 1.0 & 0 & 0 & 0 & 0 & 0 \\ 0 & 0 & 1.0 & 0 & 0 & 0 \\ 0 & 0 & 0 & 1.0 & 0 & 0 \\ 0 & 1.0 & 0 & 0 & 0 & 0 \\ 0 & 0 & 0 & 0 & 1.0 & 0 \\ 0 & 0 & 0 & 0 & 0 & 1.0 \end{bmatrix} \quad (21.26)$$

Each of these two components is taken to extend from the left to the right. The angle θ for these components is therefore equal to 0 degrees.

The post—the vertical member in the middle—is assumed to go from the bottom to the top. As such, this member is in the structure at an angle of 90 degrees. Its stiffness and transformation matrices are

$$[k]_3 = \begin{bmatrix} 222 & -222 & 0 & 0 & 0 & 0 \\ -222 & 222 & 0 & 0 & 0 & 0 \\ 0 & 0 & 0.686 & 24.7 & -0.686 & 24.7 \\ 0 & 0 & 24.7 & 1185 & -24.7 & 592 \\ 0 & 0 & -0.686 & -24.7 & 0.686 & -24.7 \\ 0 & 0 & 24.7 & 592 & -24.7 & 1185 \end{bmatrix} \quad (21.27)$$

$$[\beta]_3^T = \begin{bmatrix} 0 & 0 & -1 & 0 & 0 & 0 \\ 1 & 0 & 0 & 0 & 0 & 0 \\ 0 & 0 & 0 & 1 & 0 & 0 \\ 0 & 0 & 0 & 0 & -1 & 0 \\ 0 & 1 & 0 & 0 & 0 & 0 \\ 0 & 0 & 0 & 0 & 0 & 1 \end{bmatrix} \quad (21.28)$$

The stiffness matrices for the two rods in the structure are determined using Equation 21.3. They are the same and are

$$[k]_{4,6} = \begin{bmatrix} 224 & -224 \\ -224 & 224 \end{bmatrix} \tag{21.29}$$

The transformation matrices for the two rods are different because their orientation in the structure is different. If both rods are assumed to extend from the left side to the right side, the orientation of the left rod is -21.8 degrees, while that of the right rod is 21.8 degrees. The resulting transformation matrices, found using Equation 21.5, are

$$[\beta]_4^T = \begin{bmatrix} 0.928 & 0 \\ 0.371 & 0 \\ 0 & 0.928 \\ 0 & 0.371 \end{bmatrix} \quad [\beta]_6^T = \begin{bmatrix} 0.928 & 0 \\ -0.371 & 0 \\ 0 & 0.928 \\ 0 & -0.371 \end{bmatrix} \tag{21.30}$$

Prior to assembling the system matrices, we need to compute the equivalent loads on the global structure caused by the uniformly distributed load on Component 1. These equivalent loads, which can be computed using Equation 21.21, are

$$\{f_g^0\}_1 = \begin{bmatrix} 1 & 0 & 0 & 0 & 0 & 0 \\ 0 & 0 & 1 & 0 & 0 & 0 \\ 0 & 0 & 0 & 1 & 0 & 0 \\ 0 & 1 & 0 & 0 & 0 & 0 \\ 0 & 0 & 0 & 0 & 1 & 0 \\ 0 & 0 & 0 & 0 & 0 & 1 \end{bmatrix} \begin{Bmatrix} 0 \\ 0 \\ 0.36 \\ 10.8 \\ 0.36 \\ -10.8 \end{Bmatrix} = \begin{Bmatrix} 0 \\ 0.36 \\ 10.8 \\ 0 \\ 0.36 \\ -10.8 \end{Bmatrix} \tag{21.31}$$

The necessary information is now available to assemble the complete system equation in global coordinates. This is accomplished using the superposition procedure discussed in Section 21.7 for stiffness and in Section 21.8 for the loads. Using these procedures, the final stiffness equation for this structure is shown in Figure 21.10 on page 571.

The partitions in Figure 21.10 are between the constrained and the unconstrained degrees-of-freedom. This equation can be solved using the procedures discussed in Chapter 20. The resulting unknown forces and displacements are

$$\begin{Bmatrix} F_{10} \\ F_{11} \\ F_{12} \end{Bmatrix} = \begin{Bmatrix} 0 \\ 0.54 \\ 0.18 \end{Bmatrix} \begin{Bmatrix} X_1 \\ X_2 \\ X_3 \\ X_4 \\ X_5 \\ X_6 \\ X_7 \\ X_8 \\ X_9 \end{Bmatrix} = \begin{Bmatrix} -0.00938 \\ -0.00443 \\ -0.0198 \\ 0.00493 \\ -0.00841 \\ -0.00230 \\ -0.00405 \\ -0.0178 \\ -0.00246 \end{Bmatrix} \tag{21.32}$$

$$
\begin{bmatrix}
0.148 & 13.3 & -0.148 & 13.30 & 0 & 0 & 0 & 0 & 0 & 0 & 0 & 0 \\
13.3 & 2419 & -146.3 & 824.7 & 0 & 0 & -0.686 & 0 & 24.7 & 0 & 0 & 0 \\
-0.148 & -146.3 & 355.1 & -13.3 & 0 & 0 & 0 & -222 & 0 & 0 & 0 & 0 \\
13.30 & 824.7 & -13.3 & 2785 & 13.3 & 13.3 & -24.7 & 0 & 592 & 0 & 0 & -0.148 \\
0 & 0 & 13.3 & 1793 & 800 & -192.9 & 77.12 & 0 & 0 & 0 & 0 & -90.42 \\
0 & 0 & 133.3 & 800 & 1600 & 0 & 0 & 0 & 0 & 0 & 0 & -13.3 \\
0 & -0.686 & 0 & -24.7 & -192.9 & 386.5 & 0 & -24.7 & -192.9 & -77.12 & -77.12 & 77.12 \\
0 & 0 & -222 & 0 & 77.12 & 0 & 252.8 & 0 & 0 & -77.12 & -30.83 & -30.83 \\
0 & 24.7 & 0 & 592 & 0 & 0 & -24.7 & 0 & 1185 & 0 & 0 & 0 \\
0 & 0 & 0 & 0 & 0 & 0 & -192.9 & -77.12 & 325.9 & -210.1 & 0 & 0 \\
0 & 0 & 0 & 0 & 0 & 0 & -77.12 & -30.83 & -210.1 & 163.8 & 0 & 0 \\
0 & 0 & 0 & -0.148 & -90.42 & -13.3 & 77.12 & -30.83 & 0 & 0 & 0 & 30.98
\end{bmatrix}
\begin{Bmatrix}
X_1 \\ X_2 \\ X_3 \\ X_4 \\ X_5 \\ X_6 \\ X_7 \\ X_8 \\ X_9 \\ 0 \\ 0 \\ 0
\end{Bmatrix}
=
\begin{Bmatrix}
-10.8 \\ 0 \\ -0.36 \\ 10.8 \\ 0 \\ 0 \\ 0 \\ 0 \\ 0 \\ F_{10} \\ F_{11} \\ F_{12}
\end{Bmatrix}
$$

Figure 21.10 Final system equation.

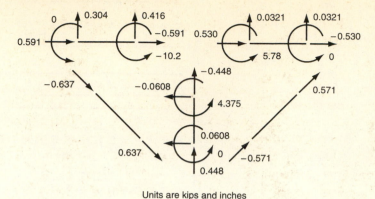

Units are kips and inches

Figure 21.11

The reaction force at Coordinate 11 has to be treated a little differently. There is a force applied at this coordinate from the member. Because this coordinate is constrained to zero displacement, all force applied to this coordinate goes directly to the reaction. But there is also a reaction force at this coordinate resulting from analysis of Equation 21.31. The final reaction force at Coordinate 11, that shown in Equation 21.32, is $0.18 + 0.36 = 0.54$. There is a 0.18 kips reaction from the structure and a 0.36 kips reaction from the load applied by the left member.

The member end forces resulting from the analysis can be found using Equation 21.24. For the first member, these forces are computed from

$$\{f\}_1 = [k]_1[\beta]_1 \begin{Bmatrix} 0 \\ 0 \\ -0.00938 \\ -0.00443 \\ -0.0198 \\ 0.00493 \end{Bmatrix} + \begin{Bmatrix} 0 \\ 0 \\ -0.360 \\ 10.8 \\ -0.360 \\ -10.8 \end{Bmatrix} = \begin{Bmatrix} 0.591 \\ -0.591 \\ 0.304 \\ 0 \\ 0.416 \\ -10.2 \end{Bmatrix} \qquad (21.33)$$

Member end forces for the other components are computed in the same manner as the first member. The final member end forces for all of the components are shown in Figure 21.11.

PROBLEMS

For problems 21.1 through 21.5 determine the displacements and forces of reaction for the structure and the member end forces for each member.

21.1   Repeat Problem 20.1

21.2   Repeat Problem 20.2 (*Ans.* $R_{AH} = -17.68^k$, $R_{DV} = 17.68^k$)

21.3   Repeat Problem 20.3

21.4 Repeat Problem 20.4 (*Ans.* $\delta_{AH} = -0.188$ m $\delta_{AV} = 0.0684$ in.)

21.5 Repeat Problem 20.5

For Problems 21.6 through 21.10 determine the displacements and forces of reaction for the structure and the member end forces for each member. For these problems, if the moment of inertia is specified as being constant, assume that $I = 150$ in^4. If the cross-sectional area of the member is not specified, assume that the area is equal to 10 percent of the moment of inertia.

21.6 Repeat Problem 13.21 (*Ans.* $R_{AV} = -8.98^k$, $M_A = -717$ in.-k)

21.7 Repeat Problem 16.4

21.8 Repeat Problem 19.8 (*Ans.* $M_A = 101.4^{lk}$, $M_D = 179.9^{lk}$)

21.9 Repeat Problem 19.20

21.10 Repeat Problem 19.26 (*Ans.* $M_B = 87.2^{lk}$, $M_A = 132.7^{lk}$)

The Catenary Equation

Some civil engineering structures include cables as part of the load-carrying system. Among these are suspension bridges and offshore platform installations that include mooring lines, such as tension leg platforms. Because the cables are not weightless, as we often assume when designing guy-lines for small towers, the tension in the cable changes along its length and the cable sags along its length when in use. These cables assume the shape of a catenary. Developed in this appendix are the fundamental relationships necessary to evaluate the force in catenary cables.

The basic geometry of the catenary cable that we will be considering is shown in Figure A.1. Notice that the bottom of the cable is located at coordinate $(0, 0)$ and that it is tangent to the horizontal coordinate axis at that point. The top of the cable is located at coordinate (X, Y).

To develop the governing equations for a cable, consider the segment of length Δs shown in Figure A.2. The equations of equilibrium for this segment of the cable are

$$\Sigma F_x = (T + \Delta T) \cos(\theta + \Delta\theta) - T \cos\theta = 0$$
$$\Sigma F_y = (T + \Delta T) \sin(\theta + \Delta\theta) - T \sin\theta - w \Delta s = 0 \qquad (A.1)$$

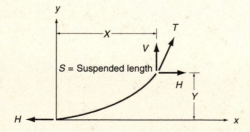

Figure A.1   Geometry of the cable.

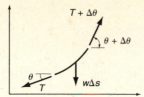

Figure A.2   A segment of the cable.

The term w in the above equation is the weight of the cable per unit length. For convenience in mathematical manipulation, let us introduce the function:

$$H(T, \theta) = T \cos \theta \tag{A.2}$$

which is the horizontal component of force in the cable. If we substitute Equation A.2 into the first of the equations in Equation A.1, that equation becomes

$$H(T + \Delta T, \theta + \Delta\theta) - H(T, \theta) = \Delta H = 0 \tag{A.3}$$

Along the segment Δs, then, the rate of change of the horizontal force is

$$\frac{dH}{ds} = \lim_{\Delta s \to 0} \left(\frac{\Delta H}{\Delta s} \right) = \lim_{\Delta s \to 0} \left(\frac{0}{\Delta s} \right) = 0 \tag{A.4}$$

Because the function H, as defined in Equation A.2, is the horizontal component of force in the cable, we can conclude from Equation A.4 that the horizontal component of force is constant along the length of the cable, namely

$$H = T \cos \theta \tag{A.5}$$

From this equation it can be seen that for H to be constant the force in the cable must change along its length.

Now let us examine the second of the equations in Equation A.1. If we rearrange that equation and divide both sides by $\Delta\theta$ we can obtain

$$\frac{(T + \Delta T) \sin(\theta + \Delta\theta) - T \sin \theta}{\Delta\theta} = \frac{w \, \Delta s}{\Delta\theta} \tag{A.6}$$

Taking the limit of this equation as $\Delta\theta$ approaches zero

$$\lim_{\Delta\theta \to 0} \left[\frac{(T + \Delta T) \sin(\theta + \Delta\theta) - T \sin \theta}{\Delta\theta} \right] = \lim_{\Delta\theta \to 0} \left[\frac{w \, \Delta s}{\Delta\theta} \right] \tag{A.7}$$

which becomes

$$\frac{d}{d\theta} [T \sin \theta] = w \frac{ds}{d\theta} \tag{A.8}$$

But from Equation A.5 it is known that

$$T = \frac{H}{\cos \theta} \tag{A.9}$$

so Equation A.8 becomes

$$\frac{d}{d\theta}[H \tan \theta] = H \sec^2 \theta = w \frac{ds}{d\theta} \qquad (A.10)$$

and can be rearranged to the form

$$\frac{ds}{d\theta} = \frac{H}{w} \sec^2 \theta \qquad (A.11)$$

Upon integrating Equation A.11, the suspended length of the cable is found to be

$$S = \int_0^s ds = \frac{H}{w} \int_0^\theta \sec^2 \theta \, d\theta = \frac{H}{w} \tan \theta \qquad (A.12)$$

From calculus it is known that

$$\cos \theta = \lim_{\Delta s \to 0} \left[\frac{\Delta x}{\Delta s} \right] = \frac{dx}{ds}$$

$$\sin \theta = \lim_{\Delta s \to 0} \left[\frac{\Delta y}{\Delta s} \right] = \frac{dy}{ds} \qquad (A.13)$$

By applying the chain rule with Equations A.11 and A.13

$$\frac{dx}{d\theta} = \left[\frac{H}{w} \right] \sec^2 \theta \cos \theta = \left[\frac{H}{w} \right] \sec \theta$$

$$\frac{dy}{d\theta} = \left[\frac{H}{w} \right] \sec^2 \theta \sin \theta \qquad (A.14)$$

Integrating the first of these equations

$$X = \int_0^X dx = \frac{H}{w} \int_0^\theta \sec \theta \, d\theta \qquad (A.15)$$

which reduces to

$$X = \left[\frac{H}{w} \right] \ln(\sec \theta + \tan \theta) \qquad (A.16)$$

This is the x projection of the cable, the horizontal distance from one end to the other end. The y projection can be found by working with the second equation in Equation A.14 and performing integration as was done to obtain the x projection. The result is

$$Y = \left[\frac{H}{w} \right] (\sec \theta - 1) \qquad (A.17)$$

These last two equations are expressed in terms of the angle θ, which probably is not known. However, the angle can be expressed in terms of other parameters of cable.

From Equation A.16

$$e^{(wX/H)} = \sec \theta + \tan \theta \qquad (A.18)$$

and from trigonometry we know that

$$\sec^2 \theta - \tan^2 \theta = (\sec \theta + \tan \theta)(\sec \theta - \tan \theta) = 1 \qquad (A.19)$$

If Equation A.18 is substituted into Equation A.19 the equation obtained is

$$\sec \theta - \tan \theta = e^{(-wX/H)} \qquad (A.20)$$

Then, by adding Equations A.18 and A.20, $\sec \theta$ is found to be

$$\sec \theta = \tfrac{1}{2}[e^{(wX/H)} + e^{(-wX/H)}] = \cosh\left(\frac{wX}{H}\right) \qquad (A.21)$$

and by subtracting Equation A.20 from Equation A.18 $\tan \theta$ is found to be

$$\tan \theta = \tfrac{1}{2}[e^{(wX/H)} - e^{(-wX/H)}] = \sinh\left(\frac{wX}{H}\right) \qquad (A.22)$$

The length of the cable is found by substituting Equation A.22 into Equation A.12 which yields

$$S = \left[\frac{H}{w}\right] \sinh\left(\frac{wX}{H}\right) \qquad (A.23)$$

By substituting Equation A.21 into Equation A.17 we find that the y projection of the cable is

$$Y = \frac{H}{w}\left[\cosh\left(\frac{wX}{H}\right) - 1\right] \qquad (A.24)$$

Lastly, using Equations A.21, A.22, and A.24, the x projection of the cable is found to be

$$X = \frac{H}{w} \cosh^{-1}\left(\frac{Y + \dfrac{H}{w}}{\dfrac{H}{w}}\right) \qquad (A.25)$$

or if Equations A.21, A.22, and A.23 had been used X would be of the form

$$X = \frac{H}{w} \sinh^{-1}\left(\frac{Sw}{H}\right) \qquad (A.26)$$

The last quantity needed is the tension in the cable. This can be obtained by using Equations A.5, A.13, and A.26. The result is

$$T = \sqrt{(wS)^2 + H^2} \qquad (A.27)$$

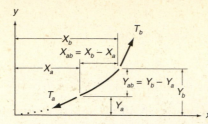

Figure A.3   Geometry of a cable segment not passing through the origin.

From the form of this last equation, we recognize that T is computed using Pythagorean's theorem applied to the vertical and horizontal components of force in the cable. As such, the vertical component of force in the cable is

$$V = wS \tag{A.28}$$

The equations developed are sufficient to determine the forces in a cable and its profile. They can be used to evaluate a cable that does not pass though the origin, as was shown in Figure A.1. To do so, a cable passing through the origin, which contains cable AB, is determined as shown in Figure A.3. The forces and projections at points A and B are then computed. They represent the forces and profile in the cable AB.

An Example   The anchor chain on a mobile offshore drilling unit is 2,000 feet long and has a submerged weight of 108 pounds per foot. It is tangent to the seafloor at the end with the anchor. The other end of the chain connects to the vessel at a point 800 feet above the seafloor. What is the tension in the chain and the distance from the vessel to the anchor?

By substituting Equation A.25 into Equation A.23 we obtain the equation

$$S = \frac{H}{W}\sinh\left[\cosh^{-1}\left(\frac{Y + \dfrac{H}{W}}{\dfrac{H}{W}}\right)\right]$$

which can be solved for H, the horizontal component of force in the chain. The result is

$$H = \tfrac{1}{2}(S^2 - Y^2)\left(\frac{W}{Y}\right) = \tfrac{1}{2}(2000^2 - 800^2)\left(\frac{108}{800}\right) = 226{,}800 \text{ lbs}$$

We then can substitute this value of H, and the other known values, into Equation A.25 to obtain X, the horizontal distance from the end of the chain to the anchor.

$$X = \frac{226{,}800}{108}\cosh^{-1}\left(\frac{800 + \dfrac{226{,}800}{108}}{\dfrac{226{,}800}{108}}\right) = 1{,}780 \text{ ft}$$

The tension in the chain can be computed using Equation A.27. The result is

$$T = \sqrt{(108\cdot 2000)^2 + 226{,}800^2} = 313{,}200 \text{ lbs}$$

Matrix Algebra

B.1 INTRODUCTION

An introduction to the basic rules of matrix algebra is presented here. The material is intended to provide introductory background for a reader of this book who may not be sufficiently familiar with the subject. The focus of the material is on applications that may be of interest primarily to structural engineers. Other excellent references exist that will provide an interested reader a more comprehensive treatment of the subject.[1,2]

B.2 MATRIX DEFINITIONS AND PROPERTIES

A matrix is defined as an ordered arrangement of numbers in rows and columns, as

$$[A] = \begin{bmatrix} a_{1,1} & a_{1,2} & \ldots & a_{1,n} \\ a_{2,1} & a_{2,2} & \ldots & a_{2,n} \\ a_{m,1} & a_{m,2} & \ldots & a_{m,n} \end{bmatrix} \qquad \text{(B.1)}$$

A representative matrix $[A]$, such as shown in Equation B.1, consists of m rows and n columns of numbers enclosed in brackets. Matrix $[A]$ is said to be of order $m \times n$. The elements $a_{i,j}$ in the array are identified by two subscripts. The first subscript designates the row and the second designates the column in which the element is located. Thus $a_{3,5}$ is the element located at the intersection of the third row and the fifth column of the matrix. Elements with repeated subscripts, for example, $a_{i,i}$ are located on the main diagonal of the matrix.

Although the number of rows and columns of a matrix may vary from problem to problem, two special cases deserve mention. When $m = 1$, the matrix consists of

---

[1] F. Ayres, *Theory and Problems of Matrices,* Schaum Outline Series (New York: McGraw-Hill, 1962).

[2] S. D. Conte and L. deBoor, *Elementary Numerical Analysis,* 2d ed. (New York: McGraw-Hill, 1972).

only one *row* of elements and is called a row matrix. It is written as

$$\lfloor A \rfloor = \lfloor a_{1,1} \quad a_{1,2} \quad \ldots \quad a_{1,n} \rfloor \tag{B.2}$$

When $n = 1$, the matrix consists of only one *column* and is called a column matrix or a vector. It is written as

$$\{A\} = \begin{Bmatrix} a_{1,1} \\ a_{2,1} \\ \cdot \\ \cdot \\ \cdot \\ a_{m,1} \end{Bmatrix} \tag{B.3}$$

Although all matrices may be considered to be row, column, or rectangular matrices, there are special types of rectangular matrices that deserve attention. These are discussed in the following section.

B.3 SPECIAL MATRIX TYPES

Square Matrix

When the number of rows m and the number of columns n are equal, the matrix is said to be *square*. As an example, if $m = n = 3$, square matrix $[A]$ may appear as

$$[A] = \begin{bmatrix} a_{1,1} & a_{1,2} & a_{1,3} \\ a_{2,1} & a_{2,2} & a_{2,3} \\ a_{3,1} & a_{3,2} & a_{3,3} \end{bmatrix} \tag{B.4}$$

Symmetric Matrix

When the elements of a square matrix obey the rule $a_{i,j} = a_{j,i}$ the matrix is said to be *symmetric*. The elements are arranged symmetrically about the main diagonal. An example of a symmetric 3×3 matrix is shown below:

$$[A] = \begin{bmatrix} 3 & -2 & 5 \\ -2 & 4 & 7 \\ 5 & 7 & 6 \end{bmatrix} \tag{B.5}$$

Symmetric matrices occur frequently in structural theory and play a key role in the matrix manipulations that develop the theory.

Identity Matrix

When all the elements on the main diagonal of a square matrix have equal unity and all the other elements are equal to zero, the matrix is called an *identity* matrix (it is also sometimes called a *unit* matrix) and is given the symbol $[I]$. An example of a 3×3 identity matrix is shown below:

$$[I] = \begin{bmatrix} 1 & 0 & 0 \\ 0 & 1 & 0 \\ 0 & 0 & 1 \end{bmatrix} \tag{B.6}$$

Transposed Matrix

When the elements of a given matrix are reordered so that the columns of the original matrix become the corresponding rows of the new matrix, the new matrix is said to be the *transpose* of the original matrix. In this book, the transpose of matrix $[A]$ is given the symbol $[A]^T$. An example of a specific matrix and its transpose is shown below:

$$[A] = \begin{bmatrix} 1 & 3 & -2 \\ 5 & 4 & 7 \\ 8 & 2 & 6 \end{bmatrix} \tag{B.7}$$

$$[A]^T = \begin{bmatrix} 1 & 5 & 8 \\ 3 & 4 & 2 \\ -2 & 7 & 6 \end{bmatrix} \tag{B.8}$$

The reader should note that if matrix $[A]$ is symmetric then $[A]^T = [A]$. This property is used frequently in the development of structural theory using matrices. As special cases, the transpose of a row matrix becomes a column matrix and vice versa. Thus

$$\lfloor A \rfloor^T = \{A\} \tag{B.9}$$

and

$$\{A\}^T = \lfloor A \rfloor \tag{B.10}$$

B.4 DETERMINANT OF A SQUARE MATRIX

A determinant of a square matrix $[A]$ is given by the symbol $|A|$ and, in its expanded form, is written as

$$|A| = \begin{vmatrix} a_{1,1} & a_{1,2} & \cdots & a_{1,m} \\ a_{2,1} & a_{2,2} & \cdots & a_{2,m} \\ \vdots & \vdots & & \vdots \\ a_{m,1} & a_{m,2} & \cdots & a_{m,m} \end{vmatrix} \tag{B.11}$$

This determinant is said to be of order m. Unlike matrix $[A]$, which has no single value, the determinant $|A|$ does have a single numerical value. The value of $|A|$ is easily found for a 2×2 array of numbers, for example

$$\begin{vmatrix} 2 & 1 \\ 4 & 5 \end{vmatrix} = 2 \cdot 5 - 4 \cdot 1 = 6 \tag{B.12}$$

The value of $|A|$ in the above example was determined by multiplying the numbers on the main diagonal and subtracting from this product the product of the numbers on the other diagonal. Unfortunately, this simple procedure does not work for determinants of order greater than 2.

A general procedure for finding the value of a determinant sometimes is called "expansion by minors." The first *minor* of array $[A]$, corresponding to the element

$a_{i,j}$, is defined as the determinant of a reduced matrix obtained by eliminating the ith row and the jth column from matrix $[A]$. The minor is a specific number, like any other determinant. As an illustration, several first minors are shown below for a specific example matrix $[A]$:

$$[A] = \begin{bmatrix} 1 & 2 & 3 \\ -2 & 3 & 4 \\ 1 & 5 & 2 \end{bmatrix}$$

$$\text{minor of } a_{1,1} = \begin{vmatrix} 3 & 4 \\ 5 & 2 \end{vmatrix} = 6 - 20 = -14$$

$$\text{minor of } a_{1,2} = \begin{vmatrix} -2 & 4 \\ 1 & 2 \end{vmatrix} = -4 - 4 = -8$$

$$\text{minor of } a_{2,3} = \begin{vmatrix} 1 & 2 \\ 1 & 5 \end{vmatrix} = 5 - 2 = 3 \tag{B.13}$$

When the proper sign is attached to a minor, the result is called a *cofactor,* and is given the symbol $A_{i,j}$. The sign of a minor is determined by multiplying the minor by $(-1)^{i+j}$. Several cofactors for the example matrix $[A]$ are shown below:

$$A_{1,1} = (-1)^2 \times \text{minor of } a_{1,1} = 1(-14) = -14$$
$$A_{1,2} = (-1)^3 \times \text{minor of } a_{1,2} = (-1)(-8) = 8$$
$$A_{2,3} = (-1)^5 \times \text{minor of } a_{2,3} = (-1)(3) = -3 \tag{B.14}$$

Now, to obtain the value of a general determinant, we can choose any arbitrary row (i) of matrix $[A]$ and expand according to the relation

$$|A| = \sum_{j=1}^{m} a_{i,j} \cdot A_{i,j} \tag{B.15}$$

The value of a determinant can also be found by choosing an arbitrary column (j) and expanding according to the relation

$$|A| = \sum_{i=1}^{m} a_{i,j} \cdot A_{i,j} \tag{B.16}$$

If the order of the original determinant is large, the procedure described above does not appear to produce a simple solution. For example, a 15th-order determinant will still have first minors that are of order 14. However, the 14th-order minors can, in themselves, be reduced to 13th-order minors by the expansion process. The process can be repeated until the resulting minors are of order 2. These minors can then be evaluated readily using the procedure described in the initial illustration of this section.

Although the procedure for evaluating determinants may appear long and tedious, computer algorithms can be written that will perform the necessary algebraic operations. Other simplifying procedures are available, which make use of special characteristics of determinants. These will not be discussed here, but interested readers may refer to books cited previously in footnotes 1 and 2 on page 579.

B.5 ADJOINT MATRIX

A special matrix exists called an *adjoint* matrix and is given the symbol adj[A]. To find the adjoint matrix corresponding to an original matrix [A], first replace each element of [A] with its cofactor; the adjoint matrix is then the transpose of this resultant matrix. Symbolically, adjoint matrix adj[A] is written as

$$\text{adj}[A] = \begin{bmatrix} A_{1,1} & A_{1,2} & \cdots & A_{1,m} \\ A_{2,1} & A_{2,2} & \cdots & A_{2,m} \\ \vdots & \vdots & & \vdots \\ A_{m,1} & A_{m,2} & \cdots & A_{m,m} \end{bmatrix} \tag{B.17}$$

A specific numerical example of a matrix [A] and its adjoint matrix is shown below:

$$[A] = \begin{bmatrix} 1 & 2 & 3 \\ 2 & 3 & 4 \\ 1 & 5 & 3 \end{bmatrix}$$

$$A_{1,1} = -11 \quad A_{1,2} = -2 \quad A_{1,3} = 7$$
$$A_{2,1} = 9 \quad A_{2,2} = 0 \quad A_{2,3} = -3$$
$$A_{3,1} = -1 \quad A_{3,2} = 2 \quad A_{3,3} = -1$$

$$\text{adj}[A] = \begin{bmatrix} -11 & -2 & 7 \\ 9 & 0 & -3 \\ -1 & 2 & -1 \end{bmatrix}^T = \begin{bmatrix} -11 & 9 & -1 \\ -2 & 0 & 2 \\ 7 & -3 & -1 \end{bmatrix} \tag{B.18}$$

The adjoint matrix and the determinant both are used in the computation of the inverse of a matrix, a topic that is treated in a later section of this appendix.

B.6 MATRIX ARITHMETIC

Equality of Matrices

Two matrices are equal only if the corresponding elements of the two matrices are equal. Thus, equality of matrices can exist only between matrices of equal orders.

Addition and Subtraction of Matrices

Two matrices may be added or subtracted only if they have the same order. Addition of two matrices [A] and [B] is performed as

$$[A] + [B] = [C] \tag{B.19}$$

where the elements of matrix [C] are the sum of corresponding elements of [A] and [B], that is, $c_{i,j} = a_{i,j} + b_{i,j}$. An example of the addition of two matrices is shown below:

$$\begin{bmatrix} 1 & 2 \\ 3 & 4 \end{bmatrix} + \begin{bmatrix} 5 & 6 \\ 7 & 8 \end{bmatrix} = \begin{bmatrix} 6 & 8 \\ 10 & 12 \end{bmatrix} \tag{B.20}$$

Subtraction of two matrices is performed similarly by subtracting corresponding elements.

Both the commutative and the associative laws hold for the addition and subtraction of two matrices. Thus

$$[A] + [B] = [B] + [A] \tag{B.21}$$

and

$$[A] + ([B] + [C]) = ([A] + [B]) + [C] \tag{B.22}$$

Scalar Multiplication of Matrices

To multiply a matrix by a scalar, *each element* of the matrix is multiplied by the scalar. Thus

$$\alpha[A] = \begin{bmatrix} \alpha a_{1,1} & \alpha a_{1,2} & \cdots & \alpha a_{1,n} \\ \alpha a_{2,1} & \alpha a_{2,2} & \cdots & \alpha a_{2,n} \\ \vdots & \vdots & & \vdots \\ \alpha a_{m,1} & \alpha a_{m,2} & \cdots & \alpha a_{m,n} \end{bmatrix} \tag{B.23}$$

where α is a constant.

Multiplication of Matrices

The product of two matrices exists only if the matrices are *conformable*. For the matrix product $[A][B]$, conformability means that the number of columns of $[A]$ equals the number of rows of $[B]$. The two matrices shown below are conformable (in the order shown) and may be multiplied.

$$[A] = \begin{bmatrix} 1 & 3 \\ 2 & 5 \end{bmatrix} \qquad [B] = \begin{bmatrix} 2 & 6 & 5 \\ 1 & 3 & -4 \end{bmatrix} \tag{B.24}$$

However, if the order of the matrices is reversed, that is, $[B][A]$, the matrices are *not* conformable and the matrix product does not exist. Furthermore, even if the matrices $[A]$ and $[B]$ are square, and thus conformable in the order $[A][B]$ and $[B][A]$, the two matrix products are generally not the same. In general,

$$[A][B] \neq [B][A] \tag{B.25}$$

The formal definition of a matrix product between matrices that are conformable is given as follows:

$$[A]_{m \times \ell}[B]_{\ell \times n} = [C]_{m \times n} \tag{B.26}$$

where $\quad c_{i,j} = \sum_{k=1}^{\ell} a_{i,k} b_{k,j}$

Note that matrix $[A]$ is not of the same order as $[B]$, but the two matrices are conformable since the number of columns of $[A]$ equals the number of rows of $[B]$.

The order of the product matrix $[C]$ is $m \times n$. A simple illustration of a matrix product is shown below:

$$\begin{bmatrix} 1 & 2 \\ -3 & 2 \end{bmatrix}_{2\times2} \begin{bmatrix} 1 & 3 & 2 \\ 4 & 5 & 3 \end{bmatrix}_{2\times3} = \begin{bmatrix} 9 & 13 & 8 \\ 5 & 1 & 0 \end{bmatrix}_{2\times3}$$

where

$$\begin{aligned}
C_{1,1} &= (1)(1) + (2)(4) = 9 \\
C_{1,2} &= (1)(3) + (2)(5) = 13 \\
C_{1,3} &= (1)(2) + (2)(3) = 8 \\
C_{2,1} &= -(3)(1) + (2)(4) = 5 \\
C_{2,2} &= -(3)(3) + (2)(5) = 1 \\
C_{2,3} &= -(3)(2) + (2)(3) = 0
\end{aligned} \tag{B.27}$$

Although the order in which two matrices are multiplied may not be reversed, in general, without obtaining different results, both the associative and the distributive laws are valid for matrix products. Thus

$$[A][B][C] = ([A][B])[C] = [A]([B][C]) \tag{B.28}$$

and

$$[A]([B] + [C]) = [A][B] + [A][C] \tag{B.29}$$

If a matrix $[A]$ is multiplied by the identity matrix $[I]$ (assuming that the matrices are conformable) the matrix $[A]$ remains unchanged. Thus

$$\begin{bmatrix} 1 & 2 & 3 \\ -2 & 4 & 6 \\ 3 & 5 & 2 \end{bmatrix} \begin{bmatrix} 1 & 0 & 0 \\ 0 & 1 & 0 \\ 0 & 0 & 1 \end{bmatrix} = \begin{bmatrix} 1 & 2 & 3 \\ -2 & 4 & 6 \\ 3 & 5 & 2 \end{bmatrix} \tag{B.30}$$

In general,

$$[A][I] = [A]$$

and

$$[I][A] = [A] \tag{B.31}$$

Transpose of a Product

If a matrix product is transposed, the result is shown symbolically as

$$([A][B])^T = [B]^T[A]^T \tag{B.32}$$

The transpose of a triple matrix product may be found by using Equation (B.32) and the associative law in several stages, as shown

$$\begin{aligned}
([A][B][C])^T &= ([A]([B][C]))^T \\
&= ([B][C])^T[A]^T \\
&= [C]^T[B]^T[A]^T
\end{aligned} \tag{B.33}$$

Note, particularly, that in finding the transpose of a matrix product the order of multiplication changes.

Matrix Inverse

Although matrix addition, subtraction, and multiplication have been defined in the preceding sections, no mention has been made of matrix division. In fact, division by matrices in the form $[A]/[B]$ does not exist. However, a matrix operation does exist that closely parallels algebraic division. This operation makes use of a matrix *inverse*.

The inverse of a square matrix $[A]$ is given the symbol $[A]^{-1}$. It is defined such that

$$[A][A]^{-1} = [A]^{-1}[A] = [I] \tag{B.34}$$

Many techniques exist with which a matrix inverse may be determined. One formal technique is described by the following relationship:

$$[A]^{-1} = \frac{\text{adj}[A]}{|A|} \tag{B.35}$$

As an example, consider the matrix $[A]$ given by Equation (B.18). The adjoint matrix adj $[A]$ is also shown by Equation (B.18). The determinant $|A|$ may be found by using the elements and first minors of the first column of $[A]$:

$$|A| = a_{1,1}A_{1,1} + a_{2,1}A_{2,1} + a_{3,1}A_{3,1}$$
$$= (1)(-11) + (2)(9) + (1)(-1) = 6$$

Thus

$$[A]^{-1} = \frac{1}{6}\begin{bmatrix} -11 & 9 & -1 \\ -2 & 0 & 2 \\ 7 & -3 & -1 \end{bmatrix} \tag{B.36}$$

The correctness of the values given for the coefficients of $[A]^{-1}$ may be verified by forming the matrix product $[A][A]^{-1}$ and checking to see if the result is the identity matrix. For the example given:

$$[A][A]^{-1} = \begin{bmatrix} 1 & 2 & 3 \\ 2 & 3 & 4 \\ 1 & 5 & 3 \end{bmatrix}\frac{1}{6}\begin{bmatrix} -11 & 9 & -1 \\ -2 & 0 & 2 \\ 7 & -3 & -1 \end{bmatrix} = \frac{1}{6}\begin{bmatrix} 6 & 0 & 0 \\ 0 & 6 & 0 \\ 0 & 0 & 6 \end{bmatrix} = [I]$$

A special situation exists regarding the inverse of a matrix and deserves attention. If the determinant of a matrix $[A]$ equals zero ($|A| = 0$), then the division operation indicated by Equation (B.35) cannot be performed. Under these circumstances, the inverse of matrix $[A]$ does not exist and matrix $[A]$ is said to be *singular*. Singular matrices occur frequently in structural theory and a reader should be aware of the meaning of this term. An example of a singular matrix is shown:

$$[A] = \begin{bmatrix} 1 & 2 & 4 \\ -2 & 3 & 2 \\ 3 & 6 & 12 \end{bmatrix}$$

and

$$|A| = a_{1,1}A_{1,1} + a_{2,1}A_{2,1} + a_{3,1}A_{3,1} = 24 + 0 - 24 = 0$$

The inverse of a matrix product may be found using rules that are very similar to those used in finding the transpose of a matrix product (given in Equation B.32). Specifically,

$$([A][B])^{-1} = [B]^{-1}[A]^{-1} \tag{B.37}$$

and

$$([A][B][C])^{-1} = ([B][C])^{-1}[A]^{-1}$$
$$= [C]^{-1}[B]^{-1}[A]^{-1} \tag{B.38}$$

Application of the Matrix Inverse

Consider a set of algebraic equations, each of which contains a number of unknown quantities x_i

$$a_{1,1}x_1 + a_{1,2}x_2 + a_{1,3}x_3 = b_1$$
$$a_{2,1}x_1 + a_{2,2}x_2 + a_{2,3}x_3 = b_2$$
$$a_{3,1}x_1 + a_{3,2}x_2 + a_{3,3}x_3 = b_3 \tag{B.39}$$

The set of algebraic equations may be cast in matrix form as follows:

$$\begin{bmatrix} a_{1,1} & a_{1,2} & a_{1,3} \\ a_{2,1} & a_{2,2} & a_{2,3} \\ a_{3,1} & a_{3,2} & a_{3,3} \end{bmatrix} \begin{Bmatrix} x_1 \\ x_2 \\ x_3 \end{Bmatrix} = \begin{Bmatrix} b_1 \\ b_2 \\ b_3 \end{Bmatrix} \tag{B.40}$$

or, symbolically, as

$$[A]\{X\} = \{B\} \tag{B.41}$$

The solution for the unknown quantities, x_1, x_2, x_3, may be found by premultiplying both sides of Equation (B.41) by $[A]^{-1}$

$$[A]^{-1}[A]\{X\} = [A]^{-1}[B]$$
$$[I]\{X\} = [A]^{-1}\{B\}$$
$$\{X\} = [A]^{-1}\{B\} \tag{B.42}$$

Therefore if $[A]^{-1}$ is known, the solution for the x_is is determined by a simple matrix product:

$$[A]^{-1}\{B\}$$

As a numerical example, consider the following algebraic equations (the coefficients in these equations are the same as those shown in matrix $[A]$ in Equation (B.18):

$$1x_1 + 2x_2 + 3x_3 = 13.0$$
$$2x_1 + 3x_2 + 4x_3 = 19.0$$
$$1x_1 + 5x_2 + 3x_3 = 22.0 \tag{B.43}$$

The solution for x_1, x_2, and x_3 is:

$$\begin{Bmatrix} x_1 \\ x_2 \\ x_3 \end{Bmatrix} = \begin{bmatrix} 1 & 2 & 3 \\ 2 & 3 & 4 \\ 1 & 5 & 3 \end{bmatrix}^{-1} \begin{Bmatrix} 13.0 \\ 19.0 \\ 22.0 \end{Bmatrix}$$

$$= \frac{1}{6} \begin{bmatrix} -11 & 9 & -1 \\ -2 & 0 & 2 \\ 7 & -3 & -1 \end{bmatrix} \begin{Bmatrix} 13.0 \\ 19.0 \\ 22.0 \end{Bmatrix} = \begin{Bmatrix} 1.0 \\ 3.0 \\ 2.0 \end{Bmatrix} \qquad \text{(B.44)}$$

Many other techniques exist for solving algebraic equations simultaneously. The use of the matrix inverse is a special technique that may be used at the option of the analyst.

B.7 GAUSS METHOD OF SOLVING SIMULTANEOUS EQUATIONS

One of the most widely used methods for solving linear, algebraic equations simultaneously is the Gauss method. This method, or some variation of it, is used in many of the currently available computer programs that deal with structural problems. Interestingly, the method is also well adapted for calculations by hand. The fundamentals of this method are illustrated below.

Consider a set of three algebraic equations written in terms of three unknowns (x_1, x_2, and x_3):

$$3x_1 + 1x_2 - 1x_3 = 2$$
$$1x_1 + 4x_2 + 1x_3 = 12$$
$$2x_1 + 1x_2 + 2x_3 = 10 \qquad \text{(B.45)}$$

First, divide each equation of the set by its leading coefficient (so that the leading coefficient becomes equal to 1.0):

$$1.0000x_1 + 0.3333x_2 - 0.3333x_3 = 0.6667$$
$$1.0000x_1 + 4.0000x_2 + 1.0000x_3 = 12.0000$$
$$1.0000x_1 + 0.5000x_2 + 1.0000x_3 = 5.0000 \qquad \text{(B.46)}$$

Next, subtract the first equation of the resultant set from each of the other equations (so that the leading coefficient of the second and third equations equals 0.0):

$$1.0000x_1 + 0.3333x_2 - 0.3333x_3 = 0.6667$$
$$3.6667x_2 + 1.3333x_3 = 11.3333$$
$$0.1667x_2 + 1.3333x_3 = 4.3333 \qquad \text{(B.47)}$$

Repeat the two operations described above, but now start with equation 2, that is, divide equations 2 and 3 by their leading coefficient (so that the leading coefficients become equal to 1.0):

$$1.0000x_1 + 0.3333x_2 - 0.3333x_3 = 0.6667$$
$$1.0000x_2 + 0.3636x_3 = 3.0909$$
$$1.0000x_2 + 7.9996x_3 = 25.9946 \qquad \text{(B.48)}$$

Subtract equation 2 from equation 3 (so that the leading coefficient of equation 3 becomes equal to 0.0):

$$1.0000x_1 + 0.3333x_2 - 0.3333x_3 = 0.6667$$
$$1.0000x_2 + 0.3636x_3 = 3.0909$$
$$7.6360x_3 = 22.9037 \qquad \text{(B.49)}$$

Divide equation 3 by its leading coefficient. Note that this operation produces a solution for x_3:

$$1.0000x_3 = 2.9994 \qquad \text{(B.50)}$$

Substitute the derived value of x_3 into the latest form of equation 2 and solve for x_2:

$$1.0000x_2 = 3.0909 - 0.3636x_3$$
$$x_2 = 2.0003 \qquad \text{(B.51)}$$

Substitute the derived value of x_2 and x_3 into equation 1 to solve for x_1:

$$1.0000x_1 = 0.6667 - 0.3333x_2 + 0.3333x_3$$
$$x_1 = 0.9997 \qquad \text{(B.52)}$$

The solution derived by the Gauss method:

$$x_1 = 0.9997$$
$$x_2 = 2.0003$$
$$x_3 = 2.9994 \qquad \text{(B.53)}$$

compares closely to the exact solution:

$$x_1 = 1.0000$$
$$x_2 = 2.0000$$
$$x_3 = 3.0000 \qquad \text{(B.54)}$$

The inexactness of the solution shown above is a function of round-off errors. Improved accuracy is attained by using a larger number of significant figures in the solution.

B.8 SPECIAL TOPICS

Matrix Partitioning

The manipulation of matrix equations is frequently made simpler by dividing the matrices into subunits, called partitions. Partitioning is indicated in this book by the lines running horizontally and vertically between the rows and the columns of the matrices. Illustrations of matrices that have been partitioned are as follows:

$$[A] = \begin{bmatrix} a_{1,1} & a_{1,2} & a_{1,3} \\ a_{2,1} & a_{2,2} & a_{2,3} \\ \hline a_{3,1} & a_{3,2} & a_{3,3} \end{bmatrix}$$

$$[A] = \begin{bmatrix} a_1 & a_2 & a_3 \end{bmatrix}$$

$$\{A\} = \begin{Bmatrix} a_1 \\ a_2 \\ \hline a_3 \end{Bmatrix} \tag{B.55}$$

Partitioning of a matrix *equation* is shown below:

$$[A]\{X\} = \{B\}$$

$$\begin{bmatrix} a_{1,1} & a_{1,2} & a_{1,3} \\ a_{2,1} & a_{2,2} & a_{2,3} \\ \hline a_{3,1} & a_{3,2} & a_{3,3} \end{bmatrix} \begin{Bmatrix} x_1 \\ x_2 \\ \hline x_3 \end{Bmatrix} = \begin{Bmatrix} b \\ b_2 \\ \hline b_3 \end{Bmatrix} \tag{B.56}$$

Note that the horizontal partition lines in Equation (B.56) extend between the same two rows for each matrix in the equation. Furthermore, the column numbers that define the vertical partition line for the square matrix are the same as the row numbers that define the horizontal partition lines for the complete equation. Thus, if the horizontal partition lines run between the second and third rows, then the vertical partition line runs between the second and third column of the square matrix.

As an illustration of the use of partitioning, consider the same matrix equation as was described in Equation (B.43):

$$\begin{bmatrix} 1 & 2 & 3 \\ 2 & 3 & 4 \\ \hline 1 & 5 & 3 \end{bmatrix} \begin{Bmatrix} x_1 \\ x_2 \\ \hline x_3 \end{Bmatrix} = \begin{Bmatrix} 13.0 \\ 19.0 \\ \hline 22.0 \end{Bmatrix} \tag{B.57}$$

In Equation (B.57), partition lines have been drawn between the second and third rows of each matrix, and between the second and third columns of matrix $[A]$. When the matrix is partitioned, the subunits, in general, still are matrices and are manipulated as matrices (although in this problem one of the subunits is a 1×1 matrix, which can be treated as a scalar). The original matrix equation may now be written as two matrix equations:

$$\begin{bmatrix} 1 & 2 \\ 2 & 3 \end{bmatrix} \begin{Bmatrix} x_1 \\ x_2 \end{Bmatrix} + \begin{Bmatrix} 3 \\ 4 \end{Bmatrix} x_3 = \begin{Bmatrix} 13.0 \\ 19.0 \end{Bmatrix} \tag{B.58}$$

and

$$\lfloor 1 \quad 5 \rfloor \begin{Bmatrix} x_1 \\ x_2 \end{Bmatrix} + 3x_3 = 22.00 \tag{B.59}$$

Equation (B.59) may be solved for x_3 in terms of x_1 and x_2

$$3x_3 = 22.0 - \lfloor 1 \quad 5 \rfloor \begin{Bmatrix} x_1 \\ x_2 \end{Bmatrix}$$

and

$$x_3 = \frac{22.0}{3} - \frac{1}{3} \lfloor 1 \quad 5 \rfloor \begin{Bmatrix} x_1 \\ x_2 \end{Bmatrix} \tag{B.60}$$

The value of x_3 from Equation (B.60) is substituted in Equation (B.58), and the resultant equation solved for x_1 and x_2:

$$\begin{bmatrix} 1 & 2 \\ 2 & 3 \end{bmatrix} \begin{Bmatrix} x_1 \\ x_2 \end{Bmatrix} + \begin{Bmatrix} 3 \\ 4 \end{Bmatrix} \left(\frac{22.0}{3} - \frac{1}{3} \lfloor 1 \quad 5 \rfloor \begin{Bmatrix} x_1 \\ x_2 \end{Bmatrix} \right) = \begin{Bmatrix} 13.0 \\ 19.0 \end{Bmatrix}$$

$$\begin{bmatrix} 1 & 2 \\ 2 & 3 \end{bmatrix} \begin{Bmatrix} x_1 \\ x_2 \end{Bmatrix} - \frac{1}{3} \begin{bmatrix} 3 & 15 \\ 4 & 20 \end{bmatrix} \begin{Bmatrix} x_1 \\ x_2 \end{Bmatrix} = \begin{Bmatrix} 13.0 \\ 19.0 \end{Bmatrix} - \frac{1}{3} \begin{Bmatrix} 66.0 \\ 88.0 \end{Bmatrix}$$

Collecting terms:

$$\frac{1}{3} \begin{bmatrix} 0 & -9 \\ 2 & -11 \end{bmatrix} \begin{Bmatrix} x_1 \\ x_2 \end{Bmatrix} = \frac{1}{3} \begin{Bmatrix} -27.0 \\ -31.0 \end{Bmatrix}$$

Solve for x_1 and x_2:

$$\begin{Bmatrix} x_1 \\ x_2 \end{Bmatrix} = \begin{bmatrix} 0 & -9 \\ 2 & -11 \end{bmatrix}^{-1} \begin{Bmatrix} -27.0 \\ -31.0 \end{Bmatrix} = \begin{Bmatrix} 1 \\ 3 \end{Bmatrix} \tag{B.61}$$

The values of x_1 and x_2 from Equation (B.61) may now be substituted in Equation (B.60) to obtain the value of x_3:

$$x_3 = \frac{22.0}{3} = \frac{1}{3} \lfloor 1 \quad 5 \rfloor \begin{Bmatrix} 1 \\ 3 \end{Bmatrix} = \frac{1}{3} (22.0 - 16.0) = 2 \tag{B.62}$$

One obvious advantage of partitioning is that the order of the matrices for which inverses must be found is reduced. However, this advantage is offset somewhat by the fact that additional algebraic manipulations are required when using partitioning schemes. Nonetheless, partitioning of matrix equations is used extensively in developing computer solutions for structural problems, and readers of this book should be acquainted with the topic.

Differentiating and Integrating a Matrix

Previously in this appendix a matrix was defined as an ordered arrangement of numbers in rows and columns. Although the elements of a matrix generally are thought of as constants, they may also be variables, as shown by the following example:

$$[A] = \begin{bmatrix} 3x & -x^2 & 2x^4 \\ -x^2 & 5x^3 & 7x \\ 2x^4 & 7x & 2x^2 \end{bmatrix} \tag{B.63}$$

Differentiating matrix $[A]$ with respect to x is performed by differentiating each element in the matrix with respect to x. The results for the example are shown as follows:

$$\frac{d}{dx}[A] = \begin{bmatrix} 3 & -2x & 8x^3 \\ -2x & 15x^2 & 7 \\ 8x^3 & 7 & 4x \end{bmatrix} \tag{B.64}$$

If the elements of $[A]$ are functions of more than one variable (say, x and y) partial differentiation of matrix $[A]$ with respect to either x or y is performed similarly, by partially differentiating each element of the matrix with respect to that variable.

In a similar way, integration of a matrix is performed by applying the integration operator to each element of the matrix. Thus, for the matrix $[A]$ given by Equation (B.63), integration produces the following results:

$$\int_a^b [A]\, dx = \begin{bmatrix} \dfrac{3x^2}{2} & \dfrac{-x^3}{3} & \dfrac{2x^5}{5} \\ \dfrac{-x^3}{3} & \dfrac{5x^4}{4} & \dfrac{7x^2}{2} \\ \dfrac{2x^5}{5} & \dfrac{7x^2}{2} & \dfrac{2x^3}{3} \end{bmatrix}_a^b$$

$$= \begin{bmatrix} \dfrac{3}{2}(b^2 - a^2) & \dfrac{-(b^3 - a^3)}{3} & \dfrac{2}{5}(b^5 - a^5) \\ \dfrac{-1}{3}(b^3 - a^3) & \dfrac{5}{4}(b^4 - a^4) & \dfrac{7}{2}(b^2 - a^2) \\ \dfrac{2}{5}(b^5 - a^5) & \dfrac{7}{2}(b^2 - a^2) & \dfrac{2}{3}(b^3 - a^3) \end{bmatrix} \tag{B.65}$$

A special application of matrix differentiation is worth noting because of its frequent appearance in the development of structural theory. Assume that a scalar variable U is defined in terms of a matrix triple product:

$$U = \tfrac{1}{2}\lfloor x \rfloor [A]\{x\} \tag{B.66}$$

where $\{x\}$ is a column matrix consisting of n variables ($x_1, x_2 \ldots x_n$). The square matrix $[A]$ is assumed to be symmetric. If U is differentiated successively with respect to $x_1, x_2, \ldots x_n$, and the results arranged in a column matrix, the results are remarkably simple, that is,

$$\begin{Bmatrix} \dfrac{\partial U}{\partial x_1} \\ \dfrac{\partial U}{\partial x_2} \\ \cdot \\ \cdot \\ \cdot \\ \dfrac{\partial U}{\partial x_n} \end{Bmatrix} = [A]\{x\} \tag{B.67}$$

A further differentiation produces

$$\frac{\partial^2 U}{\partial x_i \partial x_j} = a_{i,j} \tag{B.68}$$

where $a_{i,j}$ are the elements of the original matrix $[A]$. Although the proof of Equations (B.67) and (B.68) is not given here, a simple example will verify their correctness. Consider the following matrix triple product:

$$U = \tfrac{1}{2}\lfloor x_1 \quad x_2 \rfloor \begin{bmatrix} 2 & 4 \\ 4 & 3 \end{bmatrix} \begin{Bmatrix} x_1 \\ x_2 \end{Bmatrix} \tag{B.69}$$

When U is expanded the result is

$$U = \tfrac{1}{2}(2x_1^2 + 8x_1x_2 + 3x^2) \tag{B.70}$$

Differentiating U successively with respect to x_1 and x_2 produces:

$$\frac{\partial U}{\partial x_1} = 2x_1 + 4x_2$$

$$\frac{\partial U}{\partial x_2} = 4x_1 + 3x_2 \tag{B.71}$$

Arrangement of these results in matrix form produces:

$$\begin{Bmatrix} \dfrac{\partial U}{\partial x_1} \\ \dfrac{\partial U}{\partial x_2} \end{Bmatrix} = \begin{bmatrix} 2 & 4 \\ 4 & 3 \end{bmatrix} \begin{Bmatrix} x_1 \\ x_2 \end{Bmatrix} \tag{B.72}$$

This result corresponds to Equation (B.67). Further differentiation produces

$$\frac{\partial^2 U}{\partial x_2^2} = 2 \qquad \frac{\partial^2 U}{\partial x_1 \partial x_2} = 4$$

$$\frac{\partial^2 U}{\partial x_2 \partial x_1} = 4 \qquad \frac{\partial^2 U}{\partial x_2^2} = 3 \tag{B.73}$$

These results agree with Equation (B.68).

The resultant form of the matrix triple product given by Equation (B.66) is sometimes called a *quadratic form* (because of the second-order appearance of the variables in the matrix product). Quadratic forms occur frequently in structural theory when strain energy is used to help derive the stiffness matrix. Although quadratic forms have not been used in this book, the topic is important in more advanced treatment of structural theory and is included here for future reference for interested readers.

Glossary

Approximate structural analysis   Analysis of structures making use of certain simplifying assumptions or "reasonable approximations."

Beam   A member that supports loads that are acting transverse to the member's axis.

Bending moment   Algebraic sum of the moments of all of the external forces to one side or the other of a particular section in a member, the moments being taken about an axis through the centroid of the section.

Braced frame   A frame that has resistance to lateral loads supplied by some type of auxiliary bracing.

Camber   The construction of a member bent or arched in one direction so that it won't look so bad when the loads bend it in the opposite direction.

Cantilever   A projecting or overhanging beam.

Cantilever construction   Two simple beams, each with overhanging or cantilevered ends

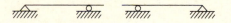

with another simple beam in between supported by the cantilevered ends.

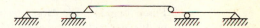

Castigliano's theorems   Energy methods for computing deformations and for analyzing statically indeterminate structures.

Cladding   The exterior covering of the structural parts of a building.

Column   A structural member whose primary function is to support compressive loads.

Concrete   A mixture of sand, gravel, crushed rock, or other aggregates held together in a rocklike mass with a paste of cement and water.

Conjugate beam   An imaginary beam that has the same length as a real beam being analyzed, and that has a set of boundary and internal continuity conditions such that the slopes and deflections in the real beam equal the shear and moment in the fictitious beam when it is loaded with the M/EI diagram.

Conservation of energy   *See* Law of conservation of energy.

Dead loads   Loads of constant magnitude that remain in one position. Examples:

weights of walls, floors, roofs, fixtures, structural frames, and so on.

Diaphragms   Those structural components that are flat plates.

Effective length   The distance between points of zero moments in a column, that is, the distance between its inflection points.

Elastic behavior   A linear relationship between stress and strain. When the external forces are removed, an elastic member will return to its original length.

Environmental loads   The loads caused by the environment in which the structure is located. Included are snow, wind, rain, and earthquakes. Strictly speaking, these also are live loads.

"Exact" structural analysis   Theoretical analysis of structures.

Fixed-end moments   The moments at the ends of loaded members when the member joints are clamped to prevent rotation.

Floor beams   The larger beams in many bridge floors that are perpendicular to the roadway of the bridge and that are used to transfer the floor loads from the stringers to the supporting girders or trusses.

Geometric instability   A situation existing when a structure has a number of reaction components equal to or greater than the number of equilibrium equations available, and yet still is unstable.

Girder   A rather loosely used term usually indicating a large beam and perhaps one into which smaller beams are framed.

Hooke's law   A statement of the linear relationship existing between force and deformation in elastic members.

Impact loads   The difference between the magnitudes of live loads actually caused and the magnitudes of those loads had they been dead loads.

Influence area   The floor area of a building that *directly influences* the forces in a particular member.

Influence line   A diagram whose ordinates show the magnitude and character of some function of a structure (shear, moment, deflection, etc.) as a unit load moves across the structure.

Joists   The closely spaced beams supporting the floors and roofs of buildings.

Law of conservation of energy   When a set of external loads is applied to a structure, the work performed by those loads equals the work performed in the elements of the structure by the internal forces.

Least work principle   The internal work accomplished by each member or each portion of a structure subjected to a set of external loads is the least possible amount necessary to maintain equilibrium in supporting the loads.

Live loads   Loads that change position and magnitude. They move or are moved. Examples: trucks, people, warehouse, materials, furniture, and so on.

Load and resistance factor design   A method of design in which the loads are multiplied by certain load or overcapacity factors (larger than 1.0) and the members are designed to have design strengths sufficient to resist these so-called factored loads.

Matrix   An ordered arrangement of numbers in rows and columns.

Maxwell's law of reciprocal deflections   The deflection at one point A in a structure due to a load applied at another point B is exactly the same as the deflection at B if the same load is applied at A. The law is applicable to members consisting of materials that follow Hooke's law.

Moment distribution   A successive correction or iteration method of analysis whereby fixed end and/or sidesway moments are balanced by a series of corrections.

Müller-Breslau's principle   The deflected shape of a structure represents to some scale the influence line for a function of the structure such as shear, moment, deflection, etc. if the function in question is allowed to act through a unit displacement.

Nodes   The locations in a structure where the elements are connected. In structures composed of beams and columns, the nodes usually are the joints.

Nominal strength   Theoretical strength.

Open-web joist   A small parallel chord truss whose members often are made from bars (hence the common name *bar joist*) or small angles or other shapes. These joists are very commonly used to support floor and roof slabs.

Plane frame   A frame that for purposes of analysis and design is assumed to lie in a single (or two-dimensional) plane.

Point of contraflexure   *See* Point of inflection.

Point of inflection (PI)   A point of zero moment. Also called *point of contraflexure*.

Ponding   A situation in which water accumulates on a roof faster than it runs off.

Principle of superposition   If a structure is linearly elastic, the forces acting on the structure may be separated or divided in any convenient fashion and the structure analyzed for the separate cases. The final results can be obtained by adding together the individual parts.

Purlins   Roof beams that span between trusses.

Qualitative influence line   A sketch of an influence line in which no numerical values are given.

Quantitative influence line   An influence line that shows numerical values.

Reinforced concrete   A combination of concrete and steel reinforcing wherein the steel provides the tensile strength lacking in the concrete. (The steel reinforcing also can be used to help the concrete resist compressive forces.)

Scuppers   Large holes or tubes in walls or parapets that enable water above a certain depth to quickly drain from roofs.

Seismic   Of or having to do with an earthquake.

Service loads   The actual loads that are assumed to be applied to a structure when it is in service (also called *working loads*).

Shear   The algebraic summation of the external forces in a member to one side or the other of a particular section that are perpendicular to the axis of the member.

Sidesway   The lateral movement of a structure caused by unsymmetrical loads and/or by an unsymmetrical arrangement of the members of the structure.

Skeleton construction   Building construction in which the loads are transferred from each floor by beams to columns and thence to the foundation.

Slenderness ratio   The ratio of the effective length of a member to its radius of gyration, both values pertaining to the same axis of bending.

Slope deflection   A classical method of analyzing statically indeterminate structures in which the moments at the ends of the members are expressed in terms of the rotations (or slopes) and deflections of the joints.

Space truss   A three-dimensional truss.

Statically determinate structures   Structures for which the equations of equilibrium are sufficient to compute all of the external reactions and internal forces.

Statically indeterminate structures   Structures for which the equations of equilibrium are insufficient for computing the external reactions and internal forces.

Steel   An alloy consisting almost entirely of iron (usually over 98 percent). It also contains small quantities of carbon, silicon, manganese, sulfur, phosphorus, and other elements.

Stringers   The beams in bridge floors that run parallel to the roadway.

Structural analysis   The computation of the forces and deformations of structures under load.

Struts Structural members that are subjected only to axial compression forces.

Superposition principle *See* Principle of superposition.

Tension coefficient The force in a truss member divided by its length.

Three-moment theorem A classical theorem that presents the relationship between the moments in the different supports of a continuous beam.

Ties Structural members that are subjected only to axial tension forces.

Tributary area The loaded area of a structure that *directly contributes* to the load applied to a particular member.

Truss A structure formed by a group of members arranged in the shape of one or more triangles.

Unbraced frame A frame whose resistance to lateral forces is provided by its members and their connections.

Unstable equilibrium A support situation whereby a structure is stable under one arrangement of loads but is not stable under other load arrangements.

Vierendeel "truss" A special type of truss (it's not really a truss by our usual definition) whose members are arranged in the shape of a set of rectangles and that requires moment-resisting joints.

Virtual displacement A fictitious displacement imposed on a structure.

Virtual work The work performed by a set of real forces during a virtual displacement.

Voussoirs The truncated wedge-shaped parts of a stone arch that are pushed together in compression.

Wichert truss A continuous statically determinate truss formerly patented by E. M. Wichert.

Working loads *See* Service loads.

Yield stress The stress at which there is a decided increase in the elongation or strain in a member without a corresponding increase in stress.

Zero-load test A procedure in which one member of a truss subjected to no external loads is given a force and the forces in the other members are computed. If all the joints balance or are in equilibrium, the structure is unstable.

Index

American Association of State Highway and Transportation Officials (AASHTO), 20, 50, 253–256
American Concrete Institute (ACI), 297, 459–460
American Institute of Steel Construction (AISC), 253, 258
American National Standards Institute (ANSI), 20
American Railway Engineering Association (AREA), 20, 256–257
American Society of Civil Engineers (ASCE), 20–50
Approximate analysis of statically indeterminate structures
  ACI coefficients, 460–461
  advantages, 455–456
  assumptions required, 456
  building frames with vertical loads, 462–465
  cantilever method, 476–479
  continuous beams, 457–460
  equivalent frame method, 460–462
  importance of, 455–456
  lateral bracing for bridges, 466–468
  lateral loads, 471–479
  mill buildings, 468–471
  moment distribution, 480
  portal method, 473–476
  portals, 465–466
  trusses, 456–457
  Vierendeel "truss," 480–482
Arches
  advantages, 86
  three-hinged, 87–93
  tied, 92–93

Archimedes, 5

Beams
  cantilever, 82–83
  continuous, 354–361, 433–439, 457–462, 495–505
  defined, 9
  fixed-ended, 284–285, 494
  floor beams, 12
  girders, 9, 55–56
  stringers, 11
Bernoulli, Johann, 309
Bertot, 416
Bowman, H. L., 184, 416, 479
Bresse, 416
Bridges
  counters, 251–253
  floor systems, 137, 142–143
  live loads, 253–256
  "see-saw," 85–86
Burgett, L. B., 27

Cables, 93–97, 574–578
Camber, 274, 595
Cantilever
  beams, 82–83, 280–282, 320–321
  method, 476–479
Cantilever erection, 347–348
Cantilevered structures, 84–86
Carryover factors, 490–492, 493
Castigliano, A., 6, 329
  first theorem, 329, 414–416, 552–554
  second theorem, 329–335, 405–413
Causey, M. L., 176, 219
Chinn, J., 27
Cladding, 30, 595
Clapeyron, B. P. E., 6, 416
Coloumb, C. A., 5
Composite structures, 410–414

Computers, 17, 187–189, 222–223, 378–379, 402, 413–414, 503–504, 530–531
Condition equations, 84–92, 94–97
Conjugate-beam method
  application to beams, 294–297
  application to frames, 297–299
  development of, 291–292
  supports, 291–292
Conservation of energy principle, 308–309
Consistent-distortion method, 353–378
Continuous beams, 354–361, 433–439, 457–462, 495–505
Cook, R. D., 415
Cooper, T., 256
Coordinate systems, 542, 544, 549–551, 557–560
Counters, 251
Critical form, 176
Cross, H., 6, 397, 488

Deflections
  beams, 275–297
  Castigliano's theorems, 330–335
  conjugate-beam method, 291–299
  elastic weight method, 286–291
  energy methods, 308–342
  frames, 297–299
  geometric methods, 271–307
  importance of, 273–274
  long-term, 297
  Maxwell's law of reciprocal deflections, 328
  moment-area method, 275–285
  reasons for computing, 273–274
  summary of beam relations, 293–294
  trusses, 311–317
  virtual work method, 309–329

Deformed shape of structures, 272
Diaphragms, 9
Displacement methods of analysis,
    430–447, 541–542
Distribution factors, 492–493
Dummy unit-load method. *See*
    Virtual work.

Earthquake loads, 47–49
Economy, 344–348
Elastic-weight method
    application of, 286–290
    derivation of, 285–286
    limitations of, 291
Energy methods, 308–342
Envelopes of forces, 68–69, 460
Environmental loads
    earthquake, 47–49
    ice, 48, 50
    rain, 27–29
    snow, 43–46
    wind, 29–46
Equations of condition, 74–75, 84,
    183–186
Equilibrium, 71, 293
Equivalent frame method, 460–462

Fairweather, V., 47, 48
Firmage, D. A., 96
Fixed-end moments, 284–285,
    493–494
Flexibility coefficients, 355
Flexibility method. *See* Force
    methods of analysis.
Floor
    beams, 12, 243
    systems for bridges, 242–244
Force envelopes, 68–69, 460
Force methods of analysis, 353–378,
    541–542
Force reversals, 249, 252, 349
Frames, 10–11, 122–125, 297–299,
    323–325, 361–362, 397,
    441–447, 465–466,
    471–482, 510–531
Free-body diagrams, 78, 149–150

Geometric instability, 76–77
Girders, 55–56
Greene, C. E., 6, 275
Grubermann, U., 136

Hooke's law, 82, 328, 354, 543

Imhotep, 4
Impact factors, 25, 257–258
Influence areas, 59–61
Influence lines
    defined, 228
    for frames, 397
    qualitative, 232–235, 394–397
    quantitative, 229–232, 235–251,
        389–394, 397–402

for statically determinate beams,
    229–240
for statically determinate trusses,
    244–253
for statically indeterminate beams,
    389–397
for statically indeterminate trusses,
    245–249
for truss reactions, 244–245
uses of, 228
Instability, 74–76, 177–181, 185
Interstate highway loading system,
    256

Jakkula, A. A., 260

Kinney, J. S., 6, 228, 299, 330, 353,
    390

Lateral bracing, 466–468
Law of conservation of energy,
    308–309
Least-work theorem, 414–416
Line diagrams, 14–15
Live loads
    highway bridges, 253–256
    impact, 25, 257–258
    railway bridges, 256–257
    reduction, 59–61
Load combinations, 61–64
Loads
    dead, 22–23
    earthquake, 47–49
    ice, 48, 50
    impact, 25, 257–258
    live, 24–27
    longitudinal, 50
    rain, 27–29
    roofs, 26–27
    snow, 43–46
    truck, 253–256
    uniform lane, 255–256
    wind, 29–43
Load and Resistance Factor Design
    (LRFD), 63–64

Manderla, H., 430
Maney, G. A., 6, 430
Marino, F. J., 27
Marshall, R. D., 30
Matrices
    algebra, 579–594
    bars, 548
    beam elements, 562–565
    coordinate systems, 557–560, 565
    defined, 579
    determinants, 581–582
    fundamentals, 543–545
    Gauss method, 588
    partitioning, 547, 590–591
    truss elements, 560–562
    types, 579–581
    uses, 540–541

Maxwell, J. C., 6, 328, 353
Maxwell's law of reciprocal
    deflections, 328, 389, 390, 392
Maxwell-Mohr method, 353
Mehta, K. C., 30
Method of joints, 149–152
Method of moments, 161–169
Method of successive approximations
    for design, 348
Method of shears, 169–172
Mill buildings, 468–471
Mohr, O., 6, 353, 430
Moment
    defined, 111
    diagrams, 113–125, 421–424,
        501–503
    influence lines for, 231–242
    maximum, 240–242
    method of, 161–169
Moment-area theorems
    application of, 278–285
    derivation of, 275–278
Moment-distribution method
    assumptions, 489
    basic relations, 490–492
    beams, 495–505
    carryover factors, 489–492
    development of, 489–495
    distribution factors, 492–493
    fixed-end moments, 493–495
    frames
        with sidesway, 512–530
        without sidesway, 510–511
        with sloping legs, 521–526
    introduction, 489–490
    modification of stiffness for simple
        ends, 499–501
    multistory frames, 526–530
    sign convention, 493
    stiffness factors, 492, 499–500
    successive corrections, 526–530
Moorman, R. B. B., 353
Morgan, N. D., 397
Morris, C. T., 526
Moving bodies, 72
Müller-Breslau, 184, 232, 235, 390,
    395
Multistory frames, 526–530

Navier, L. M. H., 5
Newton, Sir Isaac, 8, 71
Norris, C. H., 177, 463, 473

O'Brien, W. H., 297
Oliver, W. A., 297

Palladio, A., 5, 136
Parcel, J. I., 353
Perry, D. C., 30
Pitch, 26
Plane trusses. *See* Trusses.
Points of inflection, 462–464, 469
Ponding, 27–29

Portal method, 473–476
Primary forces, 138
Przemienieski, J. S., 542
Purlins, 137
Pythagoras, 5

Rain loads, 27–29
Reactions, 71–109
Rogers, G. L., 176, 219
Roofs, 26–29
Rubinstein, M. F., 542
Ruddy, J. L., 27

SABLE, 17, 187–189, 222–223,
    378–379, 402, 413–414,
    478–479, 503–504, 530–531
Scofield, W. F., 297
Scuppers, 28
Secondary forces, 138
"See-saw" bridges, 85–86
Seismic loads, 47–49
Shear
    defined, 110
    diagrams, 112–125, 421–424,
        501–503
    influence lines, 229–231, 234–235
    method of, 169–172
Sidesway, 441–447, 512–530, 597
Sign conventions, 77–78, 111–112,
    118, 124, 162, 229, 420,
    434–435, 493
Slope deflection method
    advantages of, 430
    application to continuous beams,
        433–439
    application to frames, 439–447
    derivation of equations, 431–433
    equations for, 433, 437
    frames with sloping legs, 447
    sidesway of frames, 441–447
    sign convention, 434–435
    support settlements, 437–439
Slopes
    by Castigliano's second theorem,
        335
    by conjugate-beam method,
        291–299
    by elastic-weight method,
        285–290
    by moment-area method, 276–285
    by virtual-work method, 325–329
Smith, A., 473
Southwell, R. V., 218
Snow loads, 43–46
Space trusses
    basic principles, 203–204
    computer analysis, 222–223
    defined, 202
    member forces, 209–223
    reactions, 209–223
    simultaneous equation analysis,
        218–222
    special theorems, 206
    stability, 205–206
    statics equations, 204–205

tension coefficients, 218–221
    types of supports, 207–209
Stability, 74–76, 181–183
Statically indeterminate structures
    advantages of, 346–348
    approximate analysis, 455–482
    beams, 354–367
    defined, 74, 144–147
    disadvantages of, 348–349
    displacement method, 430–447
    force methods, 353–378
    frames, 361–362
    general, 343–344
    trusses, 367–378
Statics equations, 71, 204–205
Steel, 597
Steinman, D. B., 186, 257
Stephenson, H. K., 260
Stevenson, R. L., 3
Stiffness, 492
Stiffness factor, 433, 492, 499–500
Strength design, 63–64
Stringers, 11, 243
Structural analysis, defined, 1
Structural design, defined, 1
Structural idealization, 14
Struts, 9
Successive corrections, 526–530
Superposition, or principle of, 82,
    279, 495
Supports
    for beams, 72–74
    for cantilevered structures, 84
    fixed-end, 74
    hinge, 72
    link, 74
    roller, 73
    for space trusses, 207–209
    for three-hinge arches, 87–89
Support settlement
    discussion, 348
    effect of by consistent distortions,
        365–367
Sutherland, H., 184, 416, 479
Sylvester, Pope, II, 5

Temperature changes, effect on truss
    forces, 376–378
Tension coefficients, 218–221
Three-dimensional trusses. See
    Space trusses.
Three-moment theorem
    application to continuous beams,
        419–425
    application to beams with fixed
        ends, 424–425
    derivation of, 417–419
    history, 416
Ties, 9
Timoshenko, S. P., 6, 181
Tributary areas, 26, 54–58
Trusses
    assumptions for analysis, 137
    Baltimore, 143
    bowstring, 140
    bridge, 141–143

complex, 171–181
compound, 175
computer solutions, 186–189
critical form, 176
deck, 142
defined, 11
deflections, 311–317
Fink, 140, 166–167
half-through, 142–143
Howe, 140
K, 143
lenticular, 143
notation, 138–139
Parker, 143
Petit, 143
plane, 136–201
pony, 142–143
Pratt, 140, 143
quadrangular, 140
roof, 139–140
sawtooth, 140
scissors, 140
simple, 174
stability, 181–183
statical determinancy, 144–147
through, 142
Vierendeel, 480–482
Warren, 140, 143
when assumptions not correct,
    173–174
Wichert, 184–186
Zero-force members, 172–173
Zero-load test, 175–176
Two-dimensional trusses. See
    Trusses.

Unique solution, 175–176
Unstable equilibrium, 76
Utku, S., 177, 463, 473

Van Ryzin, G., 27
Vierendeel, A., 480
Vierendeel "truss," 480–482
Virtual work
    beam deflections, 317–322
    complementary virtual work,
        309–311
    frame deflections, 322–327
    slopes or angle changes for beams
        and frames, 325–329
    truss deflections, 311–317
Voussoirs, 86–87

Wang, C. K., 271
Westergaard, H. M., 6
Whipple, S., 5
Wichert, E. M., 184
Wilbur, J. B., 177, 463, 473
Wilson, A. C., 476
Wind loads, 29–46
Winkler, E., 228

Young, D. H., 181
Young, W. C., 415

Zero-load test, 175–176